Carbonic Anhydrase

Its Inhibitors and Activators

CRC Enzyme Inhibitors Series

Series Editors
H. John Smith and Claire Simons
Cardiff Univeristy
Cardiff, UK

Carbonic Anhydrase: Its Inhibitors and Activators

Edited by Claudiu T. Supuran, Andrea Scozzafava and Janet Conway

CRC Enzyme Inhibitors Series

Carbonic Anhydrase

Its Inhibitors and Activators

Edited by

Claudiu T. Supuran
Andrea Scozzafava
Janet Conway

CRC PRESS

Boca Raton London New York Washington, D.C.

Library of Congress Cataloging-in-Publication Data

Carbonic anhydrase : its inhibitors and activators / edited by Claudiu T. Supuran, Andrea
   Scozzafava, and Janet Conway.
       p.   cm. -- (Enzyme inhibitors)
   Includes bibliographical references and index.
   ISBN 0-415-30673-6 (alk. paper)
       1. Carbonic anhydrase. I. Supuran, Claudiu T., 1962- II. Scozzafava, Andrea, 1947-
   III. Conway, Janet. IV. Series.

QP613.C37C373 2004
612′.01519—dc22   2003065063

Visit the CRC Press Web site at www.crcpress.com

© 2004 by CRC Press LLC

No claim to original U.S. Government works
International Standard Book Number 0-415-30673-6
Library of Congress Card Number 2003065063
Printed in the United States of America 1 2 3 4 5 6 7 8 9 0
Printed on acid-free paper

Preface

Carbonic anhydrases (CAs) are widespread metalloenzymes in higher vertebrates, wherein they play crucial physiological roles. Some of these isozymes are cytosolic (CA I, CA II, CA III and CA VII), others are membrane bound (CA IV, CA IX, CA XII and CA XIV), one is mitochondrial (CA V) and one is secreted in the saliva (CA VI). Three acatalytic forms, which are designated as CA-related proteins (CARPs), CARP VIII, CARP X and CARP XI, are also known. Representatives of the β- and γ-CA family are highly abundant in plants, bacteria and archaea. These enzymes are very efficient catalysts of the reversible hydration of carbon dioxide to bicarbonate, and at least the α-CAs possess a high versatility and are able to catalyze other hydrolytic processes. Recently, two new families have been discovered — the δ- and ε-CA classes of these enzymes.

Chapter 1 deals with the catalytic mechanism of CAs, which is understood in great detail at present. The active site consists of a Zn(II) ion coordinated by three histidine residues and a water molecule/hydroxide ion. Several important physiological and physiopathological functions are played by the CA isozymes present in organisms at all levels of the phylogenetic tree, such as respiration and transport of CO_2/bicarbonate between metabolizing tissues and the lungs, pH and CO_2 homeostasis, electrolyte secretion in a variety of tissues and organs, and biosynthetic reactions such as the gluconeogenesis and urea synthesis (in animals) and CO_2 fixation (in plants and algae). The presence of these ubiquitous enzymes in so many tissues and in so many different isoforms makes them useful in designing inhibitors or activators that have biomedical applications.

Chapter 2 deals with the CARPs, the CA-related proteins that do not possess catalytic activity and have been discovered only recently. Many evidences connect these proteins with many types of tumors, and these new data are thoroughly presented by one of the major contributors in the field.

Chapter 3 presents a challenging finding regarding the multiple binding modes of different CA ligands to the enzyme, based on very recent x-ray crystallographic findings. This contribution is essential for those working in drug design and tries to develop isozyme-specific inhibitors, and the proposed solutions are quite surprising.

Chapter 4 deals with the unsubstituted aromatic/heterocyclic sulfonamides — the most investigated CA inhibitors (CAIs) — that were known to inhibit CAs since the beginning of research in this field. Starting with the 1950s, potent CAIs belonging to the heterocyclic sulfonamide class have been developed, which has led to benzothiadiazine and high-ceiling diuretics and to the systemic antiglaucoma drugs acetazolamide, methazolamide, ethoxzolamide and dichlorophenamide. The discovery of these drugs has highly benefited the chemistry of sulfonamides, as thousands of derivatives belonging to the heterocyclic, aromatic and *bis*-sulfonamide classes have been synthesized and investigated for their biological activity. In the late 1980s and early 1990s, the topically effective antiglaucoma CAIs were discovered, with two such drugs — dorzolamide and brinzolamide — presently being used clinically. Both have been designed by the ring approach, i.e., by investigating many ring systems incorporating sulfamoyl moieties. More recently, the tail approach has been reported for designing antiglaucoma CAIs with topical activity, but this approach can be extended for other applications of these pharmacological agents. It consists of attaching tails that will induce the desired physicochemical properties to aromatic/heterocyclic sulfonamide scaffolds possessing derivatizable amino or hydroxy moieties. Important progress has been achieved in finding inhibitors with a higher affinity for a certain isozyme, although clear-cut isozyme-specific inhibitors are not currently available. Inhibitors selective for membrane-associated (CAs IV, IX, XII and XIV) and cytosolic isozymes are available, belonging either to macromolecular compounds or to the positively charged derivatives (of low molecular weight). CAIs possessing relevant antitumor properties were also discovered, with one such derivative – indisulam – in advanced clinical trials for treating solid tumors. CAIs with good anticonvulsant activity and several other biomedical applications have recently been reported. Some aliphatic sulfonamides show significant CA inhibitory properties, and compounds possessing zinc-binding functions different from the classical one (of aromatic/heterocyclic sulfonamide type) have been reported, incorporating, among others, sulfamate, sulfamide and hydroxamate moieties. This field is in constant progress and might lead to the discovery of very interesting pharmacological agents.

Chapter 5 presents an updated review on QSAR of sulfonamide CAIs. Based on the most modern computer-assisted drug design techniques, different mathematical models that explain the biological activity of this class of pharmacological agents are presented, together with their impact on improving the actually available drugs.

Chapter 6 reviews metal complexes of sulfonamides with CA inhibitory activity. This type of inhibitors has been studied in detail only in the past 10 years and might provide a host of important biomedical applications. The detailed chemical and crystallographic studies leading to these CA inhibitors are presented.

Chapter 7 deals with the nonsulfonamide type of CAIs, i.e., the anions and the azoles. These compounds are primarily important for understanding the inhibition mechanism of these enzymes.

Chapter 8 to Chapter 11 deal with the many different clinical applications of CAIs: in ophthalmology, as antiglaucoma and antimacular degeneration agents (Chapter 8); in oncology — a new and hot topic (Chapter 9); in gastroenterology, neurology and nephrology (Chapter 10); and in dermatology (Chapter 11). Although used for many years to treat or prevent different diseases, CAIs might still play an important role in therapy, as exemplified by these reviews of well-known contributors in the field.

Activation of CAs (Chapter 12) has been a controversial phenomenon for a long time. Recently, kinetic, spectroscopic and x-ray crystallographic data have offered a clear-cut explanation of this phenomenon, based on the catalytic mechanism of these enzymes. It has been demonstrated that molecules acting as CA activators (CAAs) bind at the entrance of the enzyme active-site cavity and participate in facilitated proton transfer processes between the active site and the reaction medium, thereby facilitating the rate-determining step of the CA catalytic cycle. In addition to CA II–activator adducts, x-ray crystallographic studies have been reported for ternary complexes of this isozyme with activators and anion inhibitors. Drug design studies have been successfully performed for obtaining strong CAAs belonging to several chemical classes, such as amines and their derivatives, azoles, amino acids and oligopeptides. Structure–activity correlations for diverse classes of activators are discussed for the isozymes for which the phenomenon has been studied. The physiological relevance of CA activation and the possible application of CAAs in Alzheimer's disease and for other memory therapies are also reviewed.

Claudiu T. Supuran
Andrea Scozzafava
Janet Conway

Contributors

Gloria Alzuet
Dipartimento Quimica Inorganica
Facultad Farmacia, Universidad
  Valencia
Valencia, Spain

Jochen Antel
Solvay Pharmaceuticals
Hannover, Germany

Mircea Desideriu Banciu
Academia Romana, Sectia Stiinte
  Chimice
Bucarest, Romania

Joaquin Borras
Dipartimento Quimica Inorganica
Facultad Farmacia, Universidad
  Valencia
Valencia, Spain

Angela Casini
Università degli Studi, Polo Scientifico
Firenze, Italy

Brian W. Clare
University of Western Australia
Department of Chemistry
Nedlands, Australia

Janet Conway
Licensing & Development
Pfizer Inc.
New York, New York

Sacramento Ferrer
Dipartemento Quimica Inorganica
Facultad Farmacia, Universidad
  Valencia
Valencia, Spain

Marc Antoniu Ilies
Texas A&M University
Galveston, Texas

Monica Ilies
Università degli Studi, Polo Scientifico
Firenze, Italy

Jyrki Kivelä
Central Military Hospital
Helsinki, Finland

Gerhard Klebe
Institute of Pharmaceutical Chemistry
Philipps University
Marburg, Germany

Antonio Mastrolorenzo
Università degli Studi
Dipartimento di Scienze
  Dermatologiche
Firenze, Italy

Luca Menabuoni
Casa di Cura Villa Donatello
Firenze, Italy

Francesco Mincione
U.O. Oculistica Az.
Pescia, Italy

Isao Nishimori
First Department of Internal Medicine
Kochi Medical School
Nankoku, Kochi, Japan

Anna-Kaisa Parkkila
Department of Neurology
Tampere University Hospital
Tampere, Finland

Seppo Parkkila
University of Tampere
Institute of Medical Technology
Tampere, Finland

Jaromír Pastorek
Institute of Virology
Slovak Academy of Sciences
Bratislava, Slovak Republic

Silvia Pastoreková
Institute of Virology
Slovak Academy of Sciences
Bratislava, Slovak Republic

Andrea Scozzafava
Università degli Studi, Polo Scientifico
Firenze, Italy

Christoph A. Sotriffer
Institute of Pharmaceutical Chemistry
Marburg, Germany

Claudiu T. Supuran
Università degli Studi, Polo Scientifico
Firenze, Italy

Alexander Weber
Institute of Pharmaceutical Chemistry
Philipps University
Marburg, Germany

Giuliano Zuccati
Università degli Studi
Dipartimento di Scienze Dermatologiche
Firenze, Italy

Table of Contents

Chapter 1   Carbonic Anhydrases: Catalytic and Inhibition Mechanisms, Distribution and Physiological Roles 1

Claudiu T. Supuran

Chapter 2   Acatalytic CAs: Carbonic Anhydrase-Related Proteins 25

Isao Nishimori

Chapter 3   Multiple Binding Modes Observed in X-Ray Structures of Carbonic Anhydrase Inhibitor Complexes and Other Systems: Consequences for Structure-Based Drug Design 45

Jochen Antel, Alexander Weber, Christoph A. Sotriffer and Gerhard Klebe

Chapter 4   Development of Sulfonamide Carbonic Anhydrase Inhibitors 67

Claudiu T. Supuran, Angela Casini and Andrea Scozzafava

Chapter 5   QSAR Studies of Sulfonamide Carbonic Anhydrase Inhibitors 149

Brian W. Clare and Claudiu T. Supuran

Chapter 6   Metal Complexes of Heterocyclic Sulfonamides as Carbonic Anhydrase Inhibitors 183

Joaquín Borrás, Gloria Alzuet, Sacramento Ferrer and Claudiu T. Supuran

Chapter 7   Nonsulfonamide Carbonic Anhydrase Inhibitors 209

Marc Antoniu Ilies and Mircea Desideriu Banciu

Chapter 8   Clinical Applications of Carbonic Anhydrase Inhibitors in Ophthalmology 243

Francesco Mincione, Luca Menabuoni and Claudiu T. Supuran

Chapter 9   Cancer-Related Carbonic Anhydrase Isozymes and Their Inhibition 255

Silvia Pastoreková and Jaromír Pastorek

Chapter 10 Role of Carbonic Anhydrase and Its Inhibitors in Biological Science Related to Gastroenterology, Neurology and Nephrology 283

Seppo Parkkila, Anna-Kaisa Parkkila and Jyrki Kivelä

Chapter 11 Carbonic Anhydrase Inhibitors in Dermatology 303

Antonio Mastrolorenzo, Giuliano Zuccati, Andrea Scozzafava and Claudiu T. Supuran

Chapter 12 Carbonic Anhydrase Activators 317

Monica Ilies, Andrea Scozzafava and Claudiu T. Supuran

Index 353

1 Carbonic Anhydrases: Catalytic and Inhibition Mechanisms, Distribution and Physiological Roles

Claudiu T. Supuran

CONTENTS

1.1   Introduction 2
1.2   Catalytic and Inhibition Mechanisms of CAs...3
        1.2.1   α-CAs...3
        1.2.2   β-CAs...12
        1.2.3   γ-CAs...15
        1.2.4   Cadmium CA 16
1.3   Distribution of CAs..16
1.4   Physiological Functions of CAs 18
References...20

At least 14 different α-carbonic anhydrase (CA, EC 4.2.1.1) isoforms have been isolated in higher vertebrates, wherein these zinc enzymes play crucial physiological roles. Some of these isozymes are cytosolic (CA I, CA II, CA III and CA VII), others are membrane bound (CA IV, CA IX, CA XII and CA XIV), one is mitochondrial (CA V) and one is secreted in the saliva (CA VI). Three acatalytic forms are also known, designated CA-related proteins (CARPs): CARP VIII, CARP X and CARP XI. Representatives of the β- and γ-CA family are highly abundant in plants, bacteria and archaea. These enzymes are very efficient catalysts for the reversible hydration of carbon dioxide to bicarbonate, and at least α-CAs possess a high versatility, being able to catalyze other different hydrolytic processes, such as the hydration of cyanate to carbamic acid or of cyanamide to urea; aldehyde hydration to *gem*-diols; hydrolysis of carboxylic or sulfonic acids esters; as well as other less investigated hydrolytic processes, such as hydrolysis of halogeno derivatives and

0-415-30673-6/04/$0.00+$1.50
© 2004 by CRC Press LLC

arylsulfonyl halides. It is not known whether the reactions catalyzed by CAs other than the hydration of CO_2/dehydration of HCO_3^- have physiological relevance in systems in which these enzymes are present. The catalytic mechanism of α-CAs is understood in great detail. The active site consists of a Zn(II) ion coordinated by three histidine residues and a water molecule/hydroxide ion. The latter is the active species, acting as a potent nucleophile. For β- and γ-CAs, the zinc hydroxide mechanism is valid too, although at least some β-class enzymes do not have water directly coordinated to the metal ion. CAs are inhibited primarily by two main classes of inhibitors: the metal-complexing inorganic anions (such as cyanide, cyanate, thiocyanate, azide and hydrogensulfide) and the unsubstituted sulfonamides possessing the general formula RSO_2NH_2 (R = aryl, hetaryl, perhaloalkyl). Several important physiological and physiopathological functions are played by the CA isozymes, which are present in organisms at all levels of the phylogenetic tree. Among these functions are respiration and transport of CO_2/bicarbonate between metabolizing tissues and lungs, pH and CO_2 homeostasis, electrolyte secretion in a variety of tissues and organs, biosynthetic reactions, such as the gluconeogenesis and urea synthesis (in animals) and CO_2 fixation (in plants and algae). The presence of these ubiquitous enzymes in so many tissues and in so many different isoforms makes them useful to design inhibitors or activators that have biomedical applications.

1.1   INTRODUCTION

The carbonic anhydrases (CAs, EC 4.2.1.1) are ubiquitous metalloenzymes present in prokaryotes and eukaryotes and encoded by three distinct evolutionarily unrelated gene families: (1) α-CAs (in vertebrates, bacteria, algae and cytoplasm of green plants), (2) β-CAs (predominantly in bacteria, algae and chloroplasts of both mono- and dicotyledons) and (3) γ-CAs (mainly in archaea and some bacteria) (Hewett-Emmett 2000; Krungkrai et al. 2000; Chirica et al. 1997; Smith and Ferry 2000; Supuran and Scozzafava 2000, 2002; Supuran et al. 2003). In higher vertebrates, including humans, 14 α-CA isozymes or CA-related proteins (CARPs) have been described (Table 1.1), with very different subcellular localizations and tissue distributions (Hewett-Emmett 2000; Supuran and Scozzafava 2000, 2002; Supuran et al. 2003). There are several cytosolic forms (CAs I–III, CA VII), four membrane-bound isozymes (CA IV, CA IX, CA XII and CA XIV), one mitochondrial form (CA V) and a secreted CA isozyme (CA VI) (Supuran and Scozzafava 2000, 2002; Supuran et al. 2003). These enzymes catalyze a very simple physiological reaction, the interconversion of the carbon dioxide and the bicarbonate ion, and are thus involved in crucial physiological processes connected with respiration and transport of CO_2/bicarbonate between metabolizing tissues and lungs, pH and CO_2 homeostasis, electrolyte secretion in a variety of tissues and organs, biosynthetic reactions (such as gluconeogenesis and lipoid and urea synthesis), bone resorption, calcification, tumorigenicity and many other physiological or pathological processes (Hewett-Emmett 2000; Supuran and Scozzafava 2000, 2002; Supuran et al. 2003). Many of these isozymes are important targets for the design of inhibitors with clinical applications.

   In addition to the physiological reaction — the reversible hydration of CO_2 to bicarbonate (Equation 1.1, Figure 1.1) — α-CAs catalyze a variety of other reactions,

TABLE 1.1
Higher-Vertebrate α-CA Isozymes, Their Relative CO$_2$ Hydrase Activity, Affinity for Sulfonamide Inhibitors and Subcellular Localization

Isozyme	Catalytic Activity (CO$_2$ Hydration)	Affinity for Sulfonamides	Subcellular Localization
CAI	Low (10% of that of CA II)	Medium	Cytosol
CA II	High	Very high	Cytosol
CA III	Very low (0.3% of that of CA II)	Very low	Cytosol
CA IV	High	High	Membrane bound
CA V	Moderate-high[a]	High	Mitochondria
CA VI	Moderate	Medium-low	Secreted into saliva
CA VII	High	Very high	Cytosol
CARP VIII	Acatalytic	[b]	Cytosol
CA IX	High	High	Transmembrane
CARP X	Acatalytic	[b]	Cytosol
CARP XI	Acatalytic	[b]	Cytosol
CA XII	Low	High	Transmembrane
CA XIII	Moderate	High	Cytosol
CA XIV	High	High	Transmembrane

[a] Moderate at pH 7.4; high at pH 8.2 or higher pH.

[b] The native CARP isozymes do not contain Zn(II), so that their affinity for the sulfonamide inhibitors has not been measured. By site-directed mutagenesis it is possible to modify these proteins and transform them into enzymes with CA-like activity, which probably are inhibited by sulfonamides, but no detailed studies on this subject are available at present (see Chapter 2 for details on CARPs).

such as hydration of cyanate to carbamic acid or of cyanamide to urea (Equation 1.2 and Equation 1.3, Figure 1.1); aldehyde hydration to *gem*-diols (Equation 1.4, Figure 1.1); hydrolysis of carboxylic or sulfonic acid esters (Equation 1.5 and Equation 1.6, Figure 1.1); as well as other less-investigated hydrolytic processes, such as those described by Equation 1.7 to Equation 1.9 in Figure 1.1 (Briganti et al. 1999; Guerri et al. 2000; Supuran et al. 1997, 2003; Supuran and Scozzafava 2000, 2002). The previously reported phosphatase activity of CA III was recently proved to be an artefact (Kim et al. 2000). It is unclear whether α-CA catalyzed reactions other than CO$_2$ hydration have physiological significance. To date, x-ray crystal structures have been determined for six α-CAs (isozymes CA I to CA V and CA XII; Stams and Christianson 2000) as well as for representatives of the β- (Mitsuhashi et al. 2000) and γ-CA families (Kisker et al. 1996; Smith and Ferry 2000; Iverson et al. 2000).

1.2 CATALYTIC AND INHIBITION MECHANISMS OF CAs

1.2.1 α-CAs

The Zn(II) ion of CAs is essential for catalysis (Lindskog and Silverman 2000; Christianson and Fierke 1996; Bertini et al. 1982; Supuran et al. 2003). X-ray

$$O{=}C{=}O + H_2O \Leftrightarrow HCO_3^- + H^+ \tag{1.1}$$

$$O{=}C{=}NH + H_2O \Leftrightarrow H_2NCOOH \tag{1.2}$$

$$HN{=}C{=}NH + H_2O \Leftrightarrow H_2NCONH_2 \tag{1.3}$$

$$RCHO + H_2O \Leftrightarrow RCH(OH)_2 \tag{1.4}$$

$$RCOOAr + H_2O \Leftrightarrow RCOOH + ArOH \tag{1.5}$$

$$RSO_3Ar + H_2O \Leftrightarrow RSO_3H + ArOH \tag{1.6}$$

$$ArF + H_2O \Leftrightarrow HF + ArOH \tag{1.7}$$

$$(Ar = 2,4\text{-dinitrophenyl})$$

$$PhCH_2OCOCl + H_2O \Leftrightarrow PhCH_2OH + CO_2 + HCl \tag{1.8}$$

$$RSO_2Cl + H_2O \Leftrightarrow RSO_3H + HCl \tag{1.9}$$

$$(R = Me; Ph)$$

FIGURE 1.1 Reactions catalyzed by α-CAs. (Reproduced from Supuran, C.T. et al. (2003) *Medicinal Research Reviews* **23**, 146–189, John Wiley & Sons. With permission.)

crystallographic data show that the metal ion is situated at the bottom of a 15-Å-deep active-site cleft (Figure 1.2), coordinated by three histidine residues (His 94, His 96 and His 119) and a water molecule/hydroxide ion (Christianson and Fierke 1996; Stams and Christianson 2000). The zinc-bound water is also engaged in hydrogen bond interactions with the hydroxyl moiety of Thr 199, which in turn is bridged to the carboxylate moiety of Glu 106. These interactions enhance the nucleophilicity of the zinc-bound water molecule and orient the substrate (CO_2) in a location favorable for nucleophilic attack (Figure 1.3; Lindskog and Silverman 2000; Stams and Christianson 2000; Supuran et al. 2003). The active form of the enzyme is the basic one, with hydroxide bound to Zn(II) (Figure 1.3A; Lindskog and Silverman 2000). This strong nucleophile attacks the CO_2 molecule bound in a hydrophobic pocket in its neighborhood (the substrate-binding site comprises residues Val 121, Val 143 and Leu 198 in human isozyme CA II — Christianson and Fierke 1996; Figure 1.3B), leading to the formation of bicarbonate coordinated to Zn(II) (Figure 1.3C). The bicarbonate ion is then displaced by a water molecule and liberated into solution, forming the acid form of the enzyme, with water coordinated to Zn(II) (Figure 1.3D), which is catalytically inactive (Lindskog and Silverman 2000; Christianson and Fierke 1996; Bertini et al. 1982; Supuran et al. 2003). To regenerate the basic form A, a proton transfer reaction from the active site to the environment occurs, which might be assisted either by active-site residues (such as His 64 — the proton shuttle in isozymes I, II, IV, VII and IX, among others; see Figure 1.2 for isozyme II) or by buffers present in the medium. The process is schematically represented by Equation 1.10 and Equation 1.11:

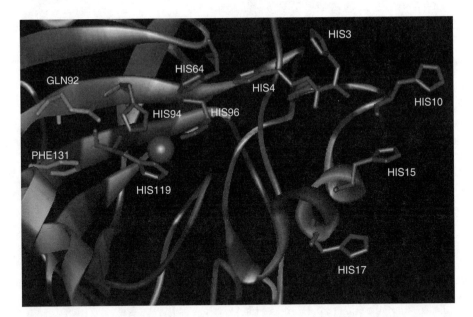

FIGURE 1.2 (See color insert following page 148.) hCA II active site. The Zn(II) ion (central pink sphere) and its three histidine ligands (in green, His 94, His 96, His 119) are shown. The histidine cluster, comprising residues His 64, His 4, His 3, His 17, His 15 and His 10, is also shown, as this is considered to play a critical role in binding activators of the types **6** to **14** reported in the chapter as well as the carboxyterminal part of the anion exchanger AE1. The figure was generated from the x-ray coordinates reported by Briganti et al. (1997) (PDB entry 4TST). (Reproduced from Scozzafava, A. and Supuran, C.T. (2002) *Biorganic Medicinal Chemistry Letters* **12**, 1177–1180. With permission from Elsevier.)

$$EZn^{2+} - OH^- + CO_2 \Leftrightarrow EZn^{2+} - HCO_3^- \overset{H_2O}{\Leftrightarrow} EZn^{2+} - OH_2 + HCO_3^- \quad (1.10)$$

$$EZn^{2+} - OH_2 \Leftrightarrow EZn^{2+} - HO^- + H^+ \quad (1.11)$$

The rate-limiting step in catalysis is the second reaction, i.e., the proton transfer that regenerates the zinc hydroxide species of the enzyme (Lindskog and Silverman 2000; Christianson and Fierke 1996; Bertini et al. 1982; Supuran et al. 2003). In the catalytically very active isozymes, such as CA II, CA IV, CA V, CA VII and CA IX, the process is assisted by a histidine residue (His 64) at the entrance of the active site, as well as by a cluster of histidines (Figure 1.2) that protrudes from the rim of the active site to the surface of the enzyme, assuring a very efficient proton transfer process for CA II, the most efficient CA isozyme (Briganti et al. 1997). This also explains why CA II is one of the most active enzymes known (with a $k_{cat}/K_m = 1.5 \times 10^8$ M^{-1}s^{-1}), approaching the limit of diffusion control (Lindskog and Silverman 2000; Christianson and Fierke 1996; Supuran et al. 2003), and also has important consequences for designing inhibitors that have clinical applications.

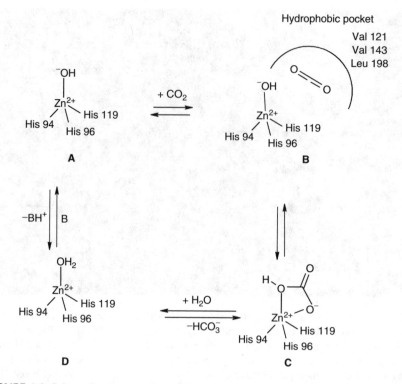

FIGURE 1.3 Schematic representation of the catalytic mechanism for the CA-catalyzed CO_2 hydration. The hypothesized hydrophobic pocket for the binding of substrates is shown schematically at Step B.

Two main classes of carbonic anhydrase inhibitors (CAIs) are known: the metal-complexing anions and the unsubstituted sulfonamides, which bind to the Zn(II) ion of the enzyme either by substituting the nonprotein zinc ligand (Equation 1.12, Figure 1.4) or add to the metal coordination sphere (Equation 1.13, Figure 1.4), generating trigonal-bipyramidal species (Bertini et al. 1982; Lindskog and Silverman 2000; Supuran et al. 2003; Supuran and Scozzafava 2000, 2002). Sulfonamides, the most important CAIs, bind in a tetrahedral geometry of the Zn(II) ion (Figure 1.4), in a deprotonated state, with the nitrogen atom of the sulfonamide moiety coordinated to Zn(II) and an extended network of hydrogen bonds involving the residues Thr 199 and Glu 106, also participating in anchoring the inhibitor molecule to the metal ion. The aromatic/heterocyclic part of the inhibitor (R) interacts with hydrophilic and hydrophobic residues of the cavity. Anions might bind either in tetrahedral geometry of the metal ion or as trigonal-bipyramidal adducts, such as the thiocyanate adduct shown in Figure 1.4B (see also Chapter 7 for a detailed description of CA inhibition by anions; Lindahl et al. 1991; Stams and Christianson 2000; Abbate et al. 2002; Supuran et al. 2003).

X-ray crystallographic structures are available for many adducts of sulfonamide inhibitors with isozymes CA I, CA II and CA IV (see also Chapter 3; Lindahl et al. 1991; Stams et al. 1996; Stams and Christianson 2000; Abbate et al. 2002, 2003;

$$E\text{-}Zn^{2+}\text{-}OH_2 + I \Leftrightarrow E\text{-}Zn^{2+}\text{-}I + H_2O \qquad \text{(substitution)} \qquad (1.12)$$

Tetrahedral adduct

$$E\text{-}Zn^{2+}\text{-}OH_2 + I \Leftrightarrow E\text{-}Zn^{2+}\text{-}OH_2(I) \qquad \text{(addition)} \qquad (1.13)$$

Trigonal-bipyramidal adduct

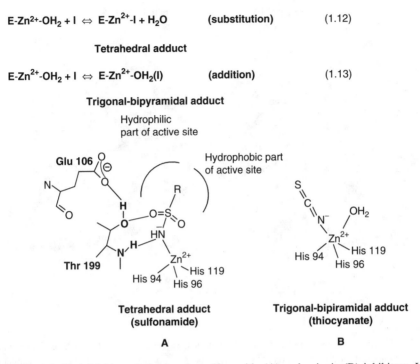

Tetrahedral adduct
(sulfonamide)

Trigonal-bipiramidal adduct
(thiocyanate)

A

B

FIGURE 1.4 CA inhibition mechanism by sulfonamide (A) and anionic (B) inhibitors. In the case of sulfonamides, in addition to the Zn(II) coordination, an extended network of hydrogen bonds ensues, involving residues Thr 199 and Glu 106, whereas the organic part of the inhibitor (R) interacts with hydrophilic and hydrophobic residues of the cavity. For anionic inhibitors such as thiocyanate (B), the interactions between inhibitor and enzyme are much simpler.

Supuran et al. 2003). In all the adducts, the deprotonated sulfonamide is coordinated to the Zn(II) ion of the enzyme and its NH moiety participates in a hydrogen bond with the Oγ of Thr 199, which in turn is engaged in another hydrogen bond to the carboxylate group of Glu 106 (Lindahl et al. 1991; Stams et al. 1996; Stams and Christianson 2000; Abbate et al. 2002; Supuran et al. 2003). One of the oxygen atoms of the SO$_2$NH moiety also participates in a hydrogen bond with the backbone NH moiety of Thr 199. Figure 1.5 shows the crystal structures of the hCA II adducts with the simplest compounds incorporating a sulfamoyl moiety (sulfamide and sulfamic acid). The binegatively charged (NH)SO$_3^{2-}$ sulfamate ion and the mono-anion of sulfamide NHSO$_2$NH$_2^-$ bind to the Zn(II) ion within the enzyme active site (Abbate et al. 2002). These two structures provide some close insights into why this functional group (the sulfonamide group) appears to have unique properties for CA inhibition: (1) it exhibits a negatively charged, most likely monoprotonated nitrogen coordinated to the Zn(II) ion; (2) simultaneously, this group forms a hydrogen bond as donor to the oxygen Oγ of the adjacent Thr 199; and (3) a hydrogen bond is formed between one of the SO$_2$ oxygens and the backbone NH of Thr 199. The basic structural elements explaining the strong affinity of the sulfonamide moiety

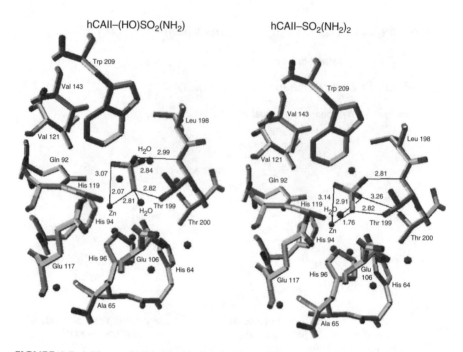

FIGURE 1.5 Adducts of hCA II with the simplest sulfonamides: sulfamic acid H_2NSO_3H, (left) and sulfamide $H_2NSO_2NH_2$ (right), determined by x-ray crystallography. (Reproduced from Supuran, C.T. et al. (2003) *Medicinal Research Reviews* **23**, 146–189, John Wiley & Sons. With permission.)

for the Zn(II) ion of CAs have been delineated in detail by using these simple compounds as prototypical CAIs (Briganti et al. 1996), without needing to analyze the interactions of the organic scaffold usually present in other inhibitors [generally belonging to the aromatic/heterocyclic sulfonamide class (Abbate et al. 2002)]. Despite important similarities in the binding of these two inhibitors to the enzyme to that of aromatic/heterocyclic sulfonamides of the type RSO_2NH_2 previously investigated, the absence of a $C–SO_2NH_2$ bond in sulfamide/sulfamic acid leads to a different hydrogen bond network in the neighborhood of the catalytical Zn(II) ion, which has been shown to be useful for the design of more potent CA inhibitors as drugs, possessing zinc-binding functions different from those of classical sulfonamides (Abbate et al. 2002).

The physiological function of the major red cell isozyme CA I (present in concentrations of up to 150 μM in the blood; Supuran et al. 2003) is unknown. Recently, the x-ray crystal structure of a natural mutant of CA I, i.e., CA I Michigan 1 (Figure 1.6), was reported. [The isozyme was isolated in three generations of a family of European Caucasians (Ferraroni et al. 2002a).] CA I Michigan 1 differs from wild-type CA I in a single amino acid residue present in the active-site cavity, i.e., Arg 67 instead of His 67 (Ferraroni et al. 2002a). This amino acid residue is located in an important region of the catalytic site, which is involved both in shuttling

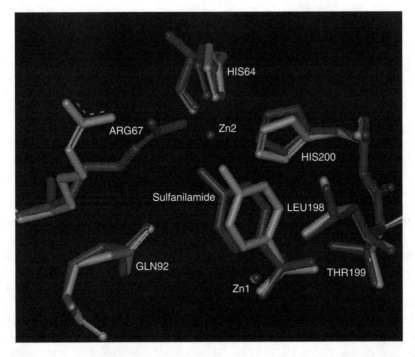

FIGURE 1.6 (See color insert.) Least-squares superimposition of the most relevant active site residues of the natural mutant CA I Michigan 1 (in yellow) and the CA I Michigan 1 $(Zn)_2$ adduct (in red) involved in sulfonamide inhibitor binding, with bound sulfanilamide, as determined by x-ray crystallography. The catalytic zinc ion is Zn1. (Reproduced from Supuran, C.T. et al. (2003) *Medicinal Research Reviews* **23**, 146–189, John Wiley & Sons. With permission.)

protons from the active cavity to the environment and in binding aromatic/hetero-cyclic sulfonamides, the classical, clinically important CAIs (Figure 1.6). The structure of the native mutant enzyme has been determined, as well as its adduct with a second zinc ion, which reveals the presence of a second metal-ion-binding site within the active cavity. Arg 67 appears to promote the binding of this second zinc ion to His 64, His 200 and itself (through one of the guanidino nitrogen atoms) (Ferraroni et al. 2002a). This second zinc ion bound to the active cavity is involved in the previously observed activation mechanism for substrate-specific α- and β-naphthyl acetate hydrolyses (Ferraroni et al. 2002a). Furthermore, this is the first example of a Zn(II) enzyme containing an arginine residue in the metal ion coordination sphere and the first CA isozyme that binds two metal ions within its active site (Ferraroni et al. 2002a). The crystal structures of sulfanilamide (4-aminobenzene sulfonamide) com-plexed to native hCA I Michigan 1 variant and to its $(Zn)_2$ adduct were also reported (Ferraroni et al. 2002b). Comparisons among these structures and the corresponding sulfonamide adduct of hCA I showed significant differences in the orientation of the inhibitor molecule and in its interactions with active-site residues such as His 200, Thr 199, Leu 198, Gln 92 and Arg/His67, which are known to play important roles in substrate or inhibitor binding and recognition (Supuran et al. 2003; Ferraroni et al. 2002a). In CA I Michigan 1 a lengthening of the Zn–N1 sulfanilamide bond distance and a corresponding shortening of the distance between the sulfamido group and Thr 199 as compared with wild-type CA I were observed. When the second Zn(II) ion was present in the active site, the p-amino group and the aromatic ring of the inhibitor molecule appeared tilted toward Gln 92 and Arg 67, moving away from His 200 and Leu 198. The structural differences in inhibitor binding between the CA I isozyme and the CA I Michigan 1 variant showed that even a point mutation within the active site of a CA isozyme might have relevant consequences on the binding of inhibitors (Ferraroni et al. 2002a). This work opens a new direction for designing isozyme-specific CAIs.

The different types of interactions by which a sulfonamide CAI achieves very high affinity (in a low nanomolar range) for the CA active site, are illustrated in Figure 1.7 for a fluoro-containing inhibitor in early clinical development, PFMZ (4-methyl-5-perfluorophenylcarboximido-δ^2-1,3,4-thiadiazoline-2-sulfonamide; Abbate et al. 2003). The ionized sulfonamide moiety of PFMZ replaces the hydroxyl ion coordinated to Zn(II) in the native enzyme (Zn–N distance 1.95 Å), with the metal ion remaining in its stable tetrahedral geometry, being coordinated in addition to the sulfonamidate nitrogen by the imidazolic nitrogens of His 94, His 96 and His 119. The proton of the coordinated sulfonamidate nitrogen atom also makes a hydrogen bond with the hydroxyl group of Thr 199, which in turn accepts a hydrogen bond from the carboxylate of Glu 106. One of the oxygen atoms of the sulfonamide moiety makes a hydrogen bond with the backbone amide of Thr 199, whereas the other is semicoordinated to the catalytic Zn(II) ion (O–Zn distance 3.0 Å). The thiadiazoline ring of the inhibitor lies in the hydrophobic part of the active-site cleft, wherein its ring atoms make van der Waals interactions with the side chains of Leu 204, Pro 202, Leu 198 and Val 135 (Figure 1.7). The carbonyl oxygen of PFMZ makes a strong hydrogen bond with the backbone amide nitrogen of Gln 92 (of 2.9 Å), an interaction also evidenced for the acetazolamide–hCA II adduct. Besides Gln 92, two other residues, Glu 69 and Asn 67, situated in the hydrophilic half of the CA active site, make van der Waals contacts with the PFMZ molecule complexed to hCA II. But the most notable and unprecedented interactions seen in this complex are the hydrogen bonds network involving the exocyclic nitrogen atom of the inhibitor, two water molecules (Wat 1194 and Wat 1199) and a fluorine atom in the *meta* position belonging to the perfluorobenzoyl tail of PFMZ (Figure 1.7). Thus, a strong hydrogen bond (of 2.9 Å) is formed between the imino nitrogen of PFMZ and Wat 1194, which in turn makes a hydrogen bond with a second water molecule of the active site, Wat 1199 (distance 2.7 Å). The second hydrogen of Wat 1194 also participates in a weaker hydrogen bond (3.3 Å) with the carbonyl oxygen of PFMZ. The other hydrogen atom of Wat 1199 makes a weak hydrogen bond with the fluorine atom in Position 3 of the perfluorobenzoyl tail of PFMZ (Figure 1.7). Finally, a very interesting interaction is observed between the perfluorophenyl ring of PFMZ and the phenyl moiety of Phe 131, a residue critical for the binding of inhibitors with

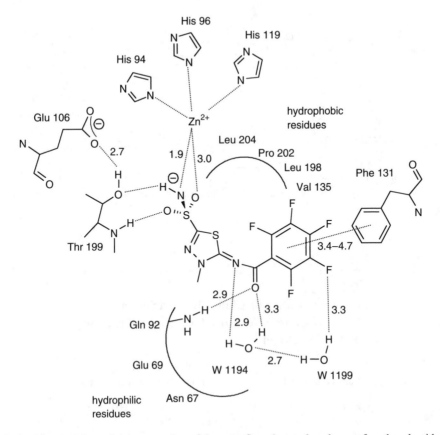

FIGURE 1.7 Schematic representation of the pentafluorobenzoyl analogue of methazolamide bound within the hCA II active site (figures represent distances in Å). (From Abbate, F. et al. (2003) *Journal of Enzyme Inhibition and Medicinal Chemistry*, **18**, 303–308. With permission.)

long tails to hCA II (Supuran et al. 2003). These two rings are almost perfectly parallel, situated at a distance of 3.4 to 4.7 Å. This type of stacking interactions has not been previously observed in a hCA II–sulfonamide adduct.

Since the report that sulfanilamide acts as a specific inhibitor of CA (Mann and Keilin 1940), four systemic sulfonamide CAIs have been developed and used clinically, mainly as antiglaucoma drugs, for some time: acetazolamide (**1.1**), methazolamide (**1.2**), ethoxzolamide (**1.3**) and dichlorophenamide (**1.4**) (Maren 1967; Supuran et al. 2003). As seen from the data in Table 1.2, Compound **1.1** to Compound **1.4** strongly inhibit several CA isozymes (such as CA I, CA II, CA IV, CA V, CA VII and CA IX), with affinities in the low nanomolar range for many of them. Recently, two new drugs have been introduced in clinical practice as topically acting sulfonamide CAIs — dorzolamide (**1.5**) and brinzolamide (**1.6**) — which also act as very potent inhibitors of most α-CA isozymes (Table 1.2). The design of sulfonamide CAIs as drugs is described in detail in Chapter 4.

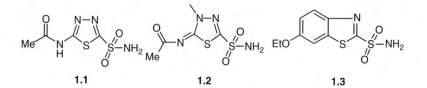

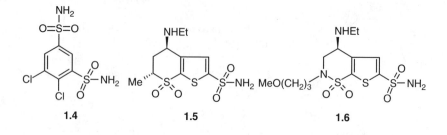

TABLE 1.2
Inhibition Data with Clinically Used Sulfonamides
1.1–1.6 against Several α-CA Isozymes of Human
or Murine Origin

Isozyme	K_I (nM)					
	1.1	1.2	1.3	1.4	1.5	1.6
hCA I	200	10	1	350	50,000	nt
hCA II	10	8	0.7	30	9	3
hCA III	3×10^5	1×10^5	5000	nt	8000	nt
hCA IV	66	56	13	120	45	45
mCA V	60	nt	5	nt	nt	nt
hCA VI	1100	560	nt	nt	nt	nt
mCA VII	16	nt	0.5	nt	nt	nt
hCA IX	25	27	34	50	52	37

Note: h = human; m: = murine isozyme; nt = not tested (no data available).

1.2.2 β-CAs

Many bacteria, some archaeas (such as *Methanobacterium thermoautotrophicum*), algae and chloroplasts of superior plants contain CAs belonging to the β-class (Smith and Ferry 1999, 2000; Kimber and Pai 2000; Mitsuhashi et al. 2000; Cronk et al. 2001). The principal difference between these enzymes and α-CAs is that β-CAs are usually oligomers, generally formed of two to six monomers of molecular weight 25 to 30 kDa each. The x-ray structures of four such β-CAs are available at the present time: (1) the enzyme isolated from the red alga *Porphyridium purpureum* (Mitsuhashi et al. 2000), (2) the enzyme from chloroplasts of *Pisum sativum* (Kimber

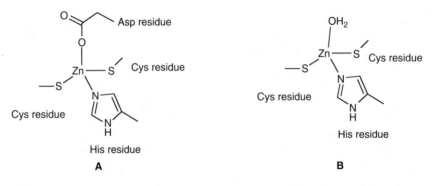

FIGURE 1.8 Schematic representation of the Zn(II) coordination sphere in β-CAs: **A:** *Porphyridium purpureum* [Mitsuhashi, S. et al. (2000) *Journal of Biological Chemistry* **275**, 5521–5526] and *Escherichia coli* [Cronk, J.D. et al. (2001) *Protein Science* **10**, 911–922] enzymes; **B:** *Pisum sativum* chloroplast and *Methanobacterium thermoautotrophicum* enzyme [Kimber, M.S., and Pai, E.F. (2000) *EMBO Journal* **15**, 2323–2330; Strop, P. et al. (2001) *Journal of Biological Chemistry* **276**, 10299–10305], as determined by x-ray crystallography.

and Pai 2000), (3) a prokaryotic enzyme isolated from *Escherichia coli* (Cronk et al. 2001) and (4) cab, an enzyme isolated from the archaeon *Methanobacterium thermoautotrophicum* (Strop et al. 2001).

The *Porphyridium purpureum* CA monomer is composed of two internally repeating structures folded as a pair of fundamentally equivalent motifs of an α/β domain and three projecting α-helices. The motif is very distinct from that of α- or γ-CAs. This homodimeric CA appeared like a tetramer with a pseudo 2-2-2 symmetry (Mitsuhashi et al. 2000). β-CAs are thus very different from the α-class enzymes. The Zn(II) ion is essential for catalysis in both families of enzymes, but its coordination is different and rather variable for β-CAs. Thus, in the prokaryotic β-CAs, the Zn(II) ion is coordinated by two cysteinate residues, an imidazole from a His residue and a carboxylate belonging to an Asp residue (Figure 1.8A), whereas the chloroplast enzyme has the Zn(II) ion coordinated by the two cysteinates, the imidazole belonging to a His residue and a water molecule (Figure 1.8B; Kimber and Pai 2000; Mitsuhashi et al. 2000; Cronk et al. 2001). The polypeptide chain folding and active-site architecture is clearly very different from those of α-CAs.

Because no water is directly coordinated to Zn(II) for some members of β-CAs (Figure 1.8A), the main question is whether the zinc hydroxide mechanism presented here for α-CAs is also valid for enzymes belonging to the β-family. A response to this question has been given by Mitsuhashi et al. (2000), who have proposed the catalytic mechanism shown in Figure 1.9.

As there are two symmetrical structural motifs in one monomer of the *Porphyridium purpureum* enzyme, resulting from two homologous repeats related to each other by a pseudo twofold axis, two Zn(II) ions are coordinated by the four amino acids mentioned. In this case, these pairs are Cys 149/Cys 403, His 205/His 459, Cys 208/Cys 462 and Asp 151/Asp 405 (Mitsuhashi et al. 2000). A water molecule is also present in the neighborhood of each metal ion, but it is not directly coordinated to it, forming a hydrogen bond with an oxygen belonging to the zinc ligand Asp

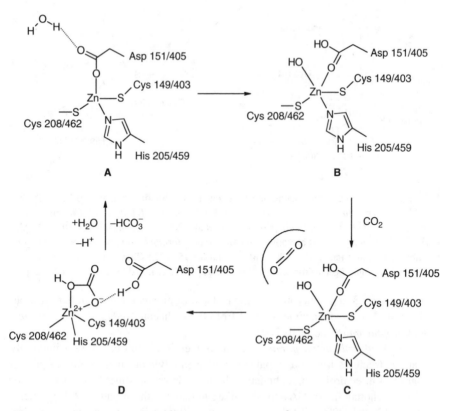

FIGURE 1.9 Proposed catalytic mechanism for prokaryotic β-CAs (*Porphyridium purpureum* enzyme numbering). (From Mitsuhashi, S. et al. (2000) *Journal of Biological Chemistry* **275**, 5521–5526.)

151/Asp 405 (Figure 1.9A). It is hypothesized that a proton transfer reaction might occur from this water molecule to the coordinated carboxylate moiety of the aspartate residue, generating a hydroxide ion, which might then be coordinated to Zn(II), which acquires a trigonal-bipyramidal geometry (Figure 1.9B). Thus, the strong nucleophile that might attack CO_2 bound within a hydrophobic pocket of the enzyme is formed (Figure 1.9C), with generation of bicarbonate bound to Zn(II) (Figure 1.9D). This intermediate is rather similar to the reaction intermediate proposed for the α-CA catalytic cycle (Figure 1.3C), except that for the β-class enzyme the aspartic acid residue originally coordinated to zinc is proposed to participate in a hydrogen bond with the coordinated bicarbonate (Figure 1.9D). In the last step, the coordinated bicarbonate is released into solution, together with a proton (no details available on this proton transfer process), the aspartate generated recoordinates the Zn(II) ion and the accompanying water molecule forms a hydrogen bond with it. The enzyme is thus ready for another cycle of catalysis.

The structure of the β-CA from the dicotyledonous plant *Pisum sativum* at a 1.93-Å resolution has also been reported (Kimber and Pai 2000). The molecule assembles as an octamer with a novel dimer of dimers of dimers arrangement. The active site is located at the interface of two monomers, with Cys 160, His 220 and Cys 223 binding the catalytic zinc ion and Asp 162 (oriented by Arg 164), Gly 224, Gln 151, Val 184, Phe 179 and Tyr 205 interacting with acetic acid. The substrate-binding groups have a one-to-one correspondence with the functional groups in the α-CA active site, with the corresponding residues closely superimposable by a mirror plane. Therefore, despite differing folds, α- and β-CAs have converged on a very similar active-site design and are likely to share a common mechanism of action (Kimber and Pai 2000).

Cab exists as a dimer with a subunit fold similar to that observed in plant-type β-CAs. The active-site zinc ion was shown to be coordinated by the amino acid residues Cys 32, His 87 and Cys 90, with the tetrahedral coordination completed by a water molecule (Strop et al. 2001). The major difference between plant- and cab-type β-CAs is in the organization of the hydrophobic pocket (except for the zinc coordination). The structure also reveals a HEPES buffer molecule bound 8 Å away from the active-site zinc, which suggests a possible proton transfer pathway from the active site to the solvent (Strop et al. 2001). No structural data are currently available on the binding of inhibitors to this type of CAs, except that acetate coordinates to the Zn(II) ion of the *Pisum sativum* enzyme (Kimber and Pai 2000).

1.2.3 γ-CAs

Cam, the prototype of the γ-class CAs, has been isolated from the methanogenic archaeon *Methanosarcina thermophila* (Iverson et al. 2000). Crystal structures of zinc-containing and cobalt-substituted Cam have been reported in the unbound form and cocrystallized with sulfate or bicarbonate (Iverson et al. 2000).

Several features differentiate Cam from α- and β-CAs. The protein fold is composed of a left-handed β-helix motif interrupted by three protruding loops and followed by short and long α-helices. The Cam monomer self-associates as a homo-trimer with an approximate molecular weight of 70 kDa (Kisker et al. 1996; Iverson et al. 2000). The Zn(II) ion within the active site is coordinated by three histidine residues, as in α-CAs, but relative to the tetrahedral coordination geometry seen at the active site of α-CAs, the active site of this γ-CA contains additional metal-bound water ligands, so that the overall coordination geometry is trigonal-bipyramidal for the zinc-containing Cam and octahedral for the cobalt-substituted enzyme. Two of the His residues coordinating the metal ion belong to one monomer (Monomer A), whereas the third is from the adjacent monomer (Monomer B). Thus, the three active sites are located at the interface of pairs of monomers (Kisker et al. 1996; Iverson et al. 2000). The catalytic mechanism of γ-CAs is proposed to be similar to that of α-class enzymes (see Section 1.2.1; Kisker et al. 1996). Still, the finding that Zn(II) is not tetracoordinated as originally reported (Kisker et al. 1996) but pentacoordi-nated (Iverson et al. 2000), with two water molecules bound to the metal ion, demonstrates that much is still to be understood about these enzymes. At present, the zinc hydroxide mechanism is accepted as being valid for γ-CAs, as it is probable that an equilibrium exists between the trigonal-bipyramidal and the tetrahedral species of the metal ion in the active site of the enzyme.

Ligands bound to the active site have been shown to make contacts with the side chain of Glu 62 in a manner that suggests that this side chain to be probably protonated. In the uncomplexed zinc-containing Cam, the side chains of Glu 62 and Glu 84 appear to share a proton; additionally, Glu 84 exhibits multiple conformations. This suggests that Glu 84 might act as a proton shuttle, which is an important aspect of the reaction mechanism of α-CAs, for which a histidine active-site residue, usually His 64, generally plays this function (see Section 1.2.1). Anions such as bicarbonate or sulfate have been shown to bind to Cam (Iverson et al. 2000), but no information is available on its inhibition by sulfonamides.

1.2.4 CADMIUM CA

X-ray absorption spectroscopy at the Zn K-edge indicates that the active site of the marine diatom *Thalassiosira weissflogii* CA (TWCA1) is strikingly similar to that of mammalian α-CAs. The zinc has three histidine ligands and a single water molecule, being quite different from the β-CAs of higher plants in which zinc is coordinated by two cysteine thiolates, one histidine and a water molecule (Cox et al. 2000). The diatom CA shows no significant sequence similarity with other CAs and probably represent an example of convergent evolution at the molecular level. In the same diatom, a rather perplexing discovery has been made — that of the first cadmium-containing enzyme, which is a CA-type protein (Lane and Morel 2000). Growth of the marine diatom *Thalassiosira weissflogii* under conditions of low zinc, typical of the marine environment, and in the presence of cadmium salts led to increased cellular CA activity, although the levels of TWCA1, the major intracellular Zn-requiring isoform of CA in *T. weissflogii*, remained low (Lane and Morel 2000). [109]Cd labeling comigrates with a protein band that shows this CA activity to be distinct from TWCA1 on native PAGE of the radiolabeled *T. weissflogii* cell lysates. The levels of the Cd protein were modulated by CO_2 in such a manner that they were consistent with the role of this enzyme in carbon acquisition. Purification of the CA-active fraction led to the isolation of a Cd-containing protein of 43 kDa, proving that *T. weissflogii* expresses a Cd-specific CA, which, particularly under conditions of Zn limitation can replace the Zn enzyme TWCA1 in its carbon-concentrating mechanism (Lane and Morel 2000).

1.3 DISTRIBUTION OF CAs

CAs were recently shown to be present in a multitude of prokaryotes, in which these enzymes play important functions, such as respiration, transport of carbon dioxide and photosynthesis (Smith and Ferry 2000). The possibility of developing CA-inhibitor-based antibiotics by inhibiting bacterial CAs present in pathogenic species raised much interest some years ago, with promising results in the use of ethoxzolamide for treating meningitis (Eickhoff and Nelson 1966; Nafi et al. 1990). This type of inhibition has also been exploited for developing selective culture media for other pathogenic bacteria, such as *Branhamella catarrhalis* (Nafi et al. 1990), in the presence of different *Neisseria* species. Some strains of *Pseudomonas, Staphylococcus,*

Streptococcus, *Serratia* and *Proteus* strongly express a gene product that was immunologically related to CA (Nafi et al. 1990). On the other hand, α-, β- and γ-CAs have been purified in many species of bacteria, such as *Neisseria* spp., *E. coli*, *Synechocystis* spp., *Acetobacterium woodi*, *Anabaena variabilis* and *Rhodospirillum rubrum*, but it is established that these enzymes are nearly ubiquitous in prokaryotes (Nafi et al. 1990, Chirica et al. 1997, 2001; Smith and Ferry 2000). Lindskog's group reported the isolation, purification and characterization of some α-CAs from pathogenic bacteria, such as *Helicobacter pylori* and *Neisseria gonorrhoeae* (Chirica et al. 1997, 2001; Elleby et al. 2001). Thus, the CA from *Helicobacter pylori* strain 26695 was toxic to *E. coli*, and therefore a modified form of the gene lacking a part that presumably encodes a cleavable signal peptide has been used for its expression. This truncated gene could be expressed in *E. coli*, yielding an active enzyme comprising 229 amino acid residues, with the amino acid sequence showing 36% identity with that of the enzyme from *N. gonorrhoeae* and 28% with that of hCA II (Elleby et al. 2001). The *H. pylori* CA was purified by sulfonamide affinity chromatography and its kinetic parameters for CO_2 hydration determined. Thiocyanate showed an uncompetitive inhibition pattern at pH 9, indicating that the maximal rate of CO_2 hydration is limited by proton transfer reactions between a zinc-bound water molecule and the reaction medium in analogy to higher-vertebrate α-CAs. The 4-nitrophenyl acetate hydrolase activity of the *H. pylori* enzyme was quite low, whereas the esterase activity against 2-nitrophenyl acetate as substrate was much better (Elleby et al. 2001). The kinetic properties of the CA isolated from *N. gonorrhoeae* (NGCA) as well as for some mutants of such enzymes have also been investigated by the same group (Chirica et al. 2001). Qualitatively, the enzyme shows the same kinetic behavior as that of the well-studied hCA II, suggesting a ping-pong mechanism with buffer as the second substrate. The ratio k_{cat}/K_m is dependent on two ionizations with pK_a values of 6.4 and 8.2, suggesting that His 66 in NGCA has the same function as a proton shuttle as does His 64 in hCA II. The kinetic defect in some NGCA mutants lacking His 66 can partially be overcome by some buffers, e.g., imidazole and 1,2-dimethylimidazole, which act as endogenous activators. The bacterial enzyme shows similar K_i values for the inhibitors cyanate, thiocyanate and azide as does hCA II, whereas cyanide and the sulfonamide ethoxzolamide are considerably weaker inhibitors of the bacterial enzyme than of hCA II (Chirica et al. 2001). Smith and Ferry (2000) recently published an excellent review on prokaryotic CAs.

The recent report on parasitic CAs by Krungkrai et al. (2000), who discovered the presence of at least two different α-CAs in *Plasmodium falciparum*, the malaria-provoking protozoan, opens new vistas to develop pharmacological agents based on CA inhibitors. Red cells infected by *Plasmodium falciparum* contain CA levels ca. twofold higher than those of normal red cells (Krungkrai et al. 2000). The three developmental forms of the asexual stages of the parasite (i.e., ring, trophozoite and schizont) were isolated from their host red cells and found to have stage-dependent CA activity. The enzyme was then purified to homogeneity and shown to have a M_r of 32 kDa, being active in monomeric form. (The human red cell enzyme was also purified for comparison with the parasite enzyme in this study; Krungkrai et al.

2000). The parasite enzyme activity was sensitive to well-known sulfonamide CAIs such as sulfanilamide and acetazolamide. The kinetic properties and the amino-terminal sequences of the purified enzymes from the parasite and host red cells were found to be different, indicating that the purified protein was a distinct enzyme, i.e., *P. falciparum* CA. In addition, the enzyme inhibitors showed antimalarial effect against *in vitro* growth of *P. falciparum*. This very important contribution shows that CAIs might be valuable drugs in the future to treat malaria.

In higher plants, algae and cyanobacteria, all members of the three CA families are present (Moroney et al. 2001). For example, analysis of the *Arabidopsis* database revealed that at least 14 different CAs are present in this plant, and six such enzymes are present in the unicellular green alga *Chlamydomonas reinhardtii* (Moroney et al. 2001). In algae, CAs were found in mitochondria, the chloroplast thylakoid, cyto-plasm and periplasmic space (Moroney et al. 2001; Park et al. 1999; Badger and Price 1994). In C_3 dicotyledons two types of CAs have been isolated, one in the chloroplast stroma and one in cytoplasm, whereas in C_4 plants these enzymes are present in the mesophyll cells, where they provide bicarbonate to phosphoenolpyru-vate (PEP) carboxylase, the first enzyme involved in fixation of CO_2 into C_4 acids (Badger and Price 1994). CAs are quite abundant in CAM (crassulacean acid metab-olism) plants, being probably present in the cytosol and very abundant in chloroplasts, where they participate in CO_2 fixation, providing bicarbonate to PEP carboxylase (Badger and Price 1994). Plant CAs have been exhaustively reviewed by Badger and Price (1994). These enzymes are highly abundant in the terrestrial vegetation and seem to be correlated with the content of atmospheric CO_2 and thus with the global warming processes (Gillon and Yakir 2001).

In animals, and more specifically vertebrates, CAs are widespread. Because this field has recently been reviewed (Parkkila 2000; Supuran et al. 2003) and several chapters of this book deal with CAs present in diverse tissues of the human body (e.g., eyes, Chapter 8; tumor tissues, Chapter 9; gastrointestinal tract, Chapter 10; CNS, Chapter 10; kidneys, Chapter 10; skin, Chapter 11), the reader should consult these chapters for a detailed overview on the distribution and function of CAs.

1.4 PHYSIOLOGICAL FUNCTIONS OF CAs

It is not clear whether the other reactions catalyzed by CAs (Figure 1.1) except for CO_2 hydration/bicarbonate dehydration have physiological relevance (Supuran et al. 2003). Thus, at present, only the reaction in Equation 1.1, Figure 1.1, is considered to be the physiological one in which these enzymes are involved.

In prokaryotes, CAs possess two general functions: (1) transport of CO_2/bicar-bonate between different tissues of the organism and (2) provide CO_2/bicarbonate for enzymatic reactions (Smith and Ferry 2000). In aquatic photosynthetic organ-isms, an additional role is that of a CO_2-concentrating mechanism, which helps overcome CO_2 limitation in the environment (Badger and Price 1994; Park et al. 1999). For example, in *Chlamydomonas reinhardtii*, this CO_2-concentrating mech-anism is maintained by the pH gradient created across the chloroplast thylakoid membranes by Photosystem II-mediated electron transport processes (Park et al. 1999).

Many nonphotosynthesizing prokaryotes catalyze reactions for which CA are expected to provide CO_2/bicarbonate in the vicinity of the active site or to remove such compounds to improve the energetics of the reaction (Smith and Ferry 2000). Smith and Ferry (2000) have reviewed many carboxylation/decarboxylation processes in which prokaryotic CAs might play such an important physiological function.

In higher organisms, including vertebrates, the physiological functions of CAs have been widely investigated over the last 70 years (Maren 1967; Chegwidden and Carter 2000; Supuran et al. 2003). Thus, isozymes I, II and IV are involved in respiration and regulation of the acid/base homeostasis (Maren 1967; Chegwidden and Carter 2000; Supuran et al. 2003). These complex processes involve both the transport of CO_2/bicarbonate between metabolizing tissues and excretion sites (lungs, kidneys), facilitated CO_2 elimination in capillaries and pulmonary microvasculature, elimination of H^+ ions in the renal tubules and collecting ducts, as well as reabsorption of bicarbonate in the brush border and thick ascending Henle loop in kidneys (Maren 1967; Chegwidden and Carter 2000; Supuran et al. 2003). Usually, isozymes I, II and IV are involved in these processes. By producing the bicarbonate-rich aqueous humor secretion (mediated by ciliary processes isozymes CA II and CA IV) within the eye, CAs are involved in vision, and their misfunctioning leads to high intraocular pressure and glaucoma (Maren 1967; Supuran et al. 2003). CA II is also involved in bone development and function, such as differentiation of osteoclasts or providing acid for bone resorption in osteoclasts (Chegwidden and Carter 2000; Supuran et al. 2003). CAs are involved in the secretion of electrolytes in many other tissues and organs, such as CSF formation, by providing bicarbonate and regulating the pH in the choroid plexus (Maren 1967; Supuran et al. 2003); saliva production in acinar and ductal cells (Parkkila 2000); gastric acid production in the stomach parietal cells (Parkkila 2000; see also Chapter 10); and bile production, pancreatic juice production, intestinal ion transport (Parkkila 2000; Maren 1967; see also Chapter 10). CAs are also involved in gustation and olfaction, protecting the gastrointestinal tract from extreme pH conditions (too acidic or too basic), regulating pH and bicarbonate concentration in the seminal fluid, muscle functions and adapting to cellular stress (Chegwidden and Carter 2000; Parkkila 2000; Supuran et al. 2003). Some isozymes, e.g., CA V, are involved in molecular signaling processes, such as insulin secretion signaling in pancreas β cells (Parkkila 2000). Isozymes II and V are involved in important metabolic processes by providing bicarbonate for gluconeogenesis, fatty acids de novo biosynthesis or pyrimidine base synthesis (Chegwidden et al. 2000). Finally, some isozymes (e.g., CA IX, CA XII, CARP VIII) are highly abundant in tumors, being involved in oncogenesis and tumor progression (Pastorek et al. 1994; Chegwidden et al. 2001; Supuran et al. 2003; see also Chapter 9).

Although the physiological function of some isozymes (CA I, CA III, CARPs) is still unclear, the data presented here helps understand the importance of CAs in a host of physiological processes, both in normal and pathological states. This might explain why inhibitors of these enzymes found a place in clinical medicine by as early as 1954, with acetazolamide (1.1) being the first nonmercurial diuretic agent used clinically (Maren 1967). At present, inhibitors of these enzymes are widely used clinically as antiglaucoma agents, diuretics, antiepileptics, to manage mountain sickness and for gastric and duodenal ulcers, neurological disorders, or osteoporosis. The development of more specific agents is required because of the high number of isozymes present in the human body as well as the isolation of many new representatives of CAs from all kingdoms. This is possible only by understanding in detail the catalytic and inhibition mechanisms of these enzymes. These enzymes and their inhibitors are indeed remarkable — after many years of intense research in this field, they continue to offer interesting opportunities to develop novel drugs and new diagnostic tools or to understanding in greater depth the fundamental processes of the life sciences.

REFERENCES

Abbate, F., Casini, A., Scozzafava, A., and Supuran, C.T., (2003) Carbonic anhydrase inhibitors: x-ray crystallographic structure of the adduct of human isozyme II with the perfluorobenzoyl analogue of methazolamide: Implications for the drug design of fluorinated inhibitors. *Journal of Enzyme Inhibition and Medicinal Chemistry*, **18**, 303–308.

Abbate, F., Supuran, C.T., Scozzafava, A., Orioli, P., Stubbs, M.T., and Klebe, G. (2002) Nonaromatic sulfonamide group as an ideal anchor for the design of potent human carbonic anhydrase inhibitors: Role of hydrogen-bonding networks in ligand binding and drug design. *Journal of Medicinal Chemistry* **45**, 3583–3587.

Badger, M.R., and Price, G.D. (1994) The role of carbonic anhydrase in photosynthesis. *Annual Reviews in Plant Physiology and Plant Molecular Biology* **45**, 369–392.

Bertini, I., Luchinat, C., and Scozzafava, A. (1982) Carbonic anhydrase: An insight into the zinc binding site and into the active cavity through metal substitution. *Structure and Bonding* **48**, 45–92.

Briganti, F., Mangani, S., Orioli, P., Scozzafava, A., Vernaglione, G., and Supuran, C.T. (1997) Carbonic anhydrase activators: x-ray crystallographic and spectroscopic investigations for the interaction of isozymes I and II with histamine. *Biochemistry* **36**, 10384–10392.

Briganti, F., Mangani, S., Scozzafava, A., Vernaglione, G., and Supuran, C.T. (1999) Carbonic anhydrase catalyzes cyanamide hydration to urea: Is it mimicking the physiological reaction? *Journal of Biological Inorganic Chemistry* **4**, 528–536.

Briganti, F., Pierattelli, A., Scozzafava, A., and Supuran, C.T. (1996) Carbonic anhydrase inhibitors. Part 37. Novel classes of carbonic anhydrase inhibitors and their interaction with the native and cobalt-substituted enzyme: kinetic and spectroscopic investigations. *European Journal of Medicinal Chemistry* **31**, 1001–1010.

Chegwidden W.R., and Carter, N. (2000) Introduction to the carbonic anhydrases. In *The Carbonic Anhydrases — New Horizons*, Chegwidden W.R., Edwards, Y., and Carter, N., Eds., Birkhäuser Verlag, Basel, pp. 14–28.

Chegwidden W.R., Dodgson, S.J., and Spencer, I.M. (2000) The roles of carbonic anhydrase in metabolism, cell growth and cancer in animals. In *The Carbonic Anhydrases — New Horizons*, Chegwidden W.R., Edwards, Y., and Carter, N., Eds., Birkhäuser Verlag, Basel, pp. 343–364.

Chegwidden, W.R., Spencer, I.M., and Supuran, C.T. (2001) The roles of carbonic anhydrase isozymes in cancer. In *Gene Families: Studies of DNA, RNA, Enzymes and Proteins*, Xue, G., Xue Y., Xu, Z., Holmes, R., Hammond, G.L., Lim, H.A., Eds., World Scientific, Singapore, pp. 157–170.

Chirica, L.C., Elleby, B., Jonsson, B.H., and Lindskog, S. (1997) The complete sequence, expression in *Escherichia coli*, purification and some properties of carbonic anhydrase from *Neisseria gonorrhoeae*. *European Journal of Biochemistry* **244**, 755–760.

Chirica, L.C., Elleby, B., and Lindskog, S. (2001) Cloning, expression and some properties of alpha-carbonic anhydrase from *Helicobacter pylori*. *Biochimica and Biophysica Acta* **1544**, 55–63.

Christianson, D.W., and Fierke, C.A. (1996) Carbonic anhydrase: evolution of the zinc binding site by nature and by design. *Accounts of Chemical Research* **29**, 331–339.

Cox, E.H., McLendon, G.L., Morel, F.M., Lane, T.W., Prince, R.C., Pickering, I.J., and George, G.N. (2000) The active site structure of *Thalassiosira weissflogii* carbonic anhydrase 1. *Biochemistry* **39**, 12128–12130.

Cronk, J.D., Endrizzi, J.A., Cronk, M.R., O'Neill, J.W., and Zhang, K.Y.J. (2001) Crystal structure of *E. coli* β-carbonic anhydrase, an enzyme with an unusual pH-dependent activity. *Protein Science* **10**, 911–922.

Eickhoff, T.C., and Nelson, M.S. (1966) *In vitro* activity of carbonic anhydrase inhibitors against *Neisseria meningitidis*. *Antimicrobial Agents and Chemotherapeutics* **6**, 389–392.

Elleby, B., Chirica, L.C., Tu, C., Zeppezauer, M., and Lindskog, S. (2001) Characterization of carbonic anhydrase from *Neisseria gonorrhoeae*. *European Journal of Biochemistry* **268**, 1613–1619.

Ferraroni, M., Briganti, F., Chegwidden, W.R., Supuran, C.T., Wiebauer, K.E., Tashian, R.E., and Scozzafava, A. (2002a) Structure of the human carbonic anhydrase (hCA) isoenzyme I Michigan 1 variant in the absence and in the presence of exogenous zinc ions: the first CA binding two zinc ions. *Biochemistry* **41**, 6237–6244.

Ferraroni, M., Briganti, F., Chegwidden, W.R., Supuran, C.T., and Scozzafava, A. (2002b) Crystal analysis of aromatic sulfonamide binding to native and (Zn)$_2$ adduct of human carbonic anhydrase I Michigan 1. *Inorganica Chimica Acta* **339**, 135–144.

Gillon, J., and Yakir, D. (2001) Influence of carbonic anhydrase activity in terrestrial vegetation on the ^{18}O content of atmospheric CO_2. *Science* **291**, 2584–2587.

Guerri, A., Briganti, F., Scozzafava, A., Supuran, C.T., and Mangani, S. (2000) Mechanism of cyanamide hydration catalyzed by carbonic anhydrase II suggested by cryogenic x-ray diffraction. *Biochemistry* **39**, 12391–12397.

Hewett-Emmett, D. (2000) Evolution and distribution of the carbonic anhydrase gene families. In *The Carbonic Anhydrases — New Horizons*, Chegwidden, W.R., Edwards, Y., and Carter, N., Eds., Birkhäuser Verlag, Basel, pp. 29–78.

Iverson, T.M., Alber, B.E., Kisker, C., Ferry, J.G., and Rees. D.C. (2000) A closer look at the active site of gamma-class carbonic anhydrases: high-resolution crystallographic studies of the carbonic anhydrase from *Methanosarcina thermophila*. *Biochemistry* **39**, 9222–9231.

Kim, G., Selengut, J., and Levine, R.L. (2000) Carbonic anhydrase III: The phosphatase activity is extrinsic. *Archives of Biochemistry and Biophysics* **377**, 334–340.

Kimber, M.S., and Pai, E.F. (2000) The active site architecture of *Pisum sativum* β-carbonic anhydrase is a mirror image of that of α-carbonic anhydrases. *EMBO Journal* **15**, 2323–2330.

Kisker, C., Schindelin, H., Alber, B.E., Ferry, J.G., and Rees, D.C. (1996) A left-hand beta-helix revealed by the crystal structure of a carbonic anhydrases from the archaeon *Methanosarcina thermophila*. *EMBO Journal* **15**, 2323–2330.

Krungkrai, S.R., Suraveratum, N., Rochanakij, S., and Krungkrai, J. (2001) Characterisation of carbonic anhydrase in *Plasmodium falciparum*. *International Journal of Parasitology* **31**, 661–668.

Lane, T.W., and Morel, F.M. (2000) A biological function for cadmium in marine diatoms. *Proceedings of the National Academy of Science of the United States of America* **97**, 4627–4631.

Lindahl, M., Vidgren, J., Eriksson, E., Habash, J., Harrop, S., Helliwell, J., Liljas, A., Lindeskog, M., and Walker, N. (1991) Crystallographic studies of carbonic anhydrase inhibition. In *Carbonic Anhydrase*, Botrè, F., Gros, G., and Storey, B.T., Eds., VCH, Weinheim, pp. 111–118.

Lindskog, S., and Silverman, D.W. (2000) The catalytic mechanism of mammalian carbonic anhydrases. In *The Carbonic Anhydrases — New Horizons*, Chegwidden W.R., Edwards, Y., and Carter, N., Eds., Birkhäuser Verlag, Basel, pp. 175–196.

Mann, T., and Keilin, D. (1940) Sulphanilamide as a specific carbonic anhydrase inhibitor. *Nature* **146**, 164–165.

Maren, T.H. (1967) Carbonic anhydrase: Chemistry, physiology and inhibition. *Physiological Reviews* **47**, 595–781.

Mitsuhashi, S., Mizushima, T., Yamashita, E., Yamamoto, M., Kumasaka, T., Moriyama, H., Ueki, T., Miyachi, S., and Tsukihara, T. (2000) X-ray structure of beta-carbonic anhydrase from the red alga, *Porphyridium purpureum*, reveals a novel catalytic site for CO_2 hydration. *Journal of Biological Chemistry* **275**, 5521–5526.

Moroney, J.V., Bartlett, S.G., and Samuelsson, G. (2001) Carbonic anhydrases in plants and algae. *Plant, Cell and Environment* **24**, 141–153.

Nafi, B.M., Miles, R.J., Butler, L.O., Carter, N.D., Kelly, C., and Jeffery, S. (1990) Expression of carbonic anhydrase in neisseriae and other heterotrophic bacteria. *Journal of Medical Microbiology* **32**, 1–7.

Park, Y., Karlsson, J., Rojdestvenski, I., Pronina, N., Klimov, V., Oquist, G., and Samuelsson, G. (1999) Role of a novel photosystem II-associated carbonic anhydrase in photosynthetic carbon assimilation in *Chlamydomonas reinhardtii*. *FEBS Letters* **444**, 102–105.

Parkkila, S. (2000) An overview of the distribution and function of carbonic anhydrase in mammals. In *The Carbonic Anhydrases — New Horizons*, Chegwidden W.R., Edwards, Y., and Carter, N., Eds., Birkhäuser Verlag, Basel, pp. 79–93.

Pastorek, J., Pastorekova, S., Callebaut, I., Mornon, J.P., Zelnik, V., Opavsky, R., Zatovicova, M., Liao, S., Portetelle, D., Stanbridge, E.J., Zavada, J., Burny, A., and Kettmann, R. (1994) Cloning and characterization of MN, a human tumor-associated protein with a domain homologous to carbonic anhydrase and a putative helix-loop-helix DNA binding segment. *Oncogene* **9**, 2877–2888.

Scozzafava, A. and Supuran, C.T. (2002) Carbonic anhydrase activators: Human isozyme II is strongly activated by oligopeptides incorporating the carboxyterminal sequence of the biocarbonate anion exchanger AE1. *Bioorganic Medicinal Chemistry Letters* **12**, 1177–1180.

Smith, K.S., and Ferry, J.G. (1999) A plant-type (β-class) carbonic anhydrase in the thermophilic methanoarchaeon *Methanobacterium thermoautotrophicum*. *Journal of Bacteriology* **181**, 6247–6253.

Smith, K.S., and Ferry, J.G. (2000) Prokaryotic carbonic anhydrases. *FEMS Microbiological Reviews* **24**, 335–366.

Stams, T., and Christianson, D.W. (2000) X-ray crystallographic studies of mammalian carbonic anhydrase isozymes. In *The Carbonic Anhydrases — New Horizons*, Chegwidden W.R., Edwards, Y., and Carter, N., Eds., Birkhäuser Verlag, Basel, pp. 159–174.

Stams, T., Nair, S.K., Okuyama, T., Waheed, A., Sly, W.S., and Christianson, D.W. (1996) Crystal structure of the secretory form of membrane-associated human carbonic anhydrase IV at 2.8 A resolution. *Proceedings of the National Academy of Science of the United States of America* **93**, 13589–13594.

Strop, P., Smith, K.S., Iverson, T.M., Ferry, J.G., and Rees, D.C. (2001) Crystal structure of the "cab"-type beta class carbonic anhydrase from the archaeon *Methanobacterium thermoautotrophicum*. *Journal of Biological Chemistry* **276**, 10299–10305.

Supuran, C.T., Conroy, C.W., and Maren, T.H. (1997) Is cyanate a carbonic anhydrase substrate? *Proteins: Structure, Function and Genetics* **27**, 272–278.

Supuran, C.T., and Scozzafava, A. (2000) Carbonic anhydrase inhibitors and their therapeutic potential. *Expert Opinion on Therapeutic Patents* **10**, 575–600.

Supuran, C.T., and Scozzafava, A. (2002) Applications of carbonic anhydrase inhibitors and activators in therapy. *Expert Opinion on Therapeutic Patents* **12**, 217–242.

Supuran, C.T., Scozzafava, A., and Casini, A. (2003) Carbonic anhydrase inhibitors. *Medicinal Research Reviews* **23**, 146–189.

2 Acatalytic CAs: Carbonic Anhydrase-Related Proteins

Isao Nishimori

CONTENTS

2.1 Introduction 26
2.2 Molecular Properties of CA-RPs...26
    2.2.1 cDNA Cloning 26
    2.2.2 Active-Site Residues and Catalytic Activity 28
2.3 Evolutionary Analysis and Subcellular Localization of CA-RPs 30
2.4 Expression of CA-RPs..31
    2.4.1 mRNA Expression 31
    2.4.2 Immunohistochemical Localization in the Brain 32
    2.4.3 Developmental Expression..34
    2.4.4 Oncogenic Expression...35
2.5 Functional Prosection of CA-RPs 35
    2.5.1 Ligand Binding to the CA-RP Domain of RPTPβ 36
    2.5.2 Interaction of CA II with Bicarbonate Transporter..........................37
2.6 Future Direction of Functional Analysis of CA-RPs 38
References..39

In the expanding gene family of carbonic anhydrases (CAs), three isoforms that are evolutionarily well conserved but lack classical CA catalytic activity have been reported and designated as carbonic anhydrase-related proteins (CA-RPs): CA-RPs VIII, X, and XI. These CA-RPs lack one or more histidine residues required to bind the zinc ion, which is essential for CO_2 hydration activity. Homologous CA-like domains without zinc-binding histidine residues have been found in extracellular parts of receptor-type protein tyrosine phosphatases, RPTPβ and RPTPγ. These CA-RPs are consistently expressed in the brain, and CA-RP VIII and RPTPγ are additionally expressed in the peripheral tissues. Similar to active transmembrane CAs IX and XII, CA-RP VIII is reported to be overexpressed in certain types of carcinoma. A CA-RP domain of RPTPβ expressed on the glial cell surface has been shown to bind to a neuronal cell-surface protein, contactin. Further, contactin forms a complex with a transmembrane protein, a contactin-associated protein (Caspr), expressed on unmyelinated axons, suggesting the presence of glial–neuronal cell interactions. Although the exact biological functions of CA-RPs are uncertain, these findings suggest a novel function of the CA-RP subfamily: they possibly bind to some other proteins and function as a protein complex.

2.1 INTRODUCTION

The α-carbonic anhydrases (CAs; EC 4.2.1.1) constitute a family of monomeric zinc metalloenzymes that catalyze the reversible hydration of CO_2 (Sly and Hu 1995). With the progress of the human genome project, the human CA gene family has been increasing in number. To date, 11 human isozymes exhibiting the characteristic enzyme activity of CA have been reported and designated CAs I–IV, VA, VB, VI, VII, IX, XII and XIV (for a review see Hewett-Emmett 2000). These isozymes vary in enzyme activity, subcellular localization and sensitivity to various types of inhibitors (details in Chapter 1). Although several expression sequence tags of CA XIII have been reported in murines (Hewett-Emmett 2000), its human homologue has not been reported. A preliminary study attempting to obtain human CA XIII cDNA by homology probing with several sets of PCR primers matching the murine sequences failed to identify any cDNA fragment for human CA XIII (Nishimori et al., unpublished data).

Along with active CA isozymes, evolutionarily conserved but acatalytic family members have been reported and designated carbonic anhydrase-related proteins (CA-RPs; Tashian et al. 2000). In the CA numbering system (the order of discovery), three isoforms, CA-RPs VIII, X and XI, have been reported (Tashian et al. 2000). CA-RPs lack one or more histidine residues required to bind the zinc ion, which is essential for CO_2 hydration activity, and are thus believed to be inactive as regards classical CA activity (Hewett-Emmett and Tashian 1996). In addition, among a family of protein tyrosine phosphatases (PTPs), two receptor-type protein tyrosine phosphatases, RPTPβ(=PTPξ) and RPTPγ, contain an N-terminal CA-like domain (Barnea et al. 1993b; Levy et al. 1993). Because of the absence of two zinc-binding histidine residues in their CA-like domain sequences, these two phosphatases also have been thought to be acatalytic isoforms. The exact biological function of these CA-RPs and CA-RP domains of RPTPs has not been elucidated (Tashian et al. 2000).

Recently, mRNA expression and immunohistochemical localization of all three CA-RPs have been reported in a panel of human organs and some cancer tissues (Fujikawa-Adachi et al. 1999; Okamoto et al. 2001; Taniuchi et al. 2000; Akisawa et al. 2003; Bellingham et al. 1998). This chapter reviews the molecular properties and tissue distribution of human and mouse CA-RPs VIII, X and XI. In addition, several lines of evidence on the biological function of CA-RPs, their possible biological roles and the future direction of functional analysis are discussed.

2.2 MOLECULAR PROPERTIES OF CA-RPs

2.2.1 cDNA CLONING

Since 1990, following the discovery of the first acatalytic isoform in the CA gene family isolated from a mouse brain cDNA library (Kato 1990), three types of CA-RP

TABLE 2.1
Molecular Properties of Human and Mouse CA-RPs

CA-RP Isoforms	CA-RP VIII		CA-RP X		CA-RP XI	
	Human	Mouse	Human	Mouse	Human	Mouse
cDNA (base)[a]	2191	3710	2926	2827	1488	1,390
Accession no.[a]	AK090655	XM_204143	AF288385	AB080741	AF067662	NM_009800
mRNA (kb)	Multiple[b] (2.4)	Multiple[b] (2.4)	2.8	2.6/3.0	1.5	1.5
Amino acid residues	290	291	328	328	328	328
Expected MW (kDa)	33.0	33.1[c]	37.6	37.6	36.2	36.1
Chromosome localization	8q11-q12	4	17q24	11	19q13.3	7

[a] The longest cDNA clone deposited in GenBank is shown with the accession numbers.
[b] Among the five transcripts of different sizes, the strongest signal was observed at 2.4 kb.
[c] Molecular size demonstrated by SDS-PAGE is 37.5 kDa [Bergenhem, N.C.H. et al. (1998) *Biochimica et Biophysica Acta* **1384**, 294–298.

Source: Adapted from Kato, K. (1990) *FEBS Letters* **271**, 137–140; Skaggs, L.A. et al. (1993) *Gene* **126**, 291–292; Okamoto, N. et al. (2001) *Biochimica et Biophysica Acta* **1518**, 311–316; Fujikawa-Adachi, K. et al. (1999) *Biochimica et Biophysica Acta* **1431**, 518–524; Taniuchi, K. et al. (2002) *Brain Research: Molecular Brain Research*, **109**, 207–215; Akisawa, Y. et al. (2003) *Virchows Archives,* **442**, 66–70; Bellingham, J. et al. (1998) *Biochemical and Biophysical Research Communications* **253**, 364–367. With permission.

cDNAs (CA-RPs VIII, X, and XI) have been reported in humans and mice (Table 2.1). Subsequently, orthoforms of these CA-RPs have been reported in a number of species, including *R. norvegicus, A. thaliana, C. elegans* and *D. melanogaster* (UniGene Cluster accession number Hs.250502 in the National Center for Biotechnology Information).

Mouse CA-RP VIII cDNA encodes 291 amino acid residues and the human homologue is one residue shorter than that of the mouse (Skaggs et al. 1993). This difference was observed in the repeat number of glutamic acid residues at the N-termini (seven repeats in mice and six in humans), and thus it might not result in a distinct difference in molecular nature between humans and mice. In both humans and mice, cDNAs of CA-RPs X and XI encode proteins with identical lengths of 328 residues (Fujikawa-Adachi et al. 1999; Okamoto et al. 2001; Taniuchi et al. 2002; Bellingham et al. 1998). Surprisingly, all three CA-RPs show very high homology in humans and mice, with 89.2 to 94.5% homology at the cDNA level and more than 97% homology at the amino acid level (Table 2.2). Although the exact function of CA-RPs is uncertain, this extremely high evolutional conservation suggests a pivotal biological function of these proteins.

The human CA-RP X cDNA contains a CCG repeat sequence in the 5′-untranslated region (Okamoto et al. 2001). This repeat sequence has also been indicated from screening analysis for genes containing CCG repeats in the human brain (Kleiderlein et al. 1998). Polymorphism in the repeat length exists in the human

TABLE 2.2
Homologies in cDNA and Amino Acid
Sequences between Human and Mouse CA-RPs

CA-RP Isoforms	cDNA (%)	Amino Acid (%)
CA-RP VIII	89.2	97.9
CA-RP X	94.5	100.0
CA-RP XI	90.3	97.0

Note: Full-length coding cDNA sequences and deduced amino
acid sequences (position 1-259) analyzed.

population: 15% for seven repeats, 5% for eight and 80% for nine (Kleiderlein et al.
1998). Four CCG repeats (actually a GCC repeat) are observed at equivalent positions of mouse CA-RP X (Okamoto et al. 2001), whereas other CA-RPs have neither
CCG repeats nor any other triplet repeats in their cDNA sequences. Expansion
mutations of trinucleotide repeats have been reported in various neuropsychiatric
disorders (Kleiderlein et al. 1998). Together with the fact that CA-RP X expression
is relatively limited to the brain, these findings prompt one to consider the possibility
that CA-RP X is a causative gene in some type of neurological disorder. However,
gene mutation in the CCG repeat length of CA-RP X has not been reported in any
pathological condition.

Expected molecular weights based on the cDNA sequences are ca. 33 kDa for
both human and mouse CA-RP VIII, 37.6 kDa for CA-RP X and 36 kDa for CA-RP
XI. Among these isozymes, only human CA-RP VIII has been experimentally studied
for the molecular size of the recombinant protein generated in the *Escherichia coli*
system: SDS-PAGE shows 37,506 g mol^{-1} and gel filtration indicates 46,919 g mol^{-1}
(Bergenhem et al. 1998). The higher molecular weight determined by gel filtration
is probably due to a stretch of acidic residues in the N-terminus of CA-RP VIII
(Bergenhem et al. 1998).

2.2.2 ACTIVE-SITE RESIDUES AND CATALYTIC ACTIVITY

Figure 2.1 shows the 36 active-site residues previously defined in catalytic CA
isozymes (Tashian 1992). All human and mouse CA-RPs are aligned with the human
CA II sequence. Seventeen residues that form an active-site hydrogen bond network
are indicated by a combination of plus signs and Zs, and three zinc-binding histidine
residues are indicated by Zs above the CA II sequence. For an easier understanding,
several amino acid residues are numbered according to the commonly used CA I
numbering system. Interestingly, all 36 active-site residues in each CA-RP are
conserved between human and mouse, except for one residue at Position 131, which
is phenylalanine in human CA-RP XI, the same as in CA II, but substituted with
leucine in mouse CA-RP XI (open box). Although 9 of 17 hydrogen bond network
residues are conserved in all CA-RPs, one or more of three zinc-binding histidine

```
                    92 94 96 119 121 131  143      200
                     \ | /  \ | /  /     |
         ++    +    +   +ZZ+++Z / /  +  ++       +  ++

hCA   II         YSNNHAFNEIQHHEHEHVFLVGWYLTTPPLCVWVNR


hCA--RP VIII     --TD-TIQYYER-----IIIIA----I--SG--L--
mCA-RP VIII      --TD-TIQYYER-----IIIIA----I--SG--L--


hCA-RP X         ---TRHVSREER----QIV--SI-M-I--YTA-I--
mCA-RP X         ---TRHVSREER----QIV--SI-M-I--YTA-I--


hCA-RP XI        ---TRHVSLSERL---QI--ISI--S---ST--L--
mCA-RP XI        ---TRHVSLSERL---QIL-ISI--S---ST--L--
```

FIGURE 2.1 An alignment of 36 active-site residues of human (h) and mouse (m) CA-RPs and human CA II. Seventeen residues that form an active-site hydrogen-bond network are indicated by a mixture of plus signs and Zs, and three zinc-binding histidine residues critical for biological activity of the catalytic CA are indicated by Zs. Several amino acid residues are numbered according to the CA I numbering system. All active-site residues of each CA-RP are conserved between human and mouse, except one residue at position 131 of CA-RP XI (open box), which is phenylalanine in human CA-RP XI, the same as in CA II, but is substituted with leucine in mouse CA-RP XI (open box). Note that CA-RPs VIII, X and XI lack one or more of the zinc-binding histidine residues (closed boxes).

residues are substituted with other amino acids. His 94 is replaced with arginine in all three CA-RPs, His 96 with leucine in CA-RP XI and His 119 with glutamine in CA-RPs X and XI. Replacements of His 94 and His 119 have been reported in the CA-RP domains of RPTPβ and RPTPδ also (Barnea et al. 1993b; Krueger et al. 1992; Levy et al. 1993; for a review see Tashian et al. 2000).

Among the CA-RP isoforms, only CA-RP VIII has been intensively studied for its catalytic activity. Recombinant proteins of human and mouse CA-RP VIII have been reported to lack not only CO_2 hydration activity but also esterase activity toward p-nitrophenyl acetate (Sjblom et al. 1996; Elleby et al. 2000). A single mutagenesis at Position 94 from arginine to histidine, similar to that for active CA isozymes, restores its CO_2 hydration activity (Elleby et al. 2000). Further mutations leading to a more CA-like active-site cavity (quintuple mutations; His-94, Gln-92, Val-121, Val-143, and Thr-200; see Figure 2.1) result in a greater increase in activity (Elleby et al. 2000). Replacement of Ile-121 or Ile-143 with a valine residue results in measurable esterase activity and introducing Thr-200 seems to stabilize the protein (Elleby et al. 2000). There is no experimental data on catalytic activity of the other two CA-RPs and the CA-RP domains of the two RPTPs. However, CA-RP VIII has no catalytic activity because of the absence of only one zinc-binding histidine residue, and other CA-RPs and CA-RP domains of RPTPs lack at least two critical histidine residues, suggesting that none of the CA-RP isoforms have catalytic activity.

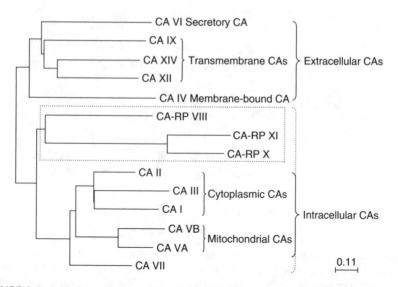

FIGURE 2.2 A phylogenetic tree of the carbonic anhydrase gene family. Full-length coding sequences of all 14 CAs and CA-RPs were translated and the resultant polypeptides (position 1-259) were analyzed. [Modified from Okamoto, N. et al. (2001) *Biochimica et Biophysica Acta* **1518**, 311–316.]

2.3 EVOLUTIONARY ANALYSIS AND SUBCELLULAR LOCALIZATION OF CA-RPs

Based on an evolutionary analysis (see phylogenetic tree, Figure 2.2), CA isozymes are subdivided into two branches: intracellular and extracellular CAs (Okamoto et al. 2001). Intracellular CAs include cytoplasmic isozymes (CA I, II and III) and mito-chondrial isozymes (CAs VA and VB). CA VII is also considered an intracellular isozyme (Montgomery et al. 1991), but its subcellular localization is undefined. Extracellular CAs include a membrane-anchored isozyme (CA IV), a secretory isozyme (CA VI) and transmembrane isozymes (CAs IX, XII and XIV). Interest-ingly, three CA-RPs form a single divergent branch that shares the same trunk as intracellular CA isozymes do, suggesting intracellular localization of CA-RPs.

Previous immunohistochemical analysis of CA-RPs expression in the human and murine central nervous system has identified their cytoplasmic localization (Taniuchi et al. 2000, 2002). Preliminary studies with confocal fluorescence micro-scopy also show cytoplasmic signals in COS-7 cells expressing a fusion protein of CA-RP VIII or X and green fluorescent protein (Nishimori et al., unpublished data). As compared with cytoplasmic CA isozymes, all human CA-RPs have N-terminal additional sequences more than 20 residues long (Fujikawa-Adachi et al. 1999; Okamoto et al. 2001; Taniuchi et al. 2002; Bellingham et al. 1998). However, there is no known motif sequence for translocation signals. Although details on the subcellular localization of CA-RPs are unclear, these findings indicate that CA-RPs are cytoplasmic proteins.

TABLE 2.3
mRNA Expression of Human and Mouse CA-RPs in Vital Organs by Northern Blot Analysis

CA-RP Isoforms	CA-RP VIII (Human[a]/Mouse[b])	CA-RP X (Human[c]/Mouse[b])	CA-RP XI (Human[d]/Mouse[b])
Heart	+/+	–/–	–/+
Brain	+/+++	+++/++	+++/++
Placenta	+/nt[e]	–[f]/nt	–/nt
Lung	–/+++	–/–	–/–
Liver	–/+++	–/–	–/–
Skeletal muscle	+/+	–/–	–/–
Kidney	++/+	+/–	–/–
Pancreas	+/nt	–/nt	–/nt
Spleen	nt (–)[g]/–	nt (–)/–	nt (–)/–
Testis	nt (+)/–	nt (–)/–	–/–

[a] Adapted from Akisawa, Y. et al. (2003) *Virchows Archives*, **442**, 66–70.

[b] Adapted from Taniuchi, K. et al. (2003) *Brain Research: Molecular Brain Research*, **109**, 207–215.

[c] Okamoto, N. et al. (2001) *Biochimica et Biophysica Acta* **1518**, 311–316.

[d] Adapted from Bellingham, J. et al. (1998) *Biochemical and Biophysical Research Communications* **253**, 364–367 and Fujikawa-Adachi, K. et al. (1999) *Biochimica et Biophysica Acta* **1431**, 518–524.

[e] nt, not tested.

[f] As compared to a normal message size for CA-RP X (2.8 kb), a shorter transcript was seen in the placenta (~2.0 kb) on Northern blot [Okamoto, N. et al. (2001) *Biochimica et Biophysica Acta* **1518**, 311–316].

[g] In parentheses are shown results from mRNA dot blot hybridization [Okamoto, N. et al. (2001) *Biochimica et Biophysica Acta* **1518**, 311–316].

2.4 EXPRESSION OF CA-RPs

2.4.1 mRNA Expression

mRNA expressions of CA-RPs VIII, X and XI have been studied in a panel of human and mouse vital organs by Northern blot analyses (Fujikawa-Adachi et al. 1999; Okamoto et al. 2001; Taniuchi et al. 2002; Akisawa et al. 2003; Bellingham et al. 1998), and the results are summarized in Table 2.3. CA-RPs X and XI show a similar expression pattern in humans and mice, with intense signals being observed in the brain. Notable differences in humans and mice are observed in the heart and kidney. Human kidney shows CA-RP X expression but mouse kidney does not, whereas mouse heart shows a positive signal for CA-RP XI but human heart does not. However, the message levels detected in these organs are much weaker than those in the brain. In this regard, mRNA expressions of CA-RPs X and XI appear to be relatively specific to the brain.

Northern blot analysis shows single transcripts of human CA-RP X (2.8 kb; Okamoto et al. 2001), human CA-RP XI (1.5 kb; Fujikawa-Adachi et al. 1999; Bellingham et al. 1998) and mouse CA-RP XI (1.5 kb; Taniuchi et al. 2003), all of which corresponded to the apparently full-length cDNAs (Table 2.1). For murine CA-RP X mRNA expression, however, two sizes of transcripts (2.6 and 3.0 kb), possibly generated by alternative splicing, are observed (Okamoto et al. 2001). Although a number of expression sequence tags of CA-RP X from the human placenta have been deposited in GenBank (Hewett-Emmett 2000), only a shorter transcript (~2.0 kb) is detected on the Northern blot, and the molecular nature of this shorter transcript remains uncertain at present (Okamoto et al. 2001).

In contrast to CA-RPs X and XI, there is much more variety in the transcript size and distribution of CA-RP VIII mRNA expression (Taniuchi et al. 2002; Akisawa et al. 2003). Both human and mouse CA-RPs VIII show broad distribution in a panel of vital organs (Table 2.3). A distinct difference between humans and mice is observed in the lung and liver: there is no detectable signal for a CA-RP VIII message in the human liver and lung, whereas mouse lung and liver as well as the brain show the strongest signals. There is no acceptable explanation for this discrepancy other than species differences.

The blots for mRNA expression of both human and mouse CA-RPs VIII show multiple transcripts: five transcripts with different sizes (2.4, 3.0, 4.0, 4.5 and 7.0 kb) for human CA-RP VIII (Akisawa et al. 2003) and six transcripts (1.4, 1.6, 2.4, 3.4, 3.8 and 4.9 kb) for mouse CA-RP VIII (Taniuchi et al. 2002). Among these transcripts, the 2.4-kb band shows the strongest intensity and is commonly observed in organs that show positive signals. Kato has also reported multiple transcripts of mouse CA-RP VIII (Kato 1990). The different-sized RNAs might be generated by alternative splicing or by the multiple poly(A) signals found in CA-RP VIII mRNA (Akisawa et al. 2003). These findings indicate that CA-RP VIII is diversely expressed not only in the brain but also in various peripheral tissues.

2.4.2 IMMUNOHISTOCHEMICAL LOCALIZATION IN THE BRAIN

Taniuchi et al. (2000, 2002) have produced monoclonal antibodies to all three human CA-RPs that are cross-reactive with mouse homologues and have studied their regional and cellular distribution in the brain by immunohistochemical analysis. Table 2.4 summarizes their results. CA-RPs VIII and XI are consistently expressed in neural cells, astrocytes and neurites of most parts of human and mouse brains (Figure 2.3, left). Lakkis et al. (1997b) have also shown mRNA expression of mouse CA-RP VIII in cerebral neurons and Purkinje cells by an *in situ* hybridization method. Epithelial cells of the choroid plexus and pia arachnoid also express CA-RPs VIII and XI.

It is of great interest that CA-RP X is expressed in the myelin sheath (Figure 2.3, right). This expression has been confirmed by using brain tissue sections under two unique pathological conditions. One is acute disseminated encephalomyelitis, a disease in which there is focal demyelinization in the human brain. In a demyelinized lesion of a patient's brain, the axons were clearly shown by lucsol fast blue stain,

TABLE 2.4
Immunohistochemical Localization of CA-RPs in Human and Mouse Brains

Region	Specific Cells	CA-RP VIII Human/Mouse	CA-RP X Human/Mouse	CA-RP XI Human/Mouse
Cerebrum	Neural cells in the cortex and medulla	+/+	–/+	+/+
	Oligodendrocytes	–/–	–/–	–/–
	Myelin sheath in the medulla	–/–	+/+	–/–
	Neural cells in the basal ganglia	+/+	–/+	+/+
Brain stem	Neural cells	+/+	+/+	+/+
	Glial cells	–/–	–/–	–/–
	Myelin sheath	–/–	+/+	–/–
Cerebellum	Neural cells in the molecular layer	+/+	–/+^{w}	–/+
	Neural cells in the granular layer	–/–	–/–	–/–
	Purkinje cells	+/+	+^{w}/+	+/+
	Neural cells in the dental nuclei	+^{w}/+	–/+	+^{w}/+
Epithelial cells in the choroid plexus		+/+	+^{w}/+^{w}	+/+
Pia arachnoid		+/+	–/+^{w}	+/+

Note: +, positive expression; +^{w}, weak expression; –, no significant expression. Previous reports of CA-RP expression in the human brain [Taniuchi, K. et al. (2000) *Neuroscience* **112**, 93–99] and the mouse brain [Taniuchi, K. et al. (2002) *Brain Research: Molecular Brain Research,* **109**, 207–215] are summarized.

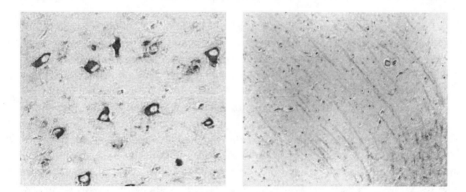

FIGURE 2.3 Immunohistochemical analysis of CA-RPs expression by using monoclonal antibodies to human CA-RPs VIII and X [Taniuchi, K. et al. (2000) *Neuroscience* **112**, 93–99]. CA-RP VIII is expressed in the cytoplasm of neural cells (left; cerebral cortex of the frontal lobe). CA-RP X is expressed in radial myelin fibers of the basal ganglia (right).

but CA-RP X expression was selectively lost (Taniuchi et al. 2000). The other is the shiverer demyelinated model mouse. In this mutant mouse, exons 3 to 7 of the myelin basic protein gene are deleted and thus myelinization is left incomplete (Mikoshiba et al. 1995). In the brain of the shiverer mouse, CA-RP X expression is almost eliminated from the myelin sheath (Taniuchi et al. 2002).

There is a notable difference in CA-RP X expression in humans and mice. In the human brain, neural cells in the cerebrum show no signal for CA-RP X expression, with only Purkinje cells and neurons in the olivary nuclei being weakly stained (Taniuchi et al. 2000). In contrast, in most parts of the murine brain, CA-RP X is expressed not only in the myelin sheath but also in the neural cell body (Taniuchi et al. 2002). This discrepancy might be caused by the difference in the postmortem time until fixation of the brain tissue. Mice brains were experimentally obtained by transcardial perfusion with 4% paraformaldehyde immediately after sacrifice, whereas human brains were obtained from necrosectomy cases in which postmortem degeneration had possibly occurred. Human CA-RP X probably disappeared from most neural cells during tissue fixation and partially remained in the Purkinje cells and neurons in the olivary nuclei. Taken together, these results suggest that CA-RP X is expressed in both the myelin sheath and neural cells.

2.4.3 Developmental Expression

Taniuchi et al. (2003) have reported mRNA expression of all three CA-RPs in whole mouse embryos at four gestational days (Days 7, 11, 15 and 17). RT-PCR and Northern blot analyses showed that CA-RPs VIII and X appeared in the middle of gestation (Day 11 by RT-PCR and Day 15 by Northern blot) and were developmentally expressed, whereas murine CA-RP XI expression was seen at the early gestational stage (Day 7 by both RT-PCR and Northern blot) and gradually decreased with the progress of gestation. Lakkis et al. (1997a) have reported that CA-RP VIII mRNA is expressed during embryonic development in the liver, lung, heart, gut and thymus.

Developmental expression of CA-RPs has also been immunohistochemically studied in the brain. Human fetal brains at five gestational periods (Days 84, 95, 121, 141 and 222) were employed for the immunohistochemical analysis (Taniuchi et al. 2000). CA-RPs VIII and XI were expressed in the neuroprogenitor cells in the subventricular zone in the early term of gestation (as early as Day 84) and subsequently detected in the neural cells migrating to the cortex on Day 95. In the epithelial cells of the choroid plexus, CA-RPs VIII and XI appeared on Day 95. Meanwhile, CA-RP X first appeared in the neural cells in the cortex on Day 141. In the mouse fetal brain, CA-RPs were expressed in the neuroprogenitor cells in the subventricular area on Day 10 of gestation and were subsequently detected in neural cells migrating to the cortex on Day 16 (Taniuchi et al. 2002). Interestingly, in the lucher mouse, which is a mutant strain with an autosomal dominant inherited disorder causing locomotor ataxia (Caddy and Biscoe 1975), CA-RP VIII has been reported to postnatally disappear from the cerebellum (Nogradi et al. 1997). Although the exact function in the developing brain is uncertain, these findings suggest certain roles of CA-RPs in the early development or differentiation of neuroprogenitor cells.

2.4.4 ONCOGENIC EXPRESSION

In 1994, a tumor-associated protein MN was found to be highly expressed in cervical tumors (Pastorekova et al. 1992; Zavada et al. 1993; Liao et al. 1994) and reported to have a CA domain and also CA activity (designated CA IX/MN; Pastorek et al. 1994; Opavsky et al. 1996). Since this discovery, CA IX expression has been immunohistochemically analyzed in various kinds of normal and tumor tissues (see details in Chapter 9). CA IX is expressed in a limited number of cell types in normal tissues, such as epithelial cells in the upper gastrointestinal tract, biliary tract, testis, ovary, and choroid plexus (Pastorekova et al. 1997; Ivanov et al. 2001). In contrast, markedly upregulated expression of CA IX has been reported in carcinoma cells in a number of organs, including the cervix, ovary, breast, kidney, gastrointestinal tract, biliary tract, pancreas and skin (Liao et al. 1994, 1997; Ivanov et al. 2001; Saarnio et al. 1998; Kivela et al. 2000b, 2001; Chia et al. 2001; Beasley et al. 2001; Wykoff et al. 2001). Subsequently, another transmembrane isozyme, CA XII, has also been reported to be overexpressed in certain tumors, including uterine carcinoma, breast cancer, renal cell tumors, gastrointestinal carcinoma and nonsmall cell carcinoma of the lung (Ivanov et al. 2001; Kivela et al. 2000a, 2001; Parkkila et al. 2000). It has been hypothesized that these transmembrane CA isozymes might contribute to the tumor microenvironment by maintaining extracellular acidic pH and thus help cancer cells grow and metastasize (Sly 2000).

Although CA-RPs have no CO_2 hydration activity, thus invalidating the previous hypothesis as applied to them, CA-RP VIII has recently been reported to be overexpressed in certain tumors, including nonsmall cell lung cancer and colorectal cancer (Akisawa et al. 2003). In normal counterparts of these two organs, CA-RP VIII expression is limited to bronchial ciliated cells and cryptal proliferating cells of the colorectal epithelium, respectively. Among a panel of nonsmall lung cancers (24 squamous cell carcinomas, 6 adenosquamous cell carcinomas and 25 adenocarcinomas), all except one case of squamous cell carcinoma showed a positive signal for CA-RP VIII expression by immunohistochemical analysis. Interestingly, the positive signals were observed in cancer cells at the front of tumor progression. Similar findings have been obtained in colorectal carcinoma (Nishimori et al., unpublished data). Furthermore, CA-RP VIII mRNA is significantly expressed in developing human lungs, and, to a much lesser extent, in normal adult lungs. These findings suggest that CA-RP VIII plays some role in cell proliferation and carcinogenesis of lung and colorectal epithelial cells.

2.5 FUNCTIONAL PROSECUTION OF CA-RPs

Several lines of evidence suggest a pivotal role of CA-RPs in cellular function: (1) amino acid sequences show high evolutional conservation (Table 2.2); (2) the secondary structure of CA-RP VIII is similar to that of CA II, suggesting that the binding interface is conserved (Bergenhem et al. 1998); and (3) replacement of an arginine residue at position 94 with histidine (Figure 2.1), which reconstructs the zinc-binding site, restores the CO_2 hydration activity of mouse CA-RP VIII (Elleby et al. 2000). These findings suggest that CA-RPs might be involved in protein– protein interactions, and, in the case of RPTPs, that CA-RP domains might function as ligand-binding domains. Although informative studies on functional properties of CA-RPs are very limited, several important findings have been reported.

2.5.1   LIGAND BINDING TO THE CA-RP DOMAIN OF RPTPβ

The most direct evidences for CA-RP function have come from the studies of RPTPs, which are type I transmembrane proteins and belong to the PTP gene family (Bixby 2001). Among the more than 30 known RPTPs, a CA-like domain has been identified in RPTPβ (= PTPξ) and RPTPγ. The extracellular part of these RPTPs possess, from the N-termini, a CA-RP domain, a fibronectin Type III (FN III) domain and a long cysteine-free region (spacer domain; Figure 2.4A; Barnea et al. 1993b; Krueger and Saito 1992; Levy et al. 1993). The intracellular part contains two catalytic domains with tyrosine phosphatase motifs (D1 and D2). Three major isoforms of RPTPβ generated from an identical gene by alternative splicing have been reported (Barnea et al. 1993a, 1994). The two transmembrane forms differ by the absence of 860 residues from the spacer domain, and one secreted form loses the transmembrane domain and a cytoplasmic tail (Figure 2.4A). The latter soluble form has been identified in rat brain as a chondroitin sulfate proteoglycan called phosphacan (Maurel et al. 1994). In addition, an alternatively spliced miniexon that encodes a seven-amino-acid residue in the juxtamembrane domain (arrows in Figure 2.4A) has been reported in both long and short transmembrane forms (Li et al. 1998). These various structural types might result in a multifunctional protein capable of interacting with other proteins under various conditions. The expression of RPTPβ is restricted to the nervous system, where it is mainly found in a glial precursor, radial glia and astrocytes (Canoll et al. 1993) as well as in certain neurons (Snyder et al. 1996), whereas RPTPγ is expressed both in the developing nervous system and in a variety of peripheral tissues in the adult rat (Barnea et al. 1993b; Canoll et al. 1993).

In 1995, a CA-RP domain of RPTPβ was reported to bind specifically to a 140-kDa protein, contactin, which is a glycosylphosphatidyl inositol (GPI)-anchored protein expressed on the surface of neuronal cells (Figure 2.4B; Peles et al. 1995). Contactin has been shown to form a *cis* complex with a 190-kDa transmembrane protein, a contactin-associated protein (Caspr; Peles et al. 1997). The cytoplasmic domain of Caspr contains a proline-rich sequence capable of binding to a subclass of SH3 domains of signaling molecules. Caspr is expressed diffusely on unmyelinated axons but becomes localized to the paranodal junctions shortly after the onset of myelination (Einheber et al. 1997), and is thus thought to promote neurite outgrowth and fasciculation and also possibly synapse formation and maintenance (Faivre-Sarrailh and Rougon 1997; Berglund et al. 1999). As expected from their molecular properties, both RPTPβ and contactin can exist as a soluble form. Along with direct glial–neuronal cell interactions, they can form a complex with each other on one side of the cell surface (Figure 2.4B). They can also transmit cytoplasmic signals and thus lead to bidirectional signals between neurons and glial cells (Peles et al. 1995). Furthermore, contactin has been shown to interact with the neuronal recognition molecules Ng-CAM and Nr-CAM (Brummendorf et al. 1993; Morales et al. 1993) and the extracellular matrix protein tenascin (Zisch et al. 1992). These

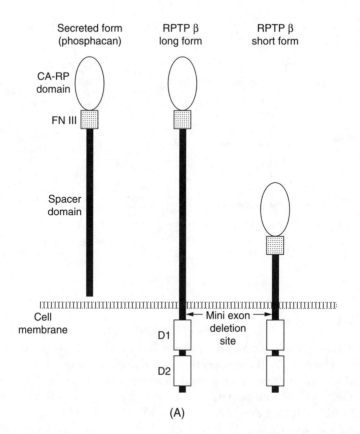

(A)

FIGURE 2.4 (A) Schematic presentation of the molecular structure of the transmembrane protein tyrosine phosphatase (RPTP). The two transmembrane forms differ by the absence of 860 residues from the spacer domain. A mini exon encoding a seven amino acid residue in the juxtamembrane domain is alternatively spliced (arrow). CA-RP, carbonic anhydrase-related protein; FN III, fibronectin type III; D1 and D2, tyrosine phosphatase domain. (B) Proposed model of the interaction among transmembrane tyrosine phosphatase (RPTPβ), contactin and contactin-associated protein (Caspr). Along with direct glial–neuronal cell interactions, RPTPβ and contactin can form a complex with each other on one side of the cell surface. GPI, glycosylphosphatidyl inositol; PI-PLC, phosphatidyl inositol-specific phospholipase C. [Based on (A) Barnea, G. et al. (1993a,b) *Cell* **46**, 205 and Barnea, G. et al. (1993b) *Molecular and Cellular Biology* **13**, 1497–1506; and (B) Peles, E. et al. (1997) *EMBO Journal* **16**, 978–988.]

findings indicate that CA-RP has ligand-binding ability and plays an important role in cell-to-cell communication between glial cells and neurons during development.

2.5.2 INTERACTION OF CA II WITH BICARBONATE TRANSPORTER

It has been reported that CA II binds to the C-terminus of a plasma membrane chloride/bicarbonate anion exchanger, AE1, and increases the rate of bicarbonate transport. This has been proved as follows: (1) the CA inhibitor acetazolamide causes

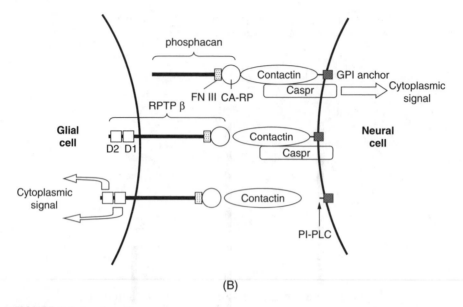

(B)

FIGURE 2.4 (continued).

a major decrease in the bicarbonate transport rate by AE1; (2) mutant AE1 lacking the ability to bind to CA II decreases transport activity as compared with wild-type AE1; and (3) overexpression of a catalytically inactive form of CA II exerts a dominant negative effect on the wild-type CA II, resulting in a loss of anion-exchange activity by AE1 (Sterling et al. 2001). Similar to the CA II–AE1 interaction, CA II has also been shown to bind and function with another type of bicarbonate transporter, the sodium bicarbonate cotransporter (kNBC1; Gross et al. 2002). The consensus C-terminal sequence in these bicarbonate transporters, Asp–Ala/Asn–Asp–Asp, binds the basic sequence at the N-terminus of CA II (Gross et al. 2002; Vince and Reithmeier 2000; Vince et al. 2000). Furthermore, five histidine residues found in the N-terminal 17 residues of CA II (underlined, MSHHWGYGKHNGPEHWH) are thought to be important for the binding to AE1 (Vince et al. 2000). Probably because of the absence of these histidine residues, human CA I fails to bind with AE1. In an alignment of known mammalian CA II sequences, interestingly, histidine residues at Positions 3, 4 and 17 are completely conserved, suggesting that these three histidine residues are more critical than the other two. Because none of the CA-RPs have histidine residues at the equivalent positions, CA-RPs seem to have no capacity to bind to the bicarbonate transporter family. However, the evidence of the interaction between CA II and membrane proteins such as certain types of bicarbonate transporters suggests a novel function of CA-RPs that can be elucidated in the future.

2.6 FUTURE DIRECTION OF FUNCTIONAL ANALYSIS OF CA-RPs

The molecular properties and expression of CA-RPs have gradually been elucidated. However, almost nothing is yet known about their biological function. The findings at present indicate that CA-RPs bind to some other proteins and function as a protein complex. Earlier efforts to obtain a putative ligand for human CA-RP VIII by using an affinity column containing a fusion protein with glutathione-*S*-transferase (GST) have failed to identify any specific ligand (Bergenhem et al. 2000). If the binding capacity is not very strong, a more sensitive method such as the yeast two-hybrid system should be employed. To study the biological role of CA-RPs in carcinogenesis, comparative screening analysis with DNA microarrays containing a panel of cancer-related genes between wild-type tumor cells and CA-RP transfectants can give informative findings. Furthermore, a gene knockout of CA-RPs in a mouse model can provide more direct evidence of the significance of CA-RPs as functional proteins. Because of the evolutionary conservation (Table 2.2) and strong expression, especially in the brain, of CA-RPs and CA-RP domains of RPTPs, there is no doubt that they have fundamental biological functions. Further functional analysis of CA-RPs can contribute to the understanding of various biological processes.

REFERENCES

Akisawa, Y., Nishimori, I., Taniuchi, K., Okamoto, N., Takeuchi, T., Sonobe, H. et al. (2003) Expression of carbonic anhydrase-related protein CA-RP VIII in non-small lung cancer. *Virchows Archives*, **442**, 66–70.

Barnea, G., Grumet, M., Milev, P., Silvennoinen, O., Levy, J.B., Sap, J., and Schlessinger, J. (1994) Receptor tyrosine phosphatase beta is expressed in the form of proteoglycan and binds to the extracellular matrix protein tenascin. *Journal of Biological Chemistry* **269**, 14349–14352.

Barnea, G., Grumet, M., Sap, J., Margolis, R.U., and Schlessinger, J. (1993a) Close similarity between receptor-linked tyrosine phosphatase and rat brain proteoglycan. *Cell* **46**, 205.

Barnea, G., Silvennoinen, O., Shaanan, B., Honegger, A.M., Canoll, P.D., D'Eustachio, P. et al. (1993b) Identification of a carbonic anhydrase-like domain in the extracellular region of RPTP gamma defines a new subfamily of receptor tyrosine phosphatases. *Molecular and Cellular Biology* **13**, 1497–1506.

Beasley, N.J., Wykoff, C.C., Watson, P.H., Leek, R., Turley, H., Gatter, K. et al. (2001) Carbonic anhydrase IX, an endogenous hypoxia marker, expression in head and neck squamous cell carcinoma and its relationship to hypoxia, necrosis, and microvessel density. *Cancer Research* **61**, 5262–5267.

Bellingham, J., Gregory-Evans, K., and Gregory-Evans, C.Y. (1998) Sequence and tissue expression of a novel human carbonic anhydrase-related protein, CARP-2, mapping to chromosome 19q13.3. *Biochemical and Biophysical Research Communications* **253**, 364–367.

Bergenhem, N.C.H., Hallberg, M., and Wisn, S. (1998) Molecular characterization of the human carbonic anhydrase-related protein (HCA-RP VIII). *Biochimica et Biophysica Acta* **1384**, 294–298.

Berglund, E.O., Murai, K.K., Fredette, B., Sekerkova, G., Marturano, B., Weber, L. et al. (1999) Ataxia and abnormal cerebellar microorganization in mice with ablated contactin gene expression. *Neuron* **24**, 739–750.

Bixby, J.L. (2001) Ligands and signaling through receptor-type tyrosine phosphatases. *IUBMB Life* **51**, 157–163.

Brummendorf, T., Hubert, M., Treubert, U., Leuschner, R., Tarnok, A., and Rathjen, F.G. (1993) The axonal recognition molecule F11 is a multifunctional protein: Specific domains mediate interactions with Ng-CAM and restrictin. *Neuron* **10**, 711–727.

Caddy, K.W., and Biscoe, T.J. (1975) Preliminary observations on the cerebellum in the mutant mouse Lurcher. *Brain Research* **91**, 276–280.

Canoll, P.D., Barnea, G., Levy, J.B., Sap, J., Ehrlich, M., Silvennoinen, O. et al. (1993) The expression of a novel receptor-type tyrosine phosphatase suggests a role in morphogenesis and plasticity of the nervous system. *Brain Research: Developmental Brain Research* **75**, 293–298.

Chia, S.K., Wykoff, C.C., Watson, P.H., Han, C., Leek, R.D., Pastorek, J. et al. (2001) Prognostic significance of a novel hypoxia-regulated marker, carbonic anhydrase IX, in invasive breast carcinoma. *Journal of Clinical Oncology* **19**, 3660–3668.

Einheber, S., Zanazzi, G., Ching, W., Scherer, S., Milner, T.A., Peles, E., and Salzer, J.L. (1997) The axonal membrane protein Caspr, a homologue of neurexin IV, is a component of the septate-like paranodal junctions that assemble during myelination. *Journal of Cellular Biochemistry* **139**, 1495–1506.

Elleby, B., Sjblom, B., Tu, C., Silverman, D.N., and Lindskog, S. (2000) Enhancement of catalytic efficiency by the combination of site-specific mutations in a carbonic anhydrase-related protein. *European Journal of Biochemistry* **267**, 5908–5915.

Faivre-Sarrailh, C., and Rougon, G. (1997) Axonal molecules of the immunoglobulin superfamily bearing a GPI anchor: their role in controlling neurite outgrowth. *Molecular and Cellular Neuroscience* **9**, 109–115.

Fujikawa-Adachi, K., Nishimori, I., Taguchi, T., Yuri, K., and Onishi, S. (1999) cDNA sequence, mRNA expression, and chromosomal localization of human carbonic anhydrase-related protein, CA-RP XI. *Biochimica et Biophysica Acta* **1431**, 518–524.

Gross, E., Pushkin, A., Abuladze, N., Fedotoff, O., and Kurtz, I. (2002) Regulation of the sodium bicarbonate cotransporter kNBC1 function: Role of Asp986, Asp988 and kNBC1-carbonic anhydrase II binding. *Journal of Physiology* **544**, 679–685.

Hewett-Emmett, D. (2000) Evolution and distribution of the carbonic anhydrase gene families. In *The Carbonic Anhydrases — New Horizons*, Chegwidden, W.R., Carter, N.D., and Edwards, Y.H., Eds., Birkhäuser Verlag, Basel, pp. 29–76.

Hewett-Emmett, D., and Tashian, R.E. (1996) Functional diversity, conservation, and convergence in the evolution of the alpha-, beta-, and gamma-carbonic anhydrase gene families. *Molecular Phylogenetics and Evolution* **5**, 50–77.

Ivanov, S., Liao, S.Y., Ivanova, A., Danilkovitch-Miagkova, A., Tarasova, N., Weirich, G. et al. (2001) Expression of hypoxia-inducible cell-surface transmembrane carbonic anhydrases in human cancer. *American Journal of Pathology* **158**, 905–919.

Kivela, A., Parkkila, S., Saarnio, J., Karttunen, T.J., Kivela, J., Parkkila, A.K. et al. (2000a) Expression of a novel transmembrane carbonic anhydrase isozyme XII in normal human gut and colorectal tumors. *American Journal of Pathology* **156**, 577–584.

Kivela, A.J., Parkkila, S., Saarnio, J., Karttunen, T.J., Kivela, J., Parkkila, A.K. et al. (2000b) Expression of transmembrane carbonic anhydrase isoenzymes IX and XII in normal human pancreas and pancreatic tumors. *Histochemistry and Cell Biology* **114**, 197–204.

Kivela, A.J., Saarnio, J., Karttunen, T.J., Kivela, J., Parkkila, A.K., Pastorekova, S. et al. (2001) Differential expression of cytoplasmic carbonic anhydrases, CA I and II, and membrane-associated isozymes, CA IX and XII, in normal mucosa of large intestine and in colorectal tumors. *Digestive Diseases and Sciences* **46**, 2179–2186.

Kleiderlein, J.J., Nisson, P.E., Jessee, J., Li, W.B., Becker, K.G., Derby, M.L. et al. (1998) CCG repeats in cDNAs from human brain. *Human Genetics* **103**, 666–673.

Tashian, R.E., Hewett-Emmett, D., Carter, N.D., and Bergenhem, N.C.H. (2000) Carbonic anhydrase (CA)-related proteins (CA-RPs), and transmembrane proteins with CA or CA-RP domains. In *The Carbonic Anhydrases — New Horizons*, Chegwidden, W.R., Carter, N.D., and Edwards, Y.H., Eds., Birkhäuser Verlag, Basel, pp. 105–120.

Vince, J.W., Carlsson, U., and Reithmeier, R.A. (2000) Localization of the Cl^-/HCO_3^- anion exchanger binding site to the amino-terminal region of carbonic anhydrase II. *Biochemistry* **39**, 13344–13349.

Vince, J.W., and Reithmeier, R.A. (2000) Identification of the carbonic anhydrase II binding site in the Cl–/HCO3– anion exchanger AE1. *Biochemistry* **39**, 5527–5533.

Wykoff, C.C., Beasley, N., Watson, P.H., Campo, L., Chia, S.K., English, R. et al. (2001) Expression of the hypoxia-inducible and tumor-associated carbonic anhydrases in ductal carcinoma *in situ* of the breast. *American Journal of Pathology* **158**, 1011–1019.

Zavada, J., Zavadova, Z., Pastorekova, S., Ciampor, F., Pastorek, J., and Zelnik, V. (1993) Expression of MaTu-MN protein in human tumor cultures and in clinical specimens. *International Journal of Cancer* **54**, 268–274.

Zisch, A.H., D'Alessandri, L., Ranscht, B., Falchetto, R., Winterhalter, K.H., and Vaughan, L. (1992) Neuronal cell adhesion molecule contactin/F11 binds to tenascin via its immunoglobulin-like domains. *Journal of Cell Biology* **119**, 203–213.

3 Multiple Binding Modes Observed in X-Ray Structures of Carbonic Anhydrase Inhibitor Complexes and Other Systems: Consequences for Structure-Based Drug Design

Jochen Antel, Alexander Weber, Christoph A. Sotriffer and Gerhard Klebe

CONTENTS

3.1 Introduction 46
3.2 Binding of Sugar Sulfamates to CAs 51
3.3 Conclusions 58
References 59

Recent years have seen a tremendous increase in the generation and deposition of x-ray structures of carbonic anhydrase inhibitor (CAI) complexes. For more than 45 years, carbonic anhydrase (CA) isozymes have been a well-known class of interesting targets for pharmaceutical research. So far, most rationally designed CAIs have been based on anchors, as zinc-binding groups, directly linked to a cyclic carbon moiety, but it is known that the more general motif $X-SO_2NH_2$ as present in sulfamides (X = N) and sulfamates (X = O) binds to the catalytical Zn(II) ion of CA isozymes as well. The increasing body of crystallographically determined protein–ligand complexes provides information about the frequent occurrence of alternative

0-415-30673-6/04/$0.00+$1.50
© 2004 by CRC Press LLC

or multiple binding modes. This chapter discusses two structures: topiramate and RWJ-37947. Both possess similar topology, but have slightly distinct substitution patterns. One of the diisopropylidene moieties of topiramate is replaced by a cyclic sulfate group in RWJ-37947. Solely based on bonding topology, it is tempting to assume similar binding properties of both resulting in similar binding modes. However, experiments show the opposite: RWJ-37947 adopts a totally different ligand pose in the active site of hCA II compared to topiramate. This chapter presents an automated docking procedure that generates conformations of CAIs close to the crystal structure. It is surprising that using as a selection criterion the frequency of how often a particular docking solution is generated, current docking algorithms such as AutoDock can reproduce correctly the observed binding modes for topiramate and RWJ-37947 in their respective crystallographically determined enzyme–inhibitor complexes.

3.1 INTRODUCTION

There has been a tremendous increase in the generation and deposition of x-ray structures of carbonic anhydrase inhibitor (CAI) complexes. Searching for *carbonic anhydrase* in the Protein Data Bank (Berman et al. 2000; http://www.PDB.org) yields 185 responses. However, this chapter is not intended to offer a comprehensive review of x-ray crystallographic studies of CA isoenzymes, but rather to highlight some recent observations on the binding modes of cocrystallized CAIs with topiramate (Casini et al. 2003; Figure 3.1) and RWJ-37947 (Maryanoff et al. 1998; Recacha et al. 2002; Figure 3.2), a related derivative. The discussion of the binding modes as well as of trials to reproduce these experimental observations by docking studies will serve as an example of more general interest and impact drug design considerations. Recent comprehensive reviews about x-ray crystallographic studies of CA isozymes and more general views on structure and function of this class of enzymes are published elsewhere (Kim et al. 2002; Liljas et al. 1994; Lindskog 1997; Stams and Christianson 2000).

For more than 45 years, CA isozymes have been a well-known class of interesting targets for pharmaceutical research (Supuran and Scozzafava 2002; Tripp et al. 2001). However, there is still room to discover new (Hebebrand et al. 2002), specific and important physiological roles (Breton 2001) that might lead to a renaissance of CAI research wherein potent and selective inhibitors with a desired, tailored inhibition profile and an acceptable safety margin can be discovered. CAIs have been considered, and are actually widely applied, for treating a variety of disease states (Supuran and Scozzafava 2000, 2002).

So far most rationally designed CAIs were based on anchors, as zinc-binding groups, directly linked to a cyclic carbon moiety (Christianson and Cox 1999; Cox et al. 2000; Liljas et al. 1994; Lindskog 1997), but it is also known that the more general motif $X\text{-}SO_2NH_2$ as present in sulfamides (X = N) and sulfamates (X = O) binds to the catalytical Zn(II) ion of CA isozymes as well (Abbate et al. 2002). It exhibits a somewhat different hydrogen-bonding network than that of the classical $C\text{-}SO_2NH_2$ inhibitors of the sulfonamide type.

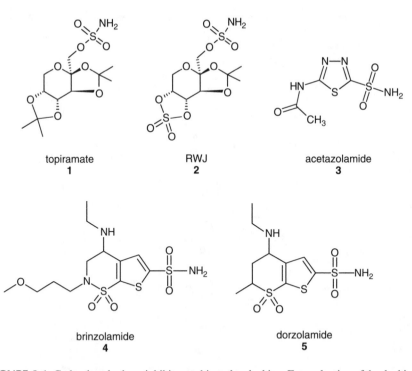

FIGURE 3.1 Carbonic anhydrase inhibitors subjected to docking. For evaluation of the docking results, x-ray crystal structures of these inhibitors in complex with CA II are used as reference.

To date, mainly sulfonamides directly attached to an aromatic or heterocyclic ring moiety [such as acetazolamide (Figure 3.1), methazolamide, ethoxozolamide and dichlorophenamide] have been used as drugs. Most notable is their successful application in treating glaucoma, e.g., dorzolamide and brinzolamide (Figure 3.1; Martens-Lobenhoffer and Banditt 2002; Sugrue 2000; Willis et al. 2002), and their use as early diuretics (Preisig et al. 1987; Seely and Dirks 1977). Among the various other disorders that can be treated with CAIs are sleep apnea (Inoue et al. 1999; Philippi et al. 2001), gastroduodenal ulcers (Erdei et al. 1990; Puscas 1984), hydrocephalus (Carrion et al. 2001; Miner 1986), some neurological (Watling and Cairncross 2002) and neuromuscular (Koller et al. 2000; Osterman et al. 1985) disorders, essential tremor (Busenbark et al. 1992) and Parkinson's diseases (Cowen et al. 1997), epilepsy (Aribi and Stringer 2002; Leniger et al. 2002; Takeoka et al. 2002), COPD (Jones and Greenstone 2001), vasodilation (Pickkers et al. 2001) and cancer (Casini et al. 2002; Olive et al. 2001). However, several mechanisms of action are being pursued in most of these areas and CAIs need to compete as regards safety and efficacy. On the other hand, several CAIs suffer not only from a lack of selectivity with respect to CA isozymes but also from mixed activity profiles owing to their interaction with non-CAI mechanisms. This makes it more difficult to assess the relative interference with one mechanism of action, resulting in an overt pharmacological effect.

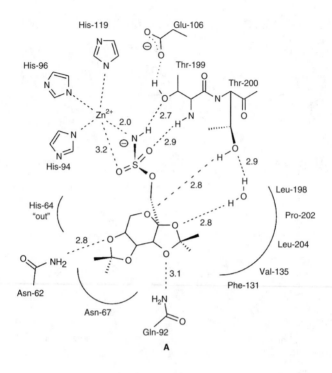

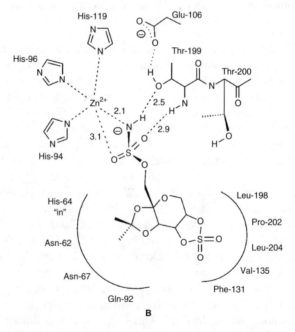

FIGURE 3.2 Schematic representation of the binding mode of topiramate (A) and RWJ-37947 (B). Hydrogen bonds between enzyme and inhibitor are shown as dotted lines. Distances (Å) are measured between the corresponding nonhydrogen atoms.

An interesting example in this respect is topiramate, a sugar (fructose) sulfamate derivative (Maryanoff et al. 1987) with a pronounced antiepileptic activity (Perucca 1997; Rosenfeld 1997; Shank et al. 1994). The mechanism of action of this drug is not yet fully understood (Shank et al. 2000; Sills et al. 2000). Several mechanisms of action have been proposed thus far (Rosenfeld 1997), such as enhancing GABA-ergic transmission (Herrero et al. 2002; Kuzniecky et al. 2002; Reis et al. 2002), antagonizing kainate/AMPA receptors (Gibbs et al. 2000; Skradski and White 2000; Zullino et al. 2003) and inhibiting the creation of action potentials in neurons by antagonizing the activation of Na^+ channels (McLean et al. 2000; Taverna et al. 1999). Another mechanism of action of topiramate that until now has not been viewed as critical (Perucca 1997), but is considered here in detail, is the inhibition of different CA isozymes (Casini et al. 2003; Dodgson et al. 2000; Masereel et al. 2002). It is hypothesized that the anticonvulsant effects of topiramate or related sulfonamides can be due to CO_2 retention following inhibition of red blood cell and brain enzymes (Masereel et al. 2002).

Reported CA inhibition data appear rather controversial. At first, topiramate was classified as a very weak (millimolar range) CA inhibitor (Maryanoff et al. 1987), but more recently the same group has reported, in a different assay that uses enzymes from an alternative source, different results (Dodgson et al. 2000) showing that topiramate is a much stronger CA inhibitor (micromolar range). Nevertheless, even in the latter report, the efficacy against a series of CA isozymes was 10 times lower than that of acetazolamide. This contradicts our data (Casini et al. 2003), which clearly suggest that topiramate is a very potent CA inhibitor. These findings match some clinically observed side-effects of topiramate (Fakhoury et al. 2002; Kuo et al. 2002; Ribacoba Montero and Salas Puig 2002), which agree with the typical pharmacological profiles of strong sulfonamide CA inhibitors used as systemic antiglaucoma agents (such as acetazolamide and methazolamide) and include paresthesias, nephrolithiasis, and weight loss (Supuran and Scozzafava 2002).

However, although a comparably strong CA isozyme inhibition profile is established, the anticonvulsant activity of topiramate appears to differ mechanistically from that of acetazolamide. This observation has been reported in some recently published *in vivo* investigations (Aribi and Stringer 2002; Stringer 2000) and agrees with former observations that derivatives of either topiramate (Maryanoff et al. 1987) or RWJ-37947 (Maryanoff et al. 1998) containing a secondary sulfamate moiety exhibit potent anticonvulsant activity but lack the inhibitory potency of CA isozymes.

Most recent studies (Shank et al. 2000) on the mechanism of action discuss the binding of topiramate to certain membrane ion channel proteins at their phosphorylation sites, resulting in an allosteric modulation of the channel conductance and subsequent inhibition of protein phosphorylation.

During the past four decades, tools for rational drug design have been developed and applied with variable success (Böhm et al. 1996). Classical quantitative structure–activity relationships (QSARs) from the early days of rational design approaches date back to original studies by Hansch and Fujita (1964). They were also widely applied in the 1970s to investigate CAIs (Hansch et al. 1985; Topliss 1977). An increase in computing power 20 years later facilitated the first structure-based drug design approaches (Babine and Bender 1997; Kubinyi 1998, 1999). The design of the CAI dorzolamide (Trusopt®) is among the first successful examples (Baldwin et al. 1989; Greer et al. 1994; Hunt et al. 1994; Smith et al. 1994). A *conditio sine qua non* for the success of the project was the availability of the protein structure. This can either be accomplished by x-ray crystallography or more recently by modern NMR techniques. Despite a significant increase in the number of solved x-ray structures, in particular of enzyme–inhibitor complexes, in the 1990s the scientific world witnessed a tremendous shift of paradigm toward high-throughput screening (HTS) and combinatorial chemistry. This subsequent stimulating progress was mainly driven by technological advances in biological sciences and chemistry. For some part, the shifted focus toward blind screening approaches in the 1990s traces back to the too limited and often untimely availability of structurally characterized targets of pharmacological relevance and the crude understanding of protein– ligand interactions. However, the initial euphoria surrounding the unlimited potential of HTS techniques has subsided because of the considerable costs involved and disappointingly low hit rates achieved. This experience, together with the required pragmatism, has paved the ground for a more integrated drug discovery approach (Antel 1999; Böhm and Stahl 2000).

In light of the current progress in structural genomics, structural information about proteins will increase dramatically and a renaissance of structure-based ligand screening approaches is expected (Klebe 2000). Structures and images of protein geometries are very suggestive; however, equally important are the thermodynamic considerations that determine affinity and selectivity. These are not simply read-off from spatial coordinates. Our present understanding of induced steric and dielectric fits and the role of water, which influence binding modes and binding energy and entropy, is rudimentary. A deeper insight into these aspects will be an essential prerequisite for the success of any structure-based approach.

Any comparative modeling technique, such as the Comparative Molecular Field Analysis (CoMFA) (Cramer et al. 1989) or Comparative Molecular Similarity Indices Analysis (CoMSIA) (Klebe et al. 1994) method, requires making assumptions about the binding mode of ligands, at least in terms of a relative mutual alignment. These aspects and the consequences of correct and arbitrarily assumed binding modes have been discussed controversially in literature (Klebe and Abraham 1993; Matter et al. 2002; Pajeva and Wiese 1998). The increasing body of crystallographically determined protein–ligand complexes provides information about the frequent occurrence of alternative or multiple binding modes (Böhm and Klebe 1996; Martinelli et al. 1989; Pavone et al. 1998; Pickens et al. 2002). In particular, low-affinity ligands seem to adopt multiple orientations in the binding pocket. Under different crystallization conditions or in distinct crystal forms, examples have been reported in which the same ligand is accommodated in different binding modes. A potent factor Xa inhibitor from Zeneca adopts reversed binding modes in trypsin crystallized under different pH conditions (Stubbs et al. 2002). The example discussed here concerns two structures: topiramate (Casini et al. 2003; Maryanoff et al. 1987; Masereel et al. 2002) and RWJ-37947 (Maryanoff et al. 1998; Recacha et al. 2002), which possess similar topology but slightly distinct substitution patterns. One of the diisopropylidene moieties of topiramate is replaced by a cyclic sulfate group in RWJ-37947 (see Figure 3.2). Based solely on the bonding topology it is tempting to assume similar binding properties of both, resulting in comparable binding modes. However, experiments show the opposite — two alternative orientations are clearly adopted.

The x-ray structure of the complex of topiramate with hCA II has been solved. The inhibitor reveals tight binding within the active site, experienced by Zn(II) ion coordination through the deprotonated sulfamate moiety.

RWJ-37947 adopts a totally different ligand pose in the active site of hCA II compared with that of topiramate. Its cyclic sulfate moiety points toward the hydrophobic pocket of the enzyme, interacting with residues Phe-131, Pro-202 and Leu-198, whereas the isopropylidene moiety is oriented toward the side-chain amide groups of Asn-67, Asn-62 and His-94 (see later for details).

In recent years, several docking and *de novo* design projects (Chazalette et al. 2001; DesJarlais et al. 1988; Esposito et al. 2000; Grzybowski et al. 2002a, 2002b; Hindle et al. 2002; Rotstein and Murcko 1993a, 1993b; Vicker et al. 2003) have been performed on CA isozymes (Gruneberg et al. 2002), and report the discovery of novel picomolar ligands (Grzybowski et al. 2002a) for CA II. Docking approaches in structure-based drug design generally face two major problems: accurate prediction of protein–ligand interaction geometries (the docking problem) and correct evaluation of the binding affinities (the scoring problem). This chapter presents an automated docking procedure that generates conformations of CAIs close to the crystal structure. Related docking studies (Sotriffer et al. 2002) have revealed that due to inaccuracies in the evaluation of the scores and the scoring functions themselves, degenerate docking solutions within a certain threshold of the top rank should be considered for analyzing the results. Furthermore, with stochastic procedures acting on an energy hypersurface, the frequency of occurrence of a certain result should also be taken into account (Sotriffer et al. 1996). Accordingly, special attention should be paid to conformations occurring more frequently, and therefore the largest cluster of results should also be analyzed when evaluating docking runs.

3.2   BINDING OF SUGAR SULFAMATES TO CAs

This study concentrates on the observed binding modes of topiramate and RWJ-37947, tries to develop a theoretical rational for this observation and investigates whether current docking algorithms will reproduce these experimental findings. This should indicate (1) whether at all virtual screening or docking attempts will result in a selection of potential inhibitors and (2) whether virtually selected inhibitors will be correctly ranked according to their potency.

Two x-ray crystal structures of CAIs with a sugar sulfamate moiety in complex with CA II have recently been solved (Casini et al. 2003; Recacha et al. 2002). Topiramate (**1**) and RWJ-37947 (**2**; Figure 3.1) inhibit CA II in the nanomolar range (Masereel et al. 2002; Recacha et al. 2002). Topiramate is clinically used as an antiepileptic drug and RWJ-37947 possesses anticonvulsant activity. The difference between the two compounds is substitution of an isopropylidene moiety in topiramate by a cyclic sulfate group in RWJ-37947 (Figure 3.1). Both sugar sulfamate derivatives **1** and **2** bind with their sulfamate moiety to zinc, resulting in a tetrahedral coordination (Figure 3.2). As generally known from NMR spectroscopy, CA inhibitors coordinate with a deprotonated anchoring group to the zinc ion (Liljas et al. 1994). In addition, both compounds make hydrogen bonds with the side chain oxygen atom of Thr-199 and the backbone nitrogen atom of the same residue. The hydroxy group of Thr-199 forms an additional hydrogen bond with Glu-106, such that the Thr-199 hydroxyl acts as a hydrogen-bond acceptor for inhibitor binding.

Despite the similarity in anchoring to the zinc according to a well-known interaction pattern (Liljas et al. 1994), a surprising difference is observed in the binding mode of the two inhibitors with respect to the attached ring system. Topiramate forms hydrogen bonds with amino acid side chains in a hydrophilic binding pocket (Asn-67, Gln-92) and to a water molecule that donates a hydrogen bond to Thr-200 (Figure 3.2A). In addition, this water interacts with the oxygen atom of the ligand's six-membered ring. RWJ-37947, instead, shows a different binding mode in which the ring system is rotated by ca. 180° (mirrored), although an orientation as observed for topiramate would appear feasible (Figure 3.3). Therefore, surprisingly, the cyclic sulfate group points to a more hydrophobic pocket (Leu-198, Pro-202, Phe-131), and except for the sulfamate anchoring group, no further hydrogen bonds are observed (Figure 3.2B).

Given the apparent topological similarity of the two compounds and the surprising difference in their binding modes, the question arises as to whether such a behavior is predictable, and, in particular, whether computational tools can highlight this difference.

Here, the grid-based automated docking program AutoDock (Goodsell and Olson 1990; Morris et al. 1996) was used to address this question and to predict the binding mode of topiramate, RWJ-37947 and three additional commonly used CA inhibitors (see Figure 3.1: acetazolamide **3**, brinzolamide **4** and dorzolamide **5**). For four compounds, x-ray crystal structures were extracted from the Protein Data Bank (PDB codes 1a42, 1azm, 1cil, 1eou). The crystal structure of CA II in complex with topiramate was kindly provided by Dr. C.T. Supuran, University of Florence. All five protein structures were superimposed with Sybyl® 6.8 (Tripos, Inc., St. Louis, MO), based on all atoms of 20 conserved amino acids in the binding pocket (Ser-29, Pro-30, Asn-61, Gly-63, Gln-92, His-94, His-96, Trp-97, Glu-106, Glu-117, His-119, His-122, Ala-142, Val-143, Ser-197, Thr-199, Pro-201, Glu-205, Trp-209, Asn-244). Comparison of the amino acids in the binding pocket of the superimposed structures indicates slight conformational flexibility for His-64 only (Figure 3.4). This is further corroborated by superpositioning 55 complex structures of hCA II with Relibase (Gunther et al. 2003; Hendlich et al. 2003), which reveals that CA II has a fairly invariant binding site, except for the amino acid at position 64 (Klebe 2003). The biochemical reason for this observation is most likely related to the functional role of His-64 in catalysis, wherein it acts as a proton shuttle in regenerating the zinc-bound hydroxide (Duda et al. 2001, 2003). In general, two main conformations of His-64 are observed: the *in* conformation, in which His-64 points to the active site, and the *out* conformation, in which His-64 points to the solvent. In the five complexes mentioned, His-64 adopts both the *in* (1azm, 1eou) and the *out* (1a42, 1cil, crystal structure of topiramate) conformations.

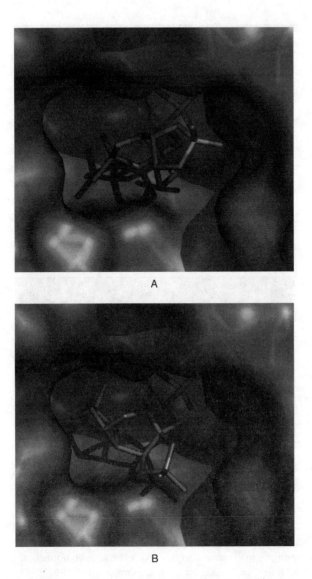

FIGURE 3.3 (See color insert following page 148.) The binding mode of RWJ-37947 observed in the crystal structure (A) and manually rotated by 220° around the single bond connecting the ring system with the sulfamate anchor (B). (B) reveals that a binding mode as observed in the topiramate complex is sterically possible. The solvent accessible surface of the binding pocket of CAII hosting topiramate is shown (A, B).

To allow as much space as possible for ligand placement, 1cil with His-64 in the *out* conformation was used to dock all five ligands. The structure was constructed for docking by removing all water and ligand atoms, adding polar hydrogens and assigning Amber atomic charges and solvation parameters as required by the

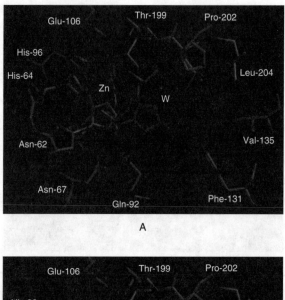

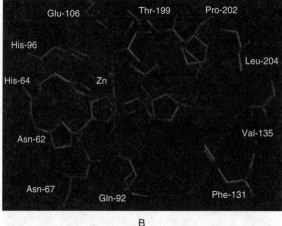

FIGURE 3.4 (See color insert.) Binding mode of topiramate (A) and RWJ-37947 (B) in CA II. All amino acids in the binding site are highly conserved, except His 64, which adopts the *out* conformation in the topiramate complex (A) and the *in* conformation in complex with RWJ-37947 (B). All molecule representations are drawn with PyMOL. (DeLano, 2002.)

AutoDock program (Morris et al. 1998). AutoDock was then allowed to dock ligands flexibly into the rigid binding pocket of CA II. For this purpose, the protein was first embedded in a three-dimensional grid, a probe atom was placed at each grid point and the interaction energy of the probe atom with the protein was assigned to the grid point. Hence, an affinity grid was calculated for each atom type in the ligand and the grids were then used as protein representation to speed up the docking process. For ligand preparation, Gasteiger–Marsili charges were assigned to the ligand atoms and rotatable bonds were explicitly defined. The largest ring system of the inhibitor was chosen as root fragment, and all rotatable bonds were kept

TABLE 3.1
Docking Results for CA Inhibitors

Inhibitor	Cluster[a] Scoring Rank	Cluster Size	Best Energy in Cluster (kcal/mol]	RMSD (x-ray structure)
		Dorzolamide (1cil)		
Top cluster	1	6	−8.61	3.55
Largest cluster	4	60	−8.45	0.54
		Brinzolamide (1a42)		
Top cluster	1	1	−9.86	3.87
Largest cluster	2	32	−9.71	1.34
		Acetazolamide (1azm)		
Top cluster	1	11	−7.30	4.38
Largest cluster	7	50	−6.90	1.23
		Topiramate		
Top cluster	1	71	−8.46	1.34
Largest cluster	1	71	−8.46	1.34
		RWJ-37947 (1eou)		
Top cluster	1	1	−8.40	4.20
largest cluster	5	81	−7.95	0.54

Note: Ligands were docked into CA II (PDB code 1cil).

[a] The total number of clusters obtained in each case is 5 (dorzolamide), 14 (brinzolamide), 13 (acetazolamide), 11 (topiramate) and 9 (RWJ-37947).

flexible. Docking was then carried out by using an empirical binding free energy function and the Lamarckian genetic algorithm (Morris et al. 1998). One hundred independent docking runs were performed for each ligand, applying a standard AutoDock protocol, with a grid spacing of 1 Å, an initial population of 50 randomly placed individuals, a maximum number of 1.5×10^6 energy evaluations, a mutation rate of 0.02, a crossover rate of 0.80 and an elitism value of 1. Results differing by less than 1 Å root-mean-square deviation (rmsd) were merged together in one cluster.

To evaluate the docking results, the rmsd values of the docked solutions with respect to the crystal structure and the corresponding energy scores were taken into account. Table 3.1 and Figure 3.5 show the docking results.

In Table 3.1, the docking solutions corresponding to the first rank of the top cluster and the first rank of the largest cluster are presented along with the corresponding cluster size, the scoring value and the rmsd with respect to the crystal structure. Except for topiramate, the rmsd of the top cluster is generally >3.5 Å and thus unsatisfactorily high; the corresponding cluster size, however, is rather small (≤11). Considering the first rank in the largest cluster, all rmsd values are below 1.4 Å, implying that the resulting binding modes are close to the crystal structure (Figure 3.5).

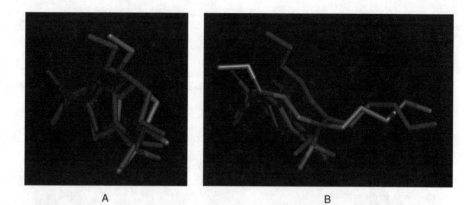

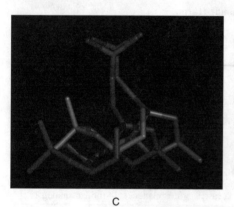

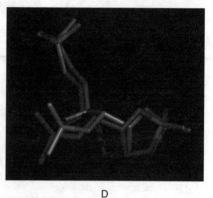

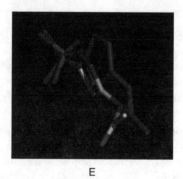

FIGURE 3.5 (See color insert.) Dorzolamide (A), brinzolamide (B), topiramate (C), RWJ-37947 (D) and acetazolamide (E) docked into the binding site of CA II (PDB code 1cil). The first rank of the largest cluster is represented. X-ray crystal structures of the corresponding ligands are shown in green.

TABLE 3.2
Observed Docking Conformations of Topiramate
and RWJ-37947

Binding Mode of the Ring System	Coordination of Sulfamate to Zinc	Topiramate (No. of Runs)	RWJ-37947 (No. of Runs)
Correct		79	90
	Correct	74	73
	Wrong	5	17
Wrong		21	10
	Correct	0	7
	Wrong	4	2
	No coordination	17	1

Note: The position of the ring systems as observed in the x-ray structure and the coordination of the sulfamate anchoring group to zinc and Thr 199 is regarded as correct.

The overall best results in terms of rmsd are observed for dorzolamide and RWJ-37947 (in both cases, 0.54 Å for the first rank of the largest cluster; Figure 3.4A and D). Comparing the scoring values between the first rank of the top cluster and the first rank of the largest cluster, the differences are small ($\leq$0.45 kcal/mol) and all within the degeneracy range identified previously for the same docking methodology (Sotriffer et al. 2002). The small differences in the scoring values indicate that, in the present case, the scoring function implemented in AutoDock does not significantly differentiate between these binding geometries.

Table 3.2 compares the docked solutions of topiramate and RWJ-37947 with respect to the conformation of the ring system and the coordination of the sulfamate anchoring group to the zinc atom. Assuming that the conformation of the ring system observed in the crystal structure represents the correct binding mode, 74 correct binding modes for topiramate and 90 correct binding modes for RWJ-37947 were generated by the 100 docking runs. As regards the zinc-coordination of the sulfamate group, an approximately equal presentation of correctly docked solutions was obtained for both cases (74 runs for topiramate and 73 runs for RWJ). Docked solutions were considered as wrong if the sulfamate group was placed far away from the zinc atom (>4 Å) or if the anchoring group interacted with Thr-200 instead of Thr-199. Regarding the wrong binding modes of the ring system of topiramate, a correct sulfamate coordination was obtained in none of the cases. In this context, the situation is different for RWJ-37947: among the 10 wrong binding modes of the ring system, 7 show correct coordination of the sulfamate group. This finding can be interpreted as the rotated binding mode of the RWJ ring system being, in principle, possible and compatible with a correct coordination of the sulfamate anchor. Nevertheless, the binding mode observed in the crystal structure is clearly preferred (73 correctly docked solutions with respect to the ring system and zinc coordination).

Also, for topiramate, the binding mode observed in the crystal structure is preferentially obtained (74 correctly docked solutions). Here, in addition, a wrongly placed ring system seems to impair the correct coordination of the sulfamate group. This finding is compatible with the tight binding mode of topiramate embedded in a well defined hydrogen bond network (Figure 3.2A).

3.3 CONCLUSIONS

It is surprising that current docking algorithms such as AutoDock yielded, in principle, the observed binding modes for topiramate as well as for RWJ-37947 in their respective crystallographically determined enzyme–inhibitor complexes. However, these binding modes were not necessarily top-ranked based on energy evaluation only, but were deemed top ranked when energy and cluster size were jointly considered. The latter criterion indirectly maps the shape of the energy surface used for docking. The most frequently detected local minimum also supposedly corresponds to the most likely geometry. More CA–inhibitor complexes have to be investigated, first to confirm the presented results on the prediction of the binding mode and second to correlate the derived scoring value with binding affinity. The implementation of a frequency factor (Sotriffer et al. 1996) that takes into account the cluster size should be beneficial to rank docked solutions and better separate cluster ranks. Other docking studies have also revealed (Sotriffer et al. 2002) that the application of knowledge-based pair potentials (Gohlke et al. 2000a, 2000b) instead of the implemented AutoDock energy function can further improve the results. Nevertheless, by incorporating the size of the cluster, the docking results presented here demonstrate that it is possible to accurately predict the binding mode of five CAIs. This observation can lead to the following conclusions: if considerably different binding modes show up as highly ranked within virtual screening or docking studies, more than one should be considered as a meaningful structural template for the design of either a single compound or of even better targeted libraries. Integration of combinatorial chemistry and structure-based drug design might be one of the most promising ways to adopt, although not the only one. Considerations of multiple binding modes will be essential for the conceptual design of the core for a chemical library aiming at high hit rates (Figure 3.6). Lead optimization will remain largely driven by intuition, because of its multidimensional character, but the quality of supporting experimental and computational tools is rapidly improving and will lead to significantly faster identification and optimization of new lead compounds.

An optimal fit in a target site neither guarantees that the desired activity of the drug will be enhanced or that undesired side effects will be diminished, nor considers the pharmacokinetics of the drug. Screening of well-designed targeted libraries with a view to pharmacodynamic and pharmacokinetic profiles will finally facilitate the discovery of optimal candidates for further development.

Where observation is concerned, chance favors only the prepared mind.

Louis Pasteur, Speech (1854)

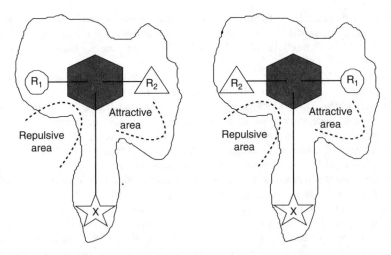

FIGURE 3.6 Consideration of multiple binding modes should have an impact on the conceptual design of compounds and compounds libraries. Although moieties responsible for specific interactions and accounting for a considerable percentage of overall intrinsic affinity can be kept constant (indicated as X), the substituent list for R_1 and R_2 might be exchangable to a certain extent.

REFERENCES

Abbate, F., Supuran, C.T., Scozzafava, A., Orioli, P., Stubbs, M.T. and Klebe, G. (2002) Nonaromatic sulfonamide group as an ideal anchor for potent human carbonic anhydrase inhibitors: Role of hydrogen-bonding networks in ligand binding and drug design. *Journal of Medicinal Chemistry* **45**, 3583–3587.

Antel, J. (1999) Integration of combinatorial chemistry and structure-based drug design. *Current Opinion in Drug Discovery and Development* **2**, 224–233.

Aribi, A.M., and Stringer, J.L. (2002) Effects of antiepileptic drugs on extracellular pH regulation in the hippocampal CA1 region in vivo. *Epilepsy Research* **49**, 143–151.

Babine, R.E., and Bender, S.L. (1997) Molecular recognition of protein–ligand complexes: Applications to drug design. *Chemical Reviews* **97**, 1359–1472.

Baldwin, J.J., Ponticello, G.S., Anderson, P.S., Christy, M.E., Murcko, M.A., Randall, W.C., Schwam, H., Sugrue, M.F., Springer, J.P., Gautheron, P. et al. (1989) Thienothiopyran-2-sulfonamides: novel topically active carbonic anhydrase inhibitors for the treatment of glaucoma. *Journal of Medicinal Chemistry* **32**, 2510–2513.

Berman, H.M., Westbrook, J., Feng, Z., Gilliland, G., Bhat, T.N., Weissig, H., Shindyalov, I.N., and Bourne, P.E. (2000) The protein data bank. *Nucleic Acids Research* **28**, 235–242.

Böhm, H.J., and Klebe, G. (1996) What can we learn from molecular recognition in protein–ligand complexes for the design of new drugs? *Angewandte Chemie International Edition in English* **35**, 2588–2614.

Böhm, H.J., Klebe, G., and Kubinyi, H. (1996) *Wirkstoffdesign*, Spektrum Akademischer Verlag, Heidelberg, Berlin, Oxford.

Bohm, H.J., and Stahl, M. (2000) Structure-based library design: molecular modelling merges with combinatorial chemistry. *Current Opinion in Chemical Biology* 4, 283–286.

Breton, S. (2001) The cellular physiology of carbonic anhydrases. *Journal of Pancreas (Online)* 2, 159–164.

Busenbark, K., Pahwa, R., Hubble, J., and Koller, W. (1992) The effect of acetazolamide on essential tremor: an open-label trial. *Neurology* 42, 1394–1395.

Carrion, E., Hertzog, J.H., Medlock, M.D., Hauser, G.J., and Dalton, H.J. (2001) Use of acetazolamide to decrease cerebrospinal fluid production in chronically ventilated patients with ventriculopleural shunts. *Archives of Disease of the Child* 84, 68–71.

Casini, A., Antel, J., Abbate, F., Scozzafava, A., David, S., Waldeck, H., Schafer, S., and Supuran, C.T. (2003) Carbonic anhydrase inhibitors: SAR and x-ray crystallographic study for the interaction of sugar sulfamates/sulfamides with isozymes I, II and IV. *Bioorganic and Medicinal Chemistry Letters* 13, 841–845.

Casini, A., Scozzafava, A., Mastrolorenzo, A., and Supuran, L.T. (2002) Sulfonamides and sulfonylated derivatives as anticancer agents. *Current Cancer Drug Targets* 2, 55–75.

Chazalette, C., Riviere-Baudet, M., Scozzafava, A., Abbate, F., Ben Maarouf, Z., and Supuran, C.T. (2001) Carbonic anhydrase inhibitors, interaction of boron derivatives with isozymes I and II: A new binding site for hydrophobic inhibitors at the entrance of the active site as shown by docking studies. *Journal of Enzyme Inhibition* 16, 125–133.

Christianson, D.W., and Cox, J.D. (1999) Catalysis by metal-activated hydroxide in zinc and manganese metalloenzymes. *Annual Reviews in Biochemistry* 68, 33–57.

Cowen, M.A., Green, M., Bertollo, D.N., and Abbott, K. (1997) A treatment for tardive dyskinesia and some other extrapyramidal symptoms. *Journal of Clinical Psychopharmacology* 17, 190–193.

Cox, J.D., Hunt, J.A., Compher, K.M., Fierke, C.A., and Christianson, D.W. (2000) Structural influence of hydrophobic core residues on metal binding and specificity in carbonic anhydrase II. *Biochemistry* 39, 13687–13694.

Cramer, R.D., III, Patterson, D.E., and Bunce, J.D. (1989) Recent advances in comparative molecular field analysis (CoMFA). *Progress in Clinical and Biological Research* 291, 161–165.

DeLano, W.L. (2002) The PyMol molecular graphics system, DeLano Scientific, San Carlos, CA, USA.

DesJarlais, R.L., Sheridan, R.P., Seibel, G.L., Dixon, J.S., Kuntz, I.D., and Venkataraghavan, R. (1988) Using shape complementarity as an initial screen in designing ligands for a receptor binding site of known three-dimensional structure. *Journal of Medicinal Chemistry* 31, 722–729.

Dodgson, S.J., Shank, R.P., and Maryanoff, B.E. (2000) Topiramate as an inhibitor of carbonic anhydrase isoenzymes. *Epilepsia* 41, S35–S39.

Duda, D., Govindasamy, L., Agbandje-McKenna, M., Tu, C., Silverman, D.N., and McKenna, R. (2003) The refined atomic structure of carbonic anhydrase II at 1.05 A resolution: implications of chemical rescue of proton transfer. *Acta Crystallografica D Biological Crystallography* 59, 93–104.

Duda, D., Tu, C., Qian, M., Laipis, P., Agbandje-McKenna, M., Silverman, D.N., and McKenna, R. (2001) Structural and kinetic analysis of the chemical rescue of the proton transfer function of carbonic anhydrase II. *Biochemistry* 40, 1741–1748.

Erdei, A., Gyori, I., Gedeon, A., and Szabo, I. (1990) Successful treatment of intractable gastric ulcers with acetazolamide. *Acta Medica Hungarica* 47, 171–178.

Esposito, E.X., Baran, K., Kelly, K., and Madura, J.D. (2000) Docking of sulfonamides to carbonic anhydrase II and IV. *Journal of Molecular Graph Model* 18, 283–289, 307–288.

Fakhoury, T., Murray, L., Seger, D., McLean, M., and Abou-Khalil, B. (2002) Topiramate overdose: Clinical and laboratory features. *Epilepsy Behavior* **3**, 185–189.

Gibbs, J.W., III, Sombati, S., DeLorenzo, R.J., and Coulter, D.A. (2000) Cellular actions of topiramate: Blockade of kainate-evoked inward currents in cultured hippocampal neurons. *Epilepsia* **41**, S10–S16.

Gohlke, H., Hendlich, M., and Klebe, G. (2000a) Knowledge-based scoring function to predict protein-ligand interactions. *Journal of Molecular Biology* **295**, 337–356.

Gohlke, H., Hendlich, M., and Klebe, G. (2000b) Predicting binding modes, binding affinities and 'hot spots' for protein-ligand complexes using a knowledge-based scoring function. *Perspectives in Drug Discovery and Design* **20**, 115–144.

Goodsell, D.S., and Olson, A. J. (1990) Automated docking of substrates to proteins by simulated annealing. *Proteins* **8**, 195–202.

Greer, J., Erickson, J.W., Baldwin, J.J., and Varney, M.D. (1994) Application of the three-dimensional structures of protein target molecules in structure-based drug design. *Journal of Medicinal Chemistry* **37**, 1035–1054.

Gruneberg, S., Stubbs, M.T., and Klebe, G. (2002) Successful virtual screening for novel inhibitors of human carbonic anhydrase: strategy and experimental confirmation. *Journal of Medicinal Chemistry* **45**, 3588–3602.

Grzybowski, B.A., Ishchenko, A.V., Kim, C.Y., Topalov, G., Chapman, R., Christianson, D.W., Whitesides, G.M., and Shakhnovich, E.I. (2002a) Combinatorial computational method gives new picomolar ligands for a known enzyme. *Proceedings of the National Academy of Sciences of the United States of America* **99**, 1270–1273.

Grzybowski, B.A., Ishchenko, A.V., Shimada, J., and Shakhnovich, E.I. (2002b) From knowledge-based potentials to combinatorial lead design in silico. *Accounts of Chemical Research* **35**, 261–269.

Gunther, J., Bergner, A., Hendlich, M., and Klebe, G. (2003) Utilising structural knowledge in drug design strategies: applications using Relibase. *Journal of Molecular Biology* **326**, 621–636.

Hansch, C., and Fujita, T. (1964) r-s-p analysis: A method for the correlation of biological activity and chemical structure. *Journal of the American Chemical Society* **86**, 1616–1626.

Hansch, C., McClarin, J., Klein, T., and Langridge, R. (1985) A quantitative structure–activity relationship and molecular graphics study of carbonic anhydrase inhibitors. *Molecular Pharmacology* **27**, 493–498.

Hebebrand, J., Antel, J., Preuschoff, U., David, S., Sann, H., and Weske, M., (2002) Drug screening method for the treatment and prophylaxis of obesity. *WO Patent* 02/07821.

Hendlich, M., Bergner, A., Gunther, J., and Klebe, G. (2003) Relibase: Design and development of a database for comprehensive analysis of protein-ligand interactions. *Journal of Molecular Biology* **326**, 607–620.

Herrero, A. I., Del Olmo, N., Gonzalez-Escalada, J. R., and Solis, J. M. (2002) Two new actions of topiramate: inhibition of depolarizing GABA(A)-mediated responses and activation of a potassium conductance. *Neuropharmacology* **42**, 210–220.

Hindle, S.A., Rarey, M., Buning, C., and Lengauer, T. (2002) Flexible docking under pharmacophore type constraints. *Journal of Computer Aided Molecular Design* **16**, 129–149.

Hunt, C.A., Mallorga, P.J., Michelson, S. R., Schwam, H., Sondey, J. M., Smith, R.L., Sugrue, M.F., and Shepard, K.L. (1994) 3-Substituted thieno[2,3-b][1,4]thiazine-6-sulfonamides. A novel class of topically active carbonic anhydrase inhibitors. *Journal of Medicinal Chemistry,* **37**, 240–247.

Inoue, Y., Takata, K., Sakamoto, I., Hazama, H., and Kawahara, R. (1999) Clinical efficacy and indication of acetazolamide treatment on sleep apnea syndrome. *Psychiatry and Clinical Neuroscience* **53**, 321–322.

Jones, P.W., and Greenstone, M. (2001) Carbonic anhydrase inhibitors for hypercapnic ventilatory failure in chronic obstructive pulmonary disease. *Cochrane Database System Reviews* **1**, CD002881.

Kim, C.Y., Whittington, D.A., Chang, J.S., Liao, J., May, J.A., and Christianson, D.W. (2002) Structural aspects of isozyme selectivity in the binding of inhibitors to carbonic anhydrases II and IV. *Journal of Medicinal Chemistry* **45**, 888–893.

Klebe, G. (2000) Recent developments in structure-based drug design. *Journal of Molecular Medicine* **78**, 269–281.

Klebe, G. (2003) From structure to recognition principles: Mining in crystal data as a prerequisite for drug design. *Ernst Schering Research Foundation Workshop* **42**, 103–126.

Klebe, G., and Abraham, U. (1993) On the prediction of binding properties of drug molecules by comparative molecular field analysis. *Journal of Medicinal Chemistry*, **36**, 70–80.

Klebe, G., Abraham, U., and Mietzner, T. (1994) Molecular similarity indices in a comparative analysis (CoMSIA) of drug molecules to correlate and predict their biological activity. *Journal of Medicinal Chemistry* **37**, 4130–4146.

Koller, W.C., Hristova, A., and Brin, M. (2000) Pharmacologic treatment of essential tremor. *Neurology* **54**, S30–S38.

Kubinyi, H. (1998) Structure-based design of enzyme inhibitors and receptor ligands. *Current Opinion in Drug Discovery and Development* **1**, 4–15.

Kubinyi, H. (1999) Chance favors the prepared mind: From serendipity to rational drug design. *Journal of Receptor and Signal Transduction Research* **19**, 15–39.

Kuo, R.L., Moran, M.E., Kim, D.H., Abrahams, H.M., White, M.D., and Lingeman, J.E. (2002) Topiramate-induced nephrolithiasis. *Journal of Endourology* **16**, 229–231.

Kuzniecky, R., Ho, S., Pan, J., Martin, R., Gilliam, F., Faught, E., and Hetherington, H. (2002) Modulation of cerebral GABA by topiramate, lamotrigine, and gabapentin in healthy adults. *Neurology* **58**, 368–372.

Leniger, T., Wiemann, M., Bingmann, D., Widman, G., Hufnagel, A. and Bonnet, U. (2002) Carbonic anhydrase inhibitor sulthiame reduces intracellular pH and epileptiform activity of hippocampal CA3 neurons. *Epilepsia* **43**, 469–474.

Liljas, A., Hakansson, K., Jonsson, B.H., and Xue, Y. (1994) Inhibition and catalysis of carbonic anhydrase: Recent crystallographic analyses. *European Journal of Biochemistry* **219**, 1–10.

Lindskog, S. (1997) Structure and mechanism of carbonic anhydrase. *Pharmacology and Therapeutics* **74**, 1–20.

Martens-Lobenhoffer, J., and Banditt, P. (2002) Clinical pharmacokinetics of dorzolamide. *Clinical Pharmacokinetics* **41**, 197–205.

Martinelli, R.A., Hanson, G.R., Thompson, J.S., Holmquist, B., Pilbrow, J.R., Auld, D.S., and Vallee, B.L. (1989) Characterization of the inhibitor complexes of cobalt carboxypeptidase A by electron paramagnetic resonance spectroscopy. *Biochemistry* **28**, 2251–2258.

Maryanoff, B.E., Costanzo, M.J., Nortey, S.O., Greco, M.N., Shank, R.P., Schupsky, J.J., Ortegon, M.P., and Vaught, J.L. (1998) Structure–activity studies on anticonvulsant sugar sulfamates related to topiramate. Enhanced potency with cyclic sulfate derivatives. *Journal of Medicinal Chemistry* **41**, 1315–1343.

Maryanoff, B.E., Nortey, S.O., Gardocki, J.F., Shank, R.P., and Dodgson, S.P. (1987) Anticonvulsant O-alkyl sulfamates. 2,3:4,5-Bis-O-(1-methylethylidene)- beta-D-fructopyranose sulfamate and related compounds. *Journal of Medicinal Chemistry* **30**, 880–887.

Masereel, B., Rolin, S., Abbate, F., Scozzafava, A., and Supuran, C.T. (2002) Carbonic anhydrase inhibitors: Anticonvulsant sulfonamides incorporating valproyl and other lipophilic moieties. *Journal of Medicinal Chemistry* **45**, 312–320.

Matter, H., Kotsonis, P., Klingler, O., Strobel, H., Frohlich, L.G., Frey, A., Pfleiderer, W., and Schmidt, H.H. (2002) Structural requirements for inhibition of the neuronal nitric oxide synthase (NOS-I): 3D-QSAR analysis of 4-oxo- and 4-amino-pteridine-based inhibitors. *Journal of Medicinal Chemistry* **45**, 2923–2941.

McLean, M.J., Bukhari, A.A., and Wamil, A.W. (2000) Effects of topiramate on sodium-dependent action-potential firing by mouse spinal cord neurons in cell culture. *Epilepsia* **41**, S21–S24.

Miner, M.E. (1986) Acetazolamide treatment of progressive hydrocephalus secondary to intraventricular hemorrhage in a preterm infant. *Childs Nervous System* **2**, 105–106.

Morris, G.M., Goodsell, D.S., Halliday, R.S., Huey, R., Hart, W.E., Belew, R.K., and Olson, A.J. (1998) Automated docking using a Lamarckian genetic algorithm and an empirical binding free energy function. *Journal of Computational Chemistry* **19**, 1639–1662.

Morris, G.M., Goodsell, D.S., Huey, R., and Olson, A. J. (1996) Distributed automated docking of flexible ligands to proteins: parallel applications of AutoDock 2.4. *Journal of Computer Aided Molecular Design* **10**, 293–304.

Olive, P.L., Aquino-Parsons, C., MacPhail, S.H., Liao, S.Y., Raleigh, J.A., Lerman, M.I., and Stanbridge, E.J. (2001) Carbonic anhydrase IX as an endogenous marker for hypoxic cells in cervical cancer. *Cancer Research* **61**, 8924–8929.

Osterman, P.O., Askmark, H., and Wistrand, P.J. (1985) Serum carbonic anhydrase III in neuromuscular disorders and in healthy persons after a long-distance run. *Journal of Neurological Science* **70**, 347–357.

Pajeva, I., and Wiese, M. (1998) Molecular modeling of phenothiazines and related drugs as multidrug resistance modifiers: a comparative molecular field analysis study. *Journal of Medicinal Chemistry* **41**, 1815–1826.

Pavone, V., De Simone, G., Nastri, F., Galdiero, S., Staiano, N., Lombardi, A., and Pedone, C. (1998) Multiple binding mode of reversible synthetic thrombin inhibitors: A comparative structural analysis. *Biological Chemistry* **379**, 987–1006.

Perucca, E. (1997) A pharmacological and clinical review on topiramate, a new antiepileptic drug. *Pharmacological Research* **35**, 241–256.

Philippi, H., Bieber, I., and Reitter, B. (2001) Acetazolamide treatment for infantile central sleep apnea. *Journal of Child Neurology* **16**, 600–603.

Pickens, J. C., Merritt, E. A., Ahn, M., Verlinde, C. L., Hol, W. G., and Fan, E. (2002) Anchor-based design of improved cholera toxin and E. coli heat-labile enterotoxin receptor binding antagonists that display multiple binding modes. *Chemical Biology* **9**, 215–224.

Pickkers, P., Hughes, A.D., Russel, F.G., Thien, T., and Smits, P. (2001) In vivo evidence for K(Ca) channel opening properties of acetazolamide in the human vasculature. *British Journal of Pharmacology* **132**, 443–450.

Preisig, P.A., Toto, R.D., and Alpern, R.J. (1987) Carbonic anhydrase inhibitors. *Renal Physiology* **10**, 136–159.

Puscas, I. (1984) Treatment of gastroduodenal ulcers with carbonic anhydrase inhibitors. *Annals of the New York Academy of Science* **429**, 587–591.

Recacha, R., Costanzo, M.J., Maryanoff, B.E., and Chattopadhyay, D. (2002) Crystal structure of human carbonic anhydrase II complexed with an anti-convulsant sugar sulphamate. *Biochemical Journal* **361**, 437–441.

Reis, J., Tergau, F., Hamer, H.M., Muller, H.H., Knake, S., Fritsch, B., Oertel, W.H., and Rosenow, F. (2002) Topiramate selectively decreases intracortical excitability in human motor cortex. *Epilepsia* **43**, 1149–1156.

Ribacoba Montero, R., and Salas Puig, X. (2002) Efficacy and tolerability of long term topiramate in drug resistant epilepsy in adults. *Reviews in Neurology* **34**, 101–105.

Rosenfeld, W.E. (1997) Topiramate: A review of preclinical, pharmacokinetic, and clinical data. *Clinical Therapy* **19**, 1294–1308.

Rotstein, S.H., and Murcko, M.A. (1993a) GenStar: a method for de novo drug design. *Journal of Computer Aided Molecular Design* **7**, 23–43.

Rotstein, S.H., and Murcko, M.A. (1993b) GroupBuild: a fragment-based method for de novo drug design. *Journal of Medicinal Chemistry* **36**, 1700–1710.

Seely, J.F., and Dirks, J.H. (1977) Site of action of diuretic drugs. *Kidney International* **11**, 1–8.

Shank, R.P., Gardocki, J.F., Streeter, A.J., and Maryanoff, B.E. (2000) An overview of the preclinical aspects of topiramate: pharmacology, pharmacokinetics, and mechanism of action. *Epilepsia*, **41**, S3–S9.

Shank, R.P., Gardocki, J.F., Vaught, J.L., Davis, C.B., Schupsky, J.J., Raffa, R. B., Dodgson, S.J., Nortey, S.O., and Maryanoff, B.E. (1994) Topiramate: preclinical evaluation of structurally novel anticonvulsant. *Epilepsia* **35**, 450–460.

Sills, G.J., Leach, J.P., Kilpatrick, W.S., Fraser, C.M., Thompson, G.G., and Brodie, M.J. (2000) Concentration-effect studies with topiramate on selected enzymes and intermediates of the GABA shunt. *Epilepsia* **41**, S30–S34.

Skradski, S., and White, H.S. (2000) Topiramate blocks kainate-evoked cobalt influx into cultured neurons. *Epilepsia* **41**, S45–S47.

Smith, G.M., Alexander, R.S., Christianson, D.W., McKeever, B.M., Ponticello, G.S., Springer, J.P., Randall, W.C., Baldwin, J.J., and Habecker, C.N. (1994) Positions of His-64 and a bound water in human carbonic anhydrase II upon binding three structurally related inhibitors. *Protein Science* **3**, 118–125.

Sotriffer, C.A., Gohlke, H., and Klebe, G. (2002) Docking into knowledge-based potential fields: A comparative evaluation of DrugScore. *Journal of Medicinal Chemistry* **45**, 1967–1970.

Sotriffer, C.A., Winger, R.H., Liedl, K.R., Rode, B.M., and Varga, J.M. (1996) Comparative docking studies on ligand binding to the multispecific antibodies IgE-La2 and IgE-Lb4. *Journal of Computer Aided Molecular Design* **10**, 305–320.

Stams, T., and Christianson, D.W. (2000) X-ray crystallographic studies of mammalian carbonic anhydrase isozymes. In *Carbonic Anhydrase — New Horizons*, Chegwidden, W.R., and Carter, N., Eds., Birkhäuser Verlag, pp. 159–174.

Stringer, J.L. (2000) A comparison of topiramate and acetazolamide on seizure duration and paired-pulse inhibition in the dentate gyrus of the rat. *Epilepsy Research* **40**, 147–153.

Stubbs, M.T., Reyda, S., Dullweber, F., Moller, M., Klebe, G., Dorsch, D., Mederski, W.W., and Wurziger, H. (2002) pH-dependent binding modes observed in trypsin crystals: lessons for structure-based drug design. *Chem-Biochem* **3**, 246–249.

Sugrue, M.F. (2000) Pharmacological and ocular hypotensive properties of topical carbonic anhydrase inhibitors. *Progress in Retinal and Eye Research* **19**, 87–112.

Supuran, C.T., and Scozzafava, A. (2000) Carbonic anhydrase inhibitors and their therapeutic potential. *Expert Opinion on Therapeutic Patents*, **10**, 575–600.

Supuran, C.T., and Scozzafava, A. (2002) Applications of carbonic anhydrase inhibitors and activators in therapy. *Expert Opinion on Therapeutic Patents* **12**, 217–242.

Takeoka, M., Riviello, J.J., Jr., Pfeifer, H., and Thiele, E.A. (2002) Concomitant treatment with topiramate and ketogenic diet in pediatric epilepsy. *Epilepsia* **43**, 1072–1075.

Taverna, S., Sancini, G., Mantegazza, M., Franceschetti, S., and Avanzini, G. (1999) Inhibition of transient and persistent Na$^+$ current fractions by the new anticonvulsant topiramate. *Journal of Pharmacology and Experimental Therapeutics* **288**, 960–968.

Topliss, J.G. (1977) A manual method for applying the Hansch approach to drug design. *Journal of Medicinal Chemistry* **20**, 463–469.

Tripp, B.C., Smith, K., and Ferry, J.G. (2001) Carbonic anhydrase: New insights for an ancient enzyme. *Journal of Biological Chemistry* **276**, 48615–48618.

Vicker, N., Ho, Y., Robinson, J., Woo, L.L., Purohit, A., Reed, M.J., and Potter, B.V. (2003) Docking studies of sulphamate inhibitors of estrone sulphatase in human carbonic anhydrase II. *Bioorganic and Medicinal Chemistry Letters* **13**, 863–865.

Watling, C.J., and Cairncross, J.G. (2002) Acetazolamide therapy for symptomatic plateau waves in patients with brain tumors. Report of three cases. *Journal of Neurosurgery* **97**, 224–226.

Willis, A.M., Diehl, K.A., and Robbin, T.E. (2002) Advances in topical glaucoma therapy. *Veterinary Ophthalmology* **5**, 9–17.

Zullino, D.F., Krenz, S., and Besson, J. (2003) AMPA blockade may be the mechanism underlying the efficacy of topiramate in PTSD. *Journal of Clinical Psychiatry* **64**, 219–220.

4 Development of Sulfonamide Carbonic Anhydrase Inhibitors

Claudiu T. Supuran, Angela Casini and
Andrea Scozzafava

CONTENTS

4.1 Introduction 68
4.2 Classical Inhibitors 69
    4.2.1 Aromatic Sulfonamides 69
    4.2.2 Heterocyclic Sulfonamides 80
    4.2.3 *Bis*-sulfonamides 89
4.3 Topical Sulfonamides as Antiglaucoma Agents 93
    4.3.1 The Ring Approach 93
    4.3.2 The Tail Approach 112
4.4 Isozyme-Specific Inhibitors 118
    4.4.1 Isozyme I 118
    4.4.2 Isozyme IV 120
    4.4.3 Isozyme III 121
4.5 Selective Inhibitors for Membrane-Associated CAs 121
4.6 Antitumor Sulfonamides 125
4.7 Sulfonamides with Modified Moieties and Other Zinc-Binding Groups 126
4.8 Antiepileptic Sulfonamides and Other Miscellaneous Inhibitors 131
4.9 Aliphatic Sulfonamides 134
4.10 Future Prospects of CAIs 136
References 138

Unsubstituted aromatic sulfonamides have been known since the beginning of research in the field to inhibit carbonic anhydrases (CAs). From the 1950s, potent carbonic anhydrase inhibitors (CAIs) belonging to the heterocyclic sulfonamide class have been developed and led to the development of benzothiadiazine and high-ceiling diuretics as well as to the systemic antiglaucoma drugs acetazolamide, methazolamide,

ethoxzolamide and dichlorophenamide. The discovery of these drugs greatly benefited the chemistry of sulfonamides, as thousands of derivatives belonging to the
heterocyclic, aromatic and *bis*-sulfonamide classes have been synthesized and investigated for their biological activity. In the late 1980s and early 1990s, the topically
effective antiglaucoma CAIs were discovered, and two such drugs — dorzolamide
and brinzolamide — are at present used clinically. Both have been designed by the
ring approach, i.e., by investigating a very large number of ring systems incorporating
sulfamoyl moieties. More recently, the tail approach has been reported for designing
antiglaucoma CAIs with topical activity, but this approach can be extended for other
applications of these pharmacological agents. It consists of attaching tails that induce
the desired physicochemical properties to aromatic/heterocyclic sulfonamide scaffolds that possess derivatizable amino or hydroxy moieties. There has been progress
in the discovery of inhibitors with higher affinity for a certain isozyme, although
clear-cut isozyme-specific inhibitors are not available at present. Inhibitors selective
for the membrane-associated (CAs IV, IX, XII and XIV) or the cytosolic isozymes
are available, being either macromolecular compounds or the positively charged
derivatives (of low molecular weight). CAIs possessing antitumor properties have
also been discovered, with one such derivative — indisulam — in advanced clinical
trials for treating solid tumors. CAIs with good anticonvulsant activity and several
other biomedical applications have recently been reported. Some aliphatic sulfonamides show significant CA inhibitory properties, and compounds that possess zinc-
binding functions different from that of the classical one (of aromatic/heterocyclic
sulfonamide type) have been reported, incorporating, among others, sulfamate, sulfamide and hydroxamate moieties. The field is in constant progress and can lead to
the discovery of interesting pharmacological agents.

4.1 INTRODUCTION

The discovery of CA inhibition with sulfanilamide **4.1** by Mann and Keilin (1940)
was the beginning of a great adventure that led to important drugs widely used to
treat or prevent a multitude of diseases. Sulfonamides constitute an important class
of drugs, with several types of pharmacological agents possessing, among others,
antibacterial, antitumor, anticarbonic anhydrase (CA), diuretic, hypoglycemic, antithyroid or protease inhibitory activity (Northey 1948; Maren 1967; Scozzafava et al.
2003; Supuran et al. 2003; Owa and Nagasu 2000). The very simple sulfanilamide
4.1 lead molecule afforded the development of all these types of pharmacological
agents that have a wide variety of biological actions, as exemplified later for the
antibacterial agent sulfathiazole, the carbonic anhydrase inhibitor acetazolamide
(clinically used for more than 45 years; Maren 1967), the widely used diuretic
furosemide, the hypoglycemic agent glibenclamide, the anticancer sulfonamide
indisulam (in advanced clinical trials), the aspartic HIV protease inhibitor
amprenavir used to treat AIDS and HIV infection and the metalloprotease (MMP)
inhibitors of the sulfonyl amino acid hydroxamate type (Scozzafava et al. 2003;
Supuran et al. 2003; Supuran 2003; Supuran and Scozzafava 2000a, 2001, 2002a,
2002b; Figure 4.1).

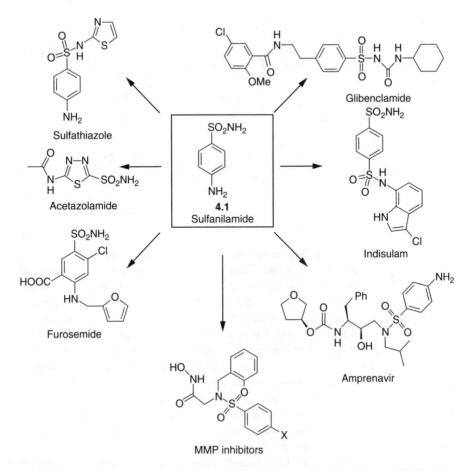

FIGURE 4.1　The main classes of pharmacolological agents developed from sulfanilamide **4.1** as lead molecule: antibacterials (such as sulfathiazole); carbonic anhydrase inhibitors (such as acetazolamide); diuretics (such as furosemide); hypoglycaemic agents (such as glibenclamide); anticancer agents (such as indisulam); anti-AIDS agents (such as the HIV protease inhibitor amprenavir) as well as MMP inhibitors.

4.2　CLASSICAL INHIBITORS

4.2.1　Aromatic Sulfonamides

After the report of Mann and Keilin (1940) that sulfanilamide **4.1** acts as a potent and specific CA inhibitor (CAI), Krebs (1948) in a classical study showed that only unsubstituted sulfonamides of the type $R\text{-}SO_2NH_2$ act as strong inhibitors, the potency of the drug being drastically affected by N-substitution at the sulfonamide moiety. Among the active structures found by Krebs were also the azodyes **4.2** and **4.3**.

Many sulfonamides derived from mono- or bicyclic aromatic hydrocarbons were then prepared and assayed for inhibitory action. The systematic study of Beasley

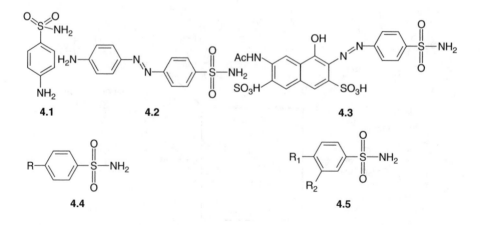

4.1   **4.2**   **4.3**

4.4   **4.5**

et al. (1958) revealed some important features for inhibitors belonging to the class of aromatic sulfonamides: (1) 4-substituted derivatives of benzenesulfonamide, of the type **4.4**, as well as of diphenyl-4-sulfonamide (**4.4b**; Table 4.1) are stronger inhibitors than are naphthalene-1- or 2-sulfonamides; (2) 4-substituents inducing good activities in compounds **4.4** include halogens, acetamido and alkoxycarbonyl, with esters of 4-sulfamoyl-benzoic acid (**4.4f–m**) needing special mention (see Table 4.1) because they are very potent CAIs (including the n-butyl, n-hexyl, n-nonyl and benzyl esters) in this class; (3) 2-substituted- and 2,4- and 3,4-disubstitued-benzenesulfonamides are generally weaker inhibitors than are 4-substituted derivatives; (4) promising activities as well as desired physicochemical properties (such as an increase in hydrosolubility) are seen for compounds possessing carboxy-, hydrazido-, ureido-, thioureido- and methylamino moieties in position 4, such as **4m–r** (Beasley et al. 1958; Supuran 1994).

Some of these compounds were recently reinvestigated. To obtain compounds for the selective radiolabeling of erythrocytes (an important tool in diagnostic nuclear medicine) via the enzyme–inhibitor approach, Singh and Wyeth (1991) reported the use of CAIs of type **4.5**. Such inhibitors possess lipophilic groups as well as reactive halogen atoms, which, by reaction with amino acid residues within the CA active site, can irreversibly inactivate the enzyme. Table 4.2 presents the comparative inhibition data for red cell isozymes (CA I and CA II) with compounds of type **4.5** incorporating reactive halogen atoms.

To obtain compounds that act as specific inhibitors for the brain enzyme, Cross et al. (1978) prepared substituted benzenedisulfonamides of type **4.8** starting from substituted sulfanilamides **4.6**, as outlined in Scheme 4.1. Such compounds showed IC$_{50}$ values of the order of magnitude 10^{-7} M, and one (**4.8b**) exhibited anticonvulsant activity in rats, with a minimal diuretic effect and a low level of metabolic acidosis as side effects.

A related approach for the preparation of diversely substituted sulfanilamide derivatives was reported by Shepard et al. (1991) for the preparation of CAIs with topical effects as antiglaucoma drugs. These researchers prepared hydroxyalkyl-sulfonylbenzenesulfonamide of types **4.12** and **4.13** (Scheme 4.2).

TABLE 4.1
Inhibition with Substituted Benzenesulfonamides

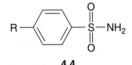

4.4

4.4	R	$IC_{50} \times 10^8$ (M)
a	H	30
b	Ph	2.6
c	Cl	19
d	Br	10
e	I	1.4
f	COOH	26
g	COOMe	1.1
h	COOEt	0.6
i	COOnBu	0.1
j	COOn-C$_6$H$_{13}$	0.5
k	COOn-C$_9$H$_{19}$	0.04
l	COOCH$_2$Ph	0.032
m	COOCH$_2$CH$_2$OH	1.0
n	CONH$_2$	24
o	CONHNH$_2$	18
p	NHCONH$_2$	27
q	NHCSNH$_2$	28
r	CH$_2$NH$_2$	10

Source: From Beasley, Y.M. et al. (1958) **10**, 696–705 and Supuran, C.T. (1994) In *Carbonic Anhydrase and Modulation of Physiologic and Pathologic Processes in the Organism*, Puscas, I., Ed., Helicon Press, Timisoara, Romania, pp. 29–111. With permission.

Because 4-chloro- and 4-bromobenzenesulfonamides hardly react with any nucleophile, the protection of the sulfonamido moiety of **4.9** was necessary, and this has been done by a sulfonilformamidino group (derivatives **4.10**). This reaction was performed by using dimethylformamide-dimethylacetal. Compounds **4.10** prepared this way are treated thereafter with a hydroxythiol in anhydrous solvents, in the presence of NaH, which by SN_{AR} reactions leads to the corresponding hydroxyalkylthio derivatives **4.11**. These derivatives are deprotected in an alkaline medium to afford the corresponding sulfonamides **4.12**. Oxidation of sulfides **4.12** to sulfones **4.13** was achieved with H_2O_2 in acetic acid, 3-chloroperoxibenzoic acid in ethyl acetate or oxone in water (Shepard et al. 1991). Among the benzenesulfonamides

TABLE 4.2
CA Inhibition against Isozymes I and II with Substituted Benzenesulfonamides 4.5 Prepared as Selective Radiolabeling Agents for Erythrocytes

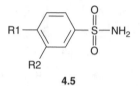

4.5

4.5	R1	R2	$IC_{50} \times 10^7$ (M)	
			CA I	CA II
a	$4\text{-I-C}_6\text{H}_4\text{NHSO}_2\text{-}$	H	0.6	0.05
b	$4\text{-I-C}_6\text{H}_4\text{SO}_2\text{NH-}$	H	1.7	2.3
c	$4\text{-I-C}_6\text{H}_4\text{S-}$	H	1.5	0.04
d	$4\text{-I-C}_6\text{H}_4\text{SO}_2\text{-}$	H	0.18	0.001
e	$\text{ClCH}_2\text{CONH-}$	H	1.1	0.9
f	$\text{ClCH}_2\text{CONH-}$	I	0.9	1.1
g	$\text{BrCH}_2\text{CONH-}$	H	1.0	0.9
h	$\text{ICH}_2\text{CONH-}$	H	0.4	0.2
i	H	$\text{ClCH}_2\text{CONH-}$	34	—
j	H	$\text{ICH}_2\text{CONH-}$	25	—
k	I	$\text{ICH}_2\text{CONH-}$	15	0.7

Source: From Singh, J., and Wyeth, P. (1991) *Journal of Enzyme Inhibition* **5**, 1–24. With permission.

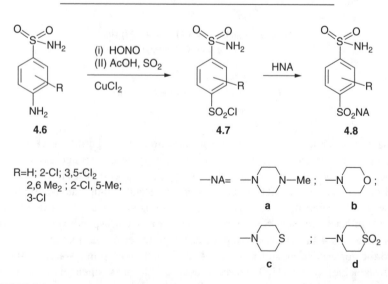

SCHEME 4.1

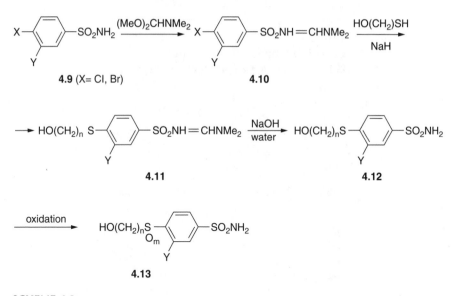

SCHEME 4.2

4.9 investigated by Shepard et al. (1991) only 3-chloro-4-nitrobenzenesulfonamide easily reacted with mercaptoalcohols in an aqueous medium (in the presence of sodium acetate) or in an alcoholic medium (EtOH and Et$_3$N), the substitution reaction being facilitated by the labilizing effect of the nitro group in the *ortho* position to the halogeno-moiety leaving group. Furthermore, in this unique case, it was not necessary to protect the SO$_2$NH$_2$ group.

Compounds **4.13** prepared by this approach (Table 4.3) were highly active CAIs. The inhibitory potency increased with the increase in the length of the alkyl chain from 2 to 5 carbon atoms. Without exception, sulfones were more potent inhibitors than were the corresponding sulfides, a phenomenon that can be correlated with the decrease of pK_a of the SO$_2$NH$_2$ protons by hydroxyalkylsulfonyl moieties (as compared to hydroxyalkylthio groups). Table 4.3 also gives the pK_a values for these sulfonamides. The presence of fluorine or chlorine atoms in position 3 of the benzene nucleus generally led to stronger CAIs, whereas moieties such as COOH and COOMe led to less potent inhibitors. Groups such as NO$_2$ or NH$_2$ in this position also led to active compounds (Shepard et al. 1991).

Shepard et al. (1991) also prepared 2-chloro-5-fluoro-4-substitued-benzene-sulfonamides of types **4.16** and **4.17** according to the strategy presented in Scheme 4.3. By treating 2-fluoro-5-chlorotoluene **4.14** with bromine in homolytic conditions, followed by reacting with sodium diethylmalonate, deprotecting the monocarboxylic acid, and reducing its ester, the hydroxypropyl derivative **4.15** was obtained, which was acetylated, chlorosulfonated and amidated. Sulfonamides **4.16** and **4.17** prepared by this procedure showed IC$_{50}$ values ca. 2 nM against hCA II, being potent CAIs, similar to the compounds **4.13** (Shepard et al. 1991).

Boddy et al. (1989) investigated erythrocyte CA inhibition with the salicylic acid derivatives **4.18**. These compounds were prepared by reacting phenyl salicylate with

TABLE 4.3
CA II Inhibition with Aromatic
Sulfonamides 4.13 and Their pKa Values

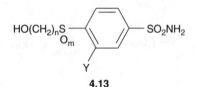

4.13

4.13	n	m	Y	IC_{50} (M)	pK_a
a	2	0	H	37	9.27
b	2	2	H	24	9.38
c	3	0	H	20	10.23
d	3	2	H	12	8.98
e	4	0	H	9	10.17
f	4	2	H	7	9.60
g	5	0	H	4.3	9.25
h	3	0	Cl	9	9.45
i	2	0	F	9	9.76
j	2	2	F	5	9.00
k	3	0	F	3.5	9.64
l	3	2	F	3.5	9.02
m	4	0	F	1.5	9.82
n	4	2	F	4	9.18
o	2	0	NO_2	21	9.42
p	3	0	NO_2	23	9.38
q	3	2	NO_2	—	8.00
r	2	2	COOMe	30	9.20
s	3	0	COOH	10^4	9.94
t	3	2	COOH	900	9.28
u	2	2	NH_2	51	9.50
v	3	2	NH_2	23	9.15

Source: From Shepard, K.L. et al. (1991) *Journal of Medicinal Chemistry* **34**, 3098–3105. With permission.

sulfanilamide ($n = 0$) and homosulfanilamide ($n = 1$). The compounds strongly inhibit human red cell CA (with K_is of 53 and 98 nM, respectively), probably because they are stronger acids than the parent unacylated sulfonamides. The same authors studied the binding to erythrocytes of these CAIs compared with that of compounds possessing $-SO_2NHMe$ and $-SO_2NH$-2-pyridyl groups (which are not CAIs). The unsubstituted sulfonamides **4.18** strongly bound to red cells, with affinities 10 to 100 times higher than the substituted derivatives incorporating *N*-modified sulfonamide moieties, whereas their affinities for CA were also 100 to 1000 times higher.

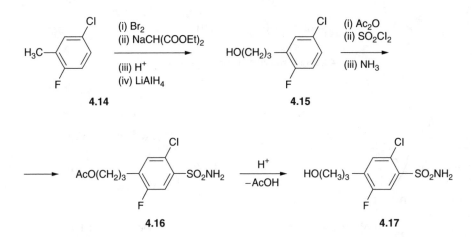

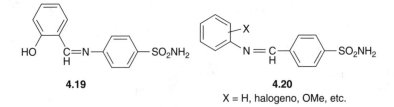

SCHEME 4.3

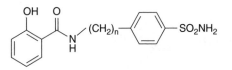

4.18 n = 0,1

Another approach used to prepare CA inhibitors from this class involved Schiff bases of aromatic sulfonamides, such as **4.19** and **4.20**. Derivative **4.19** was prepared from sulfanilamide and salicyl aldehyde, whereas **4.20**, from 4-sulfamoyl-benzaldehyde and substituted anilines, had already been reported by Beasley et al. (1958). This class of CAIs was subsequently extensively investigated by our group (Supuran et al. 1996b, 1996c, 1997b; Popescu et al. 1999; Scozzafava et al. 2000a). Initially, a large series of sulfanilamide-derived Schiff bases of type **4.21**, which incorporated a large number of aromatic and heterocyclic moieties (arising from the aldehyde component of the condensation), was prepared (Supuran et al. 1996a). These compounds were potent CAIs and also showed more affinity for the membrane-bound (CA IV) than the cytosolic (CA I and CA II) isozymes. The related derivatives **4.22** and **4.23**, prepared from aromatic/heterocyclic sulfonamides and chalcones (Supuran et al. 1996c), showed the same interesting biological activities, being generally more potent CA IV than CA I/II inhibitors. Even stronger CAIs were then reported, which

4.21: Ar = substituted phenyl; hetaryl

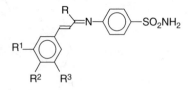

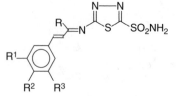

4.22   R=Me, Ph, subst. Ph
R^1-R^3=H, Cl, MeO, Me$_2$N

4.23   R=Me, Ph, subst. Ph
R^1-R^3=H, Cl, MeO, Me$_2$N

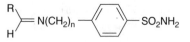

4.24 (n = 0–2)

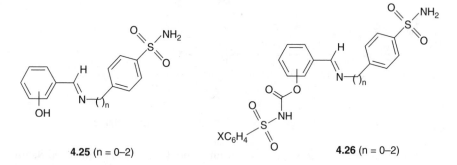

4.25 (n = 0–2)

4.26 (n = 0–2)

For **4.25** and **4.26**: OH and bulky substituents in *ortho* and *para*; X = *p*-F-, *p*-Cl-, *p*-Me-
and *o*-Me.

incorporated homosulfanilamide, *p*-aminoethylbenzenesulfonamide and different
aromatic/heterocyclic aldehyde moieties of types **4.24** to **4.26** (Popescu et al. 1999;
Scozzafava et al. 2000a). Many of these compounds showed affinities in the low
nanomolar range for the physiologically relevant isozymes CA I, II and IV, with a
slightly higher affinity for CA IV as compared to that for the cytosolic isozymes.

It is worth noting that though the compound saccharin **4.27** has a substituted
sulfonamido group, Supuran and Banciu (1991) showed that it acts as a weak CAI
(IC$_{50}$ = 97 μ*M* against hCA II). It is not known whether this has physiological
consequences in humans.

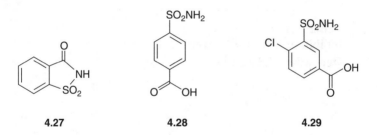

4.27   **4.28**   **4.29**

The reaction of 4-carboxy-benzenesulfonamide **4.28** or 4-chloro-3-sulfamoyl benzoic acid **4.29** with carboxy-protected amino acids/dipeptides, or aromatic/heterocyclic sulfonamides/mercaptans afforded the corresponding benzene-carboxamide derivatives **4.30** and **4.31** (Table 4.4; Mincione et al. 2001). These amides were tested as inhibitors of three isozymes, CAs I, II and IV. Some of the new derivatives showed affinity in the low nanomolar range for isozymes CA II and CA IV, which are involved in aqueous humor secretion within the eye and were tested as topically acting antiglaucoma agents in normotensive and glaucomatous rabbits. Good *in vivo* activity and prolonged duration of action have been observed for some of these derivatives, whereas some of the 4-chloro-3-sulfamoyl benzenecarboxamides **4.31** showed higher affinity for CA I than for the sulfonamide avid isozyme CA II.

Carboxamides structurally related to **4.30** were also reported by several other groups. Whitesides' group reported derivatives of 4-carboxy-benzenesulfonamide to which oligopeptidyl moieties were attached of type **4.32** (Jain et al. 1994; Boriack et al. 1995) and **4.33** (Avila et al. 1993; Gao et al. 1995). In another series of such derivatives, oligoethylene glycol units were attached to 4-carboxy-benzenesulfonamide and the terminal hydroxy moiety of the tail was derivatized by acyl amino moieties. [Six derivatives of type **4.34** were thus obtained by Gao et al. (1996).] Finally, Baldwin's group reported several structurally related inhibitors of type **4.35**, obtained again from 4-carboxy-benzenesulfonamide by attaching peptidyl moieties incorporating nipecotic acid at its carboxy group (Burbaum et al. 1995).

Jain's group reported carboxamides of type **4.36,** incorporating fluoro-substituted anilines (Doyon and Jain 1999). Some of these derivatives showed very good hCA II inhibitory properties.

Casini et al. (2000) reported an alternative approach for obtaining water-soluble, potent CAIs with putative applications as agents used to treat ocular hypertension and glaucoma. Thus, 4-isothiocyanato-benzenesulfonamide **4.37**, obtained from sulfanilamide **4.1** and thiophosgene (Scheme 4.4), was reacted with many amines, amino acids or oligopeptides, and the thioureas **4.38** thus obtained showed excellent CA inhibitory properties against isozymes I, II and IV (Table 4.5) and water solubilities either as sodium salts (for the amino acid/ologopeptide derivatives) or as hydrochlorides/triflates (in the case of the amine derivatives). The nucleophiles used in the syntheses were chosen such that they had pK_a values in the physiological range. More precisely, salts of these new inhibitors applied in the eyes of experimental animals generally possessed pH values in the range of 6.5 to 7.0. Such salts were applied topically in the eyes of normotensive or glaucomatous rabbits and produced a powerful, long-lasting reduction of intraocular pressure (IOP).

TABLE 4.4
Inhibition Data for Benzene-Carboxamide
Derivatives 4.30 and 4.31

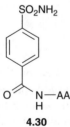

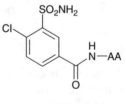

		K$_i$ (nM)		
4.30	**Inhibitor AA**	**hCA I**[a]	**hCA II**[a]	**bCA IV**[b]
a	Gly	500	85	200
b	β-Ala	435	85	190
c	GABA	415	73	120
d	Ala	425	79	75
e	Val	420	74	150
f	Leu	405	53	90
g	Ile	350	48	86
h	α-Ph-Gly	250	36	75
i	Ser	440	120	325
j	β-Ph-Ser	54	21	62
k	Thr	475	115	350
l	Cys	450	110	230
m	Met	400	66	150
n	Asp	425	73	150
o	Asn	410	72	155
p	Glu	430	74	160
q	Gln	425	75	160
r	His	270	40	125
s	Phe	560	52	130
t	Tyr	535	45	120
u	DOPA	520	43	105
v	GlyGly	350	42	50
x	β-AlaHis	270	16	21
y	HisGly	380	10	19
z	HisPhe	240	9	23
aa	AlaPhe	320	12	24
ab	LeuGly	400	9	16
ac	-C$_6$H$_4$-SO$_2$NH$_2$ (*p*)	38	10	25
ad	CH$_2$-C$_6$H$_4$-SO$_2$NH$_2$ (*p*)	40	7	16
ae	(CH$_2$)$_2$C$_6$H$_4$-SO$_2$NH$_2$ (*p*)	40	5	13
af	-1,3,4-Thiadiazole-SO$_2$NH$_2$	35	3	15
ag	-1,3,4-Thiadiazole-SH	38	5	14

TABLE 4.4 (continued)
Inhibition Data for Benzene-Carboxamide
Derivatives 4.30 and 4.31

4.31	Inhibitor AA	hCA I[a]	hCA II[a]	bCA IV[b]
			K_I (nM)	
a	Gly	210	450	600
b	β-Ala	205	450	610
c	GABA	200	400	540
d	Ala	175	420	475
e	Val	130	330	360
f	Leu	125	305	330
g	Ile	150	280	360
h	α-Ph-Gly	105	160	240
i	Ser	170	290	395
j	β-Ph-Ser	37	62	98
k	Thr	320	520	735
l	Cys	300	490	720
m	Met	240	410	650
n	Asp	285	300	355
o	Asn	240	275	330
p	Glu	305	340	400
q	Gln	250	305	410
r	His	103	210	350
s	Phe	135	250	340
t	Tyr	115	250	325
u	DOPA	100	240	330
v	GlyGly	69	85	170
x	β-AlaHis	30	36	50
y	HisGly	41	50	73
z	HisPhe	27	38	45
aa	AlaPhe	36	44	72
ab	LeuGly	39	51	80
ac	–C_6H_4-SO_2NH_2 (p)	24	30	66
ad	CH_2-C_6H_4-SO_2NH_2 (p)	24	28	61
ae	$(CH_2)_2C_6H_4$-SO_2NH_2 (p)	21	26	63
af	–1,3,4-Thiadiazole-SO_2NH_2	13	10	22
ag	–1,3,4-Thiadiazole-SH	15	11	21

[a] Human (cloned) isozymes.
[b] From bovine lung microsomes by the esterase method.

Source: From Mincione, F. et al. (2001) *Bioorganic and Medicinal Chemistry Letters* **11**, 1787–1791. With permission.

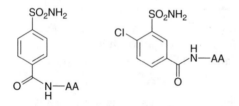

4.30   **4.31**

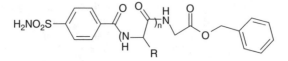

4.32: R = H, Ph; PhCH$_2$; n = 2–4

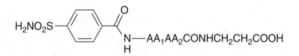

4.33: AA$_1$ = AA$_2$ = L-Leu; D-Leu; L-Thr; D-Thr; L-Ser; D-Ser; Gly

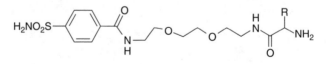

4.34: Gly; Leu; Ser; Lys; Glu; Phe derivatives

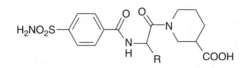

4.35: R = *i*-Pr-CH$_2$; HOOCCH$_2$; H$_2$NCOCH$_2$CH$_2$

4.2.2 HETEROCYCLIC SULFONAMIDES

Investigation of heterocyclic sulfonamides as CAIs has been fostered by the discovery of Davenport (1945) that thiophene-2-sulfonamide is 40 times more active than sulfanilamide as a CAI. Shortly thereafter, in a classical work, Roblin's group prepared a very large series of heterocyclic sulfonamides derived from the most important ring systems (imidazole, alkyl- and aryl imidazoles, benzimidazoles, benzothiazole,

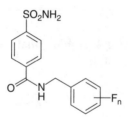

4.36: n = 1–5

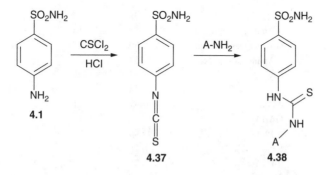

SCHEME 4.4

1,2,4-triazole, thiazole, tetrazole and alkyl/aryl tetrazoles, 1,3,4-thiadiazole, pyrimidine, pyrazine, etc.; Roblin and Clapp 1950; Miller et al. 1950).

Two strategies were used to prepare such derivatives: (1) sulfonation/chlorosulfonation of the heterocyclic compound (RH in Scheme 4.5), followed by transformation of the sulfonic acid **4.39** into the sulfonyl chloride **4.40** and subsequent amidation with formation of the desired sulfonamide **4.42**; and (2) oxidative chlorination of heterocyclic mercaptans **4.41** (generally in acetic acid as a solvent), followed by amidation of the sulfonyl chloride generated this way (Scheme 4.5) and formation of the desired heterocyclic sulfonamide **4.42**.

The second approach has largely been applied to the preparation of heterocyclic sulfonamides, due to the fact that either sulfonation/chlorosulfonation cannot be always carried out successfully for most heterocycles or because heterocyclic mercaptans are easily prepared by a variety of methods, the sulfonyl chlorides are generally obtained in very good yields and the amidation of sulfonyl chlorides **4.40** occurs in quantitative yield (Roblin and Clapp 1950; Miller et al. 1950). The large number of compounds prepared in the classical studies of Roblin's group helped establish two important facts connecting chemical structure with CA inhibitory action: (1) five-membered derivatives were more effective CAIs than six-membered ring compounds, and (2) the presence of nitrogen and sulfur atoms within the ring led to the most potent inhibitors. Thus, extremely powerful inhibitors were found to be 5-substituted-1,3,4-thiadiazole-2-sulfonamides **4.43** (R = NH$_2$) and **4.44** (R = NHAc), benzothiazole-2-sulfonamide **4.45**, as well as 1,3,4-thiadiazoline-sulfonamides

TABLE 4.5
CA Inhibition Data with 4-Isothiocyanatobenzenesulfonamide
4.37 and the Thioureas 4.38

4.38	A-NH$_2$	K_I (nM)		
		hCA I[a]	hCA II[a]	bCA IV[b]
4.37	—	5000	185	300
a	2-Pyridylmethylamine	135	45	76
b	2-Pyridylethylamine	124	40	77
c	Phenethylamine	125	42	75
d	Histamine	92	33	56
e	o-Aminobenzoic acid	55	13	29
f	m-Aminobenzoic acid	50	18	36
g	p-Aminobenzoic acid	59	16	40
h	Gly	62	20	39
i	Ala	54	12	36
j	β-Ala	47	11	35
k	GABA	45	11	33
l	α-Ph-Gly	40	6	15
m	Ser	28	3	15
n	β-Ph-Ser	24	2	9
o	Thr	30	4	16
p	Cys	25	5	10
q	Met	27	4	12
r	Val	23	5	13
s	Leu	21	5	12
t	Ile	22	4	10
u	Asp	35	8	17
v	Asn	27	6	11
w	Glu	38	10	19
x	Gln	40	9	23
y	Pro	110	25	68
z	His	32	5	15
aa	Phe	30	6	13
ab	Tyr	26	5	13
ac	DOPA	35	7	18
ad	Trp	47	11	24
ae	Lys	97	12	55
af	Arg	115	15	61
ag	GlyGly	54	15	32
ah	β-AlaHis	23	3	10
ai	HisGly	13	1	6
aj	HisPhe	21	3	9
ak	AlaPhe	23	4	13
al	LeuGly	20	5	13
am	AspAsp	50	15	32

TABLE 4.5 (continued)
CA Inhibition Data with 4-Isothiocyanatobenzenesulfonamide
4.37 and the Thioureas 4.38

an	ProGlyGly	20	3	11
ao	AspAspAspAsp	59	10	27

[a] Human (cloned) isozymes.
[b] From bovine lung microsomes by the esterase method.

Source: Reported from Casini, A. et al. (2000) *Journal of Medicinal Chemistry* **43**, 4884–4892. With permission.

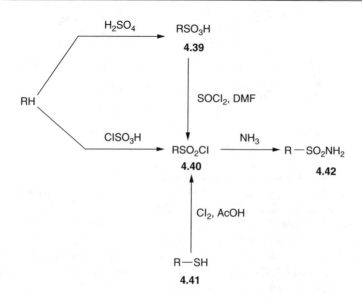

SCHEME 4.5

4.46. Compound **4.44**, acetazolamide, proved to possess excellent pharmacological properties (low toxicity, very good CA inhibitory action, excellent bioavailability) and was shortly thereafter introduced into clinical medicine as the first nonmercurial diuretic (Maren, 1967). Since 1954, acetazolamide, has been continuously used to manage glaucoma, gastroduodenal ulcers and many other disorders (Maren 1967; Supuran and Scozzafava 2000a, 2002a).

The clinical success of acetazolamide fostered a further search for active structures in this ring system. Thus, Vaughan et al. (1956) and Young et al. (1956) prepared and assayed many acetazolamide-like sulfonamides **4.47** possessing 5-acylamido and 5-arylsulfonamido moieties, as well as 2-sulfonamido-Δ^5-1,3,4-thiadiazoline derivatives of type **4.46**. Some of the most active inhibitors are presented in Table 4.6 (thiadiazoles) and Table 4.7 (thiadiazolines), with structurally related compounds investigated subsequently by other groups.

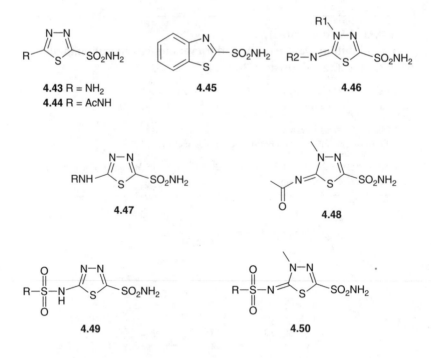

4.43 R = NH₂
4.44 R = AcNH

4.45

4.46

4.47

4.48

4.49

4.50

Data in Table 4.6 show that the acylation of amine **4.43** led to inhibitors with enhanced efficiency; acylamido- and alkyl/aryl-sulfonamido groups were equally useful in inducing this effect. CA inhibitory action is particularly higher in derivatives bearing lipophilic moieties (**4.47f, g, m, o**). On the other hand, compounds substituted with arylamido- and arylsulfonamido- groups are stronger acids than the other derivatives, which was also correlated with increased CA inhibition. (Thio)ureido-substituted derivatives **4.47p–s** recently reported are hydrosoluble, but were unfortunately not active by the topical route for the management of glaucoma (Supuran 1994).

Methazolamide (**4.48**), already reported in 1956, has been used in clinical medicine for more than 35 years. Although it has a structure similar to that of acetazolamide (**4.44**), methazolamide is more liposoluble, possesses only one acidic proton and is more diffusible in body fluids than is acetazolamide (Maren 1967). As methazolamide is relatively nontoxic and because of its pharmacological properties the same physiological responses being achieved at only one third the dose of acetazolamide, it is not surprising that many derivatives related to methazolamide, of type **4.46,** have been investigated (Table 4.7). Generally, the activity was influenced in the same way as for compounds **4.47** by changing the group in position 5 of the heterocyclic ring (Supuran 1994; see Table 4.7). The same is true for the moiety substituting the N-4 atom (although the greater majority of these compounds bear an N-Me group). Importantly, this research line led to the discovery topically active inhibitors in glaucoma. Compound **4.46g**, trifluoromethazolamide, was the first inhibitor possessing such properties (see later).

Recently, 1,3,4-thiadiazole-2-sulfonamide and the corresponding thiadiazolines bearing 5-alkyl/arylsulfonamido moieties, of types **4.49** and **4.50**, respectively, have

TABLE 4.6
Inhibition Data for hCA II with
1,3,4-Thiadiazole-2-Sulfonamide
Derivatives 4.47

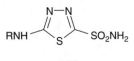

4.47

4.47	R	$IC_{50} \times 10^8$ (*M*)
a	H	60.00
b	Ac	1.20
c	HCO-	2.00
d	Me_2CHCO-	1.50
e	CF_3CO-	4.50
f	PhCO-	0.70
g	$PhCH_2OCO-$	0.50
h	MeOCO-	0.40
i	2-pyrydyl-CO	0.60
j	$n\text{-}Pr_2CHCO-$	0.10
k	$cyclo\text{-}C_4H_7CO-$	0.80
l	1-Adamantyl-CO-	1.10
m	$PhSO_2$	0.30
n	$4\text{-}H_2N\text{-}C_6H_4\text{-}SO_2-$	0.40
o	CF_3SO_2-	0.15
p	H_2NCS-	0.70
q	H_2NCO-	0.85
r	MeNHCO-	0.73
s	PhNHCO-	0.70

Source: From Roblin, R.O., Jr., and Clapp, J.W. (1950) *Journal of the American Chemical Society* **72**, 4890–4892 and Supuran, C.T. (1994) In *Carbonic Anhydrase and Modulation of Physiologic and Pathologic Processes in the Organism*, Puscas, I., Ed., Helicon Press, Timisoara, Romania, pp. 29–111. With permission.

been investigated in detail (Table 4.8 and Table 4.9; Supuran et al. 1998a). Derivatives **4.49** and **4.50** contained alkylsulfonyl-, dimethylaminosulfamoyl- or halogeno-, alkyl-, methoxy-, amino- and nitro- substituted phenylsulfonyl moieties, and were chosen to obtain compounds with different physicochemical properties (e.g., enhanced lipophilicity, or, conversely, enhanced hydrosolubility) and also those

TABLE 4.7
CA II Inhibition with Δ^5-1,3,4-Thiadiazoline-2-Sulfonamide Derivatives 4.46

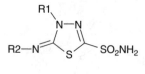

4.46

4.46	R1	R2	$IC_{50} \times 10^8$ (M)
a	Me	H	10
b	Me	Ac	0.9
c	Me	EtCO-	A
d	Me	ClCH$_2$CO-	A
e	Me	Cl$_2$CHCO-	A
f	Me	Cl$_3$CO-	A
g	Me	F$_3$CO-	A
h	Me	PhCO-	A
i	Me	n-Pr$_2$CHCO-	0.6
j	Me	4-MeNH-cyclohexylCO-	0.9
k	Me	1-Adamantylacetyl	1.2
l	Me	CF$_3$SO$_2$-	0.1
m	Me	3,4,5-(MeO)$_3$C$_6$H$_2$CO-	4.0
n	Me	4-Cl-C$_6$H$_4$-CO-	5.2
o	Me	4-I-C$_6$H$_4$-CO-	4.7
p	Et	Ac	1.1
q	Et	EtCO-	1.2
r	PhCH$_2$	Ac	1.9

Source: From Vaughan, J.R. et al. (1956) *Journal of Organic Chemistry* **21**, 700–771 and Supuran, C.T. (1994) In *Carbonic Anhydrase and Modulation of Physiologic and Pathologic Processes in the Organism*, Puscas, I., Ed., Helicon Press, Timisoara, Romania, pp. 29–111. With permission.

containing groups that would allow an easy derivatization, which can be exploited to prepare isotopically labeled compounds, envisaging a possible application of such inhibitors in PET imaging (Supuran et al. 1998a). As seen from the data in Table 4.8 and Table 4.9, most of these sulfonamides are very effective inhibitors of isozymes I, II and IV. Some 1,3,4-thiadiazole-5-sulfonamides incorporating 2-aminoacyl moieties were reported by Jayaweera et al. (1991).

Derivatives of thiophene-2-sulfonamide of the types **4.52** to **4.55** that possess different substitution patterns, such as 5-arylthio, 5-arylsulfinyl and 5-arylsulfonyl moieties, were prepared in the search for anticonvulsant and cerebrovasodilator agents (Scheme 4.6; Barnish et al. 1981). Sulfones **4.55** were generally more active

TABLE 4.8
Compounds of Type 4.49: Inhibition Data (K_I) toward hCA I, hCA II and bCA IV, Chloroform–Buffer Partition Coefficients and Solubility Data

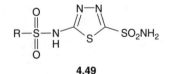

4.49

4.49	R	K_I (nM)			Log P^c	S^d (mg/100 ml)
		hCA I[a]	hCA II[a]	bCA IV[b]		
a	C_6H_5 (Benzolamide)	15	9	12	0.0001	40
b	Me	10	6	5	0.024	542
c	$PhCH_2$	7	5	6	0.104	67
d	4-Me-C_6H_4	5	4	3	0.025	2030
e	4-F-C_6H_4	4	4	7	0.019	1232
f	4-Cl-C_6H_4	4	3	5	0.027	347
g	4-Br-C_6H_4	3	2	4	0.140	295
h	4-I-C_6H_4	2	1	2	0.165	280
i	4-MeOC_6H_4	5	3	4	0.030	195
j	4-AcNHC_6H_4	10	3	8	0.024	900
k	4-H_2N-C_6H_4	6	2	5	0.015	1050
l	3-H_2N-C_6H_4	9	1	7	0.016	975
m	4-O_2N-C_6H_4	3	1	2	0.013	1700
n	3-O_2N-C_6H_4	2	0.9	1	0.014	1278
o	2-O_2N-C_6H_4	5	3	4	0.026	549
p	Me_2N	19	8	13	0.382	60
q	2-HO_2CC_6H_4	1	0.5	0.6	0.001	176
r[e]	4-Me_3Py[+]C_6H_4	18	4	10	<0.0001	475
s[f]	4-Ph_3Py[+]C_6H_4	360	110	320	0.430	320
t	2,4-$(O_2N)_2C_6H_3$	12	5	28	0.740	50
u	4-Cl-3-O_2N-C_6H_3	9	3	7	0.085	407
v	2,4,6-$Me_3C_6H_2$	15	9	12	0.540	72

[a] Human (cloned) isozymes.
[b] From bovine lung microsomes.
[c] Chloroform–buffer partition coefficient.
[d] Solubility in pH 7.40 buffer at 25°C.
[e] As perchlorate salt.
[f] As tetrafluoroborate salt.

Source: From Supuran, C.T. et al. (1998a). With permission.

than sulfoxides **4.54**, which in turn were more active than sulfides **4.53** (IC_{50} values 10^{-8} to 10^{-9} *M*). More recently, Chow et al. (1996) reported a series of 5-substituted-3-thiophenesulfonamides with good inhibitory properties.

TABLE 4.9
Compounds of Type 4.50: Inhibition toward hCA I, hCA II and bCA IV, Partition Coefficient and Solubility Data

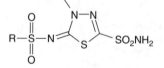

4.50

4.50	R	K_I (nM)			Log P^c	S^d (mg/100 ml)
		hCA I[a]	hCA II[a]	bCA IV[b]		
a	Me	17	4	8	0.104	240
b	PhCH$_2$	6	8	9	0.097	115
c	4-Me-C$_6$H$_4$	5	3	3	0.009	50
d	4-F-C$_6$H$_4$	8	4	7	0.144	330
e	4-Cl-C$_6$H$_4$	8	3	5	0.217	134
f	4-Br-C$_6$H$_4$	5	2	6	0.286	127
g	4-I-C$_6$H$_4$	1	0.6	1	0.310	78
h	4-MeOC$_6$H$_4$	6	3	5	0.209	64
i	4-AcNHC$_6$H$_4$	2	0.7	2	0.066	800
j	4-H$_2$N-C$_6$H$_4$	1	0.6	0.8	0.395	110
k	3-H$_2$N-C$_6$H$_4$	1	0.5	0.8	0.338	125
l	4-O$_2$N-C$_6$H$_4$	8	4	6	0.402	75
m	3-O$_2$N-C$_6$H$_4$	7	2	5	0.396	81
n	2-O$_2$N-C$_6$H$_4$	5	1	3	0.542	62
o	Me$_2$N	9	5	8	0.441	60
p	2-HO$_2$CC$_6$H$_4$	1	0.2	0.5	0.002	117
q[e]	4-Me$_3$Py$^+$C$_6$H$_4$	17	4	12	<0.0001	75
r[f]	4-Ph$_3$Py$^+$C$_6$H$_4$	455	110	180	0.345	64
s	2,4-(O$_2$N)$_2$C$_6$H$_3$	10	4	8	0.852	77
t	4-Cl-3-O$_2$N-C$_6$H$_3$	7	2	5	0.098	140
u	2,4,6-Me$_3$C$_6$H$_2$	13	7	9	0.659	38

[a] Human (cloned) isozymes.

[b] From bovine lung microsomes.

[c] Chloroform–buffer partition coefficient.

[d] Solubility in pH 7.40 buffer at 25°C.

[e] As perchlorate salt.

[f] As tetrafluoroborate salt.

Source: From Supuran, C.T. et al. (1998a). With permission

Other approaches were designed to obtain bicyclic inhibitors commencing from 5-amino-1,3,4-thiadiazole-2-sulfonamide **4.43** or 2-amino-thiazole-5-sulfonamide, which were annulated with α-bromoketones, phosgene, carbonsuboxide, ethene-sulfonylchloride and chlorosulfonyl isocyanate (Scheme 4.7; Barnish et al. 1980;

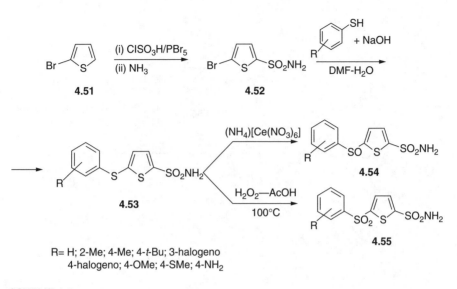

R= H; 2-Me; 4-Me; 4-*t*-Bu; 3-halogeno
4-halogeno; 4-OMe; 4-SMe; 4-NH$_2$

SCHEME 4.6

Katritzky et al. 1987). The imidazo[2,1-*b*]thiadiazole-5-sulfonamides (**4.56**) as well as the thiadiazolo[3,2-*a*]pyrimidinesulfonamides (**4.58** to **4.61**) and thiadiazolo[3,2-*a*]triazinesulfonamides (**4.62**) thus obtained were very effective inhibitors, with IC$_{50}$ values of 10^{-8} to 10^{-9} *M* (Katritzky et al. 1987).

Some thieno[2,3-*b*]pyrrole-5-sulfonamides of types **4.68** and **4.69** were prepared as outlined in Scheme 4.8, but no data have been published on their inhibitory properties toward CAs (Hartman and Halczenko 1989).

Annulation of thiophene-3-carbaldehyde with methyloxycarbonylazide afforded the thieno[2,3-*b*]pyrrole **4.64**, which was protected at the NH group by treatment with benzenesulfonyl chloride. After chlorosulfonation, amidation and conversion of the carboxymethyl group to a secondary amine, inhibitors of types **4.68** and **4.69** were obtained.

4.2.3 *Bis*-sulfonamides

Compounds that possess more than one sulfonamide group in their molecule were tested as CAIs in the search for either more active structures or to design compounds with a different biological activity, e.g., saluretics and high-ceiling diuretics (the CA inhibitory effect being subsidiary in the latter cases; Beasley et al. 1958; Maren 1967). Thus, 1,3,4-thiadiazole-2,5-disulfonamide **4.70**, already reported by Roblin and Clapp (1950), was reported to be a stronger inhibitor than acetazolamide **4.44**, but this finding was later revealed to be incorrect (Supuran et al. 1996a). This compound had constituted the lead molecule for designing important classes of other pharmacological agents, such as benzothiadiazine diuretics and high-ceiling diuretics (Maren 1967). It was then proved that **4.70** is actually quite a weak CAI ($K_I = 10$ µ*M* against hCA II), whereas 1,3-disulfamoylbenzenes of type **4.71**, prepared from it as

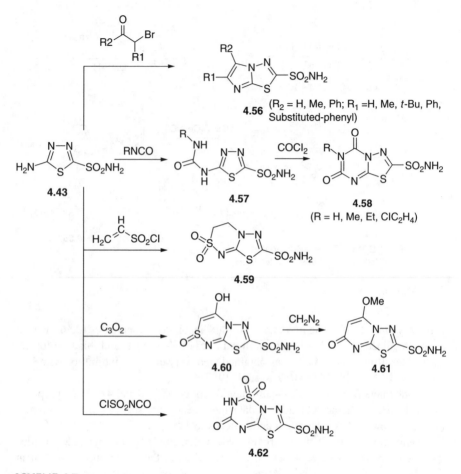

SCHEME 4.7

lead, ironically, are potent CAIs. Indeed, dichlorophenamide **4.71** has an IC_{50} of $3 \times 10^{-8} \, M$ and is still used clinically as a diuretic, antiglaucoma and antihypertensive drug (Maren 1967; Supuran and Scozzafava 2000a, 2001).

Bis-sulfonamides derived from biphenyl, biphenylether, biphenylsulfide, biphenyl-methane or containing urea or guanidine moieties as spacers, of types **4.72**, reported by Beasley et al. (1958) were generally more active than the corresponding compounds containing only one SO_2NH_2 group and possessed IC_{50} values of $1–7 \times 10^{-7} \, M$ (against unpurified blood CA, i.e., a mixture of isozymes I and II). Derivatives **4.73** containing thiophene instead of phenylene moieties, reported by Buzas et al. (1961), were again prepared in the search for more potent diuretics and possess CA inhibitory effects (IC_{50} values not specified).

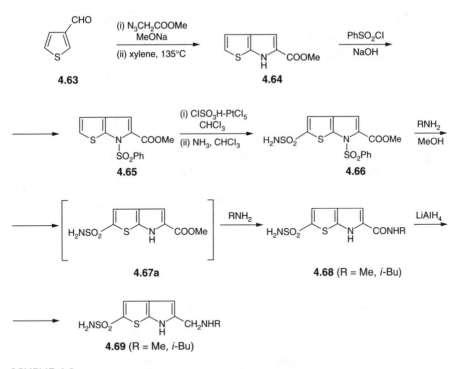

SCHEME 4.8

Bis-sulfonamides of type **4.75**, containing the classical 2-sulfonamido-1,3,4-thiadiazole-moieties were prepared by Supuran et al. (1991) as outlined in Scheme 4.9 by the bisacylation of amine **4.43** with diacyl halides (Table 4.10).

Compounds containing aliphatic, aromatic or heterocyclic groups as spacers between the two thiadiazolic moieties were very effective as CAIs, in contrast to derivatives possessing bulky spacers (such as 2,2′-biphenyl-carboxamido in **4.75e** or the macrocyclic ring in **4.75i**) which showed poor activities, probably because of the hindered access of the latter ones to the active site of the enzyme. Some of these compounds might possess a depot effect (*in vivo*) as CAIs, because of their very low solubilities in most solvents and water (Supuran et al. 1991).

The reaction of sulfanilamide **4.1** with potassium thiocyanate in the presence of hydrochloric acid, affording the corresponding thiourea **4.76** (A = 1,4-phenylene; Scheme 4.10) was described by Beasley et al. (1958). They also oxidized this thiourea with iodine but could not establish the structure of the oxidation product. By reinvestigating this reaction, it was found out that a new route to 2,5-disubstituted thiadiazoles is readily available (Scozzafava and Supuran 1998a). Reaction of several aromatic sulfonamides containing a free amino group with cyanate or thiocyanate in the presence of acid afforded the corresponding urea/thiourea derivatives, which were assayed as inhibitors of three isozymes, CA I, II and IV. Oxidation of the obtained thioureas **4.76** with iodine/KI in an acidic medium afforded symmetrical

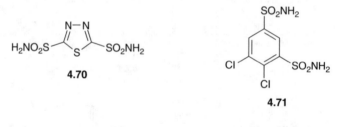

4.70

4.71

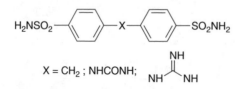

X = CH₂ ; NHCONH;

$$\underset{NH}{\overset{NH}{\underset{}{\parallel}}}\overset{}{\underset{NH}{C}}$$

4.72

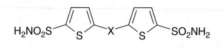

X = CO; CH₂

4.73

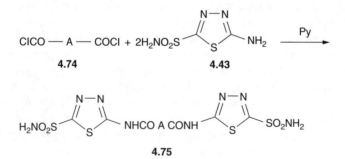

ClCO — A — COCl + 2H₂NO₂S ⟨thiadiazole⟩ NH₂ → Py

4.74   **4.43**

H₂NO₂S ⟨thiadiazole⟩ NHCO A CONH ⟨thiadiazole⟩ SO₂NH₂

4.75

SCHEME 4.9

2,5-*bis*-(substituted-phenyl)-1,3,4-thiadiazole derivatives that possessed sulfonamido groups on the aromatic ring (**4.78**) via a new synthesis of the heterocyclic
moiety (Scheme 4.10). Good inhibition of all three CA isozymes was observed with
the new compounds **4.78**, but the exciting finding was that the ureas/thioureas
reported here have an increased affinity to the slow-reacting isozyme CA I, generally
more insensitive to inhibition by sulfonamides as compared to the rapid-reacting
isozymes CA II and CA IV (Scozzafava and Supuran 1998a).

TABLE 4.10
hCA II Inhibition Data with *bis*-Sulfonamides 4.75

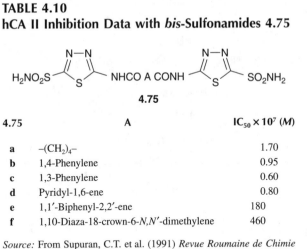

4.75

4.75	A	$IC_{50} \times 10^7$ (*M*)
a	–(CH$_2$)$_4$–	1.70
b	1,4-Phenylene	0.95
c	1,3-Phenylene	0.60
d	Pyridyl-1,6-ene	0.80
e	1,1′-Biphenyl-2,2′-ene	180
f	1,10-Diaza-18-crown-6-*N,N*′-dimethylene	460

Source: From Supuran, C.T. et al. (1991) *Revue Roumaine de Chimie* **36**, 251–255. With permission.

Recently, Arslan et al. (2002) reported the *bis*-sulfonamides **4.79** and **4.80**, obtained by condensation of 4-carboxybenzenesulfonamide with triethyleneglycol or 1,2-bis(aminoethoxy)ethane. These compounds were not very effective as CAIs.

4.3 TOPICAL SULFONAMIDES AS ANTIGLAUCOMA AGENTS

4.3.1 THE RING APPROACH

Although the treatment of glaucoma with CAIs is very effective in reducing elevated intraocular pressure (IOP), the systemic administration of drugs such as acetazolamide **4.44**, methazolamide **4.48** or dichlorophenamide **4.71** leads to unpleasant side effects due to inhibition of the enzyme present in other tissues (kidneys, red cells, stomach, etc.; Maren 1967; Supuran and Scozzafava 2000a, 2002a; Supuran et al. 2003). The possibility of topical administration of the classical drugs from this class (acetazolamide, methazolamide or dichlorophenamide) was extensively investigated by several researchers, but negative results were constantly obtained, and for more than 25 years it was considered that CA inhibitors could only be given systemically (Maren 1967; see also Chapter 8 of this book). In an elegant study, Maren et al. (1983) then showed that some polyhalogenoacetamido-thiadiazoles- and thiadiazolines (of the types **4.47** and **4.46**, see Tables 4.6 and 4.7) possess such properties, inducing an increased liposolubility and a decrease in pK_a of the sulfonamide moiety, and, in consequence, very strong CA inhibitory properties that allow their penetration through the cornea. Indeed, trifluoromethazolamide **4.46g** was found to reduce IOP in rabbits (–4.0 mmHg) when applied as a 3% aqueous solution for 25 min directly into the eye of the experimental animal (Maren et al. 1983). This was an extraordinary result and the beginning of the search for a new antiglaucoma drug. Unfortunately,

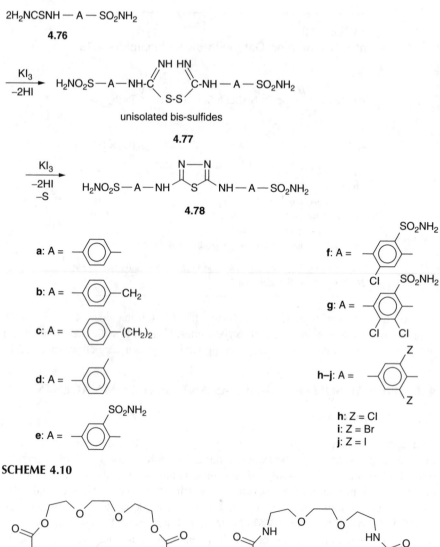

$$2H_2NCSNH—A—SO_2NH_2$$

4.76

unisolated bis-sulfides

4.77

4.78

a: A =

b: A = —CH$_2$

c: A = —(CH$_2$)$_2$

d: A =

e: A =

SO$_2$NH$_2$

f: A =

g: A =

h–j: A =

h: Z = Cl
i: Z = Br
j: Z = I

SCHEME 4.10

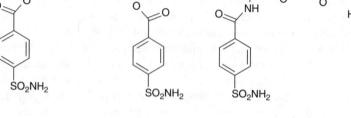

4.79   **4.80**

trifluoromethazolamide could not be developed for clinical use, its major drawbacks being its *in vivo* hydrolysis to trifluoroacetate and the imine derivative of methazolamide within a few hours, and a rather low efficacy of the drug in reducing IOP.

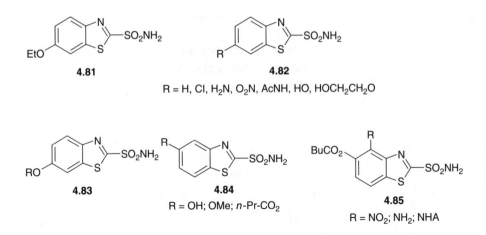

4.81

4.82

R = H, Cl, H_2N, O_2N, AcNH, HO, $HOCH_2CH_2O$

4.83

4.84

R = OH; OMe; n-Pr-CO_2

4.85

R = NO_2; NH_2; NHA

Programs to develop novel types of topical CAIs were thereafter initiated by several groups. Maren et al. (1987) showed that other homologues of methazolamide of type **4.46** (R1 = Me and R2 = EtCO, n-PrCO, n-Bu-CO) possessed an unexpected increase in water solubility (as well as an increase in liposolubility) and were weakly effective by the topical route in decreasing IOP in animal models of glaucoma. But the main advances in this field were achieved by the Merck group, and this led to the first clinically used, topically effective antiglaucoma sulfonamide dorzolamide. The approach for arriving at this compound (and also to brinzolamide, the second such clinically used pharmacological agent) is known as the ring approach, as it involved exploring a very wide range of ring systems to which sulfamoyl moieties were incorporated (Supuran and Scozzafava 2001; Supuran et al. 2003).

Roblin and Clapp (1950) had already obtained bicyclic sulfonamides, and they were reinvestigated as CAIs with putative topical activity in glaucoma because of their increased lipid solubility (Korman 1958). Ethoxozolamide **4.81** — the fourth classical drug from the family of CA inhibitors, discovered at Upjohn Laboratories — was used clinically in the same manner as acetazolamide, methazolamide or dichloro-phenamide (Korman 1958; Maren 1967) and constituted the starting point for pre-paring novel topically acting antiglaucoma derivatives.

Thus, Schoenwald et al. (1984) reported 6-substituted-benzothiazole-2-sulfona-mides of type **4.82** and developed a mathematical model relating corneal permeability to chemical structure (Eller et al. 1985). *In vivo* studies (in humans) with these inhibitors showed that the amino derivative had a weak topical activity, but these results were not reproduced in subsequent studies (Maren 1995).

O-Acylderivatives of 6-hydroxy-benzothiazole-2-sulfonamide **4.83** were pre-pared by acylation of the phenol **4.82** (R = OH) with anhydrides, mixed anhydrides (R-COOCONPh₂) or acyl halides in pyridine or acetone (Woltersdorf et al. 1989). The last procedure had minimum complications (side reactions of acylation/conden-sation at the sulfonamidic moiety) and was extensively applied. Table 4.11 shows some of the most effective inhibitors reported by Woltersdorf et al. (1989).

Some of the reported compounds **4.83** showed very potent *in vitro* CA inhibitory properties. *In vivo* studies showed that some esters of type **4.83** are hydrolyzed by

TABLE 4.11
CA II Inhibition Data with
6-Substituted-Benzothiazole-
2-Sulfonamides 4.83

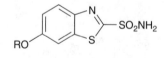

4.83

4.83	R	IC$_{50}$ (nM)
a	H	4
b	Ac	2
c	n-PrCO-	3.2
d	i-PrCO-	6.8
e	n-BuCO-	4
f	n-C$_{11}$H$_{23}$CO-	3.1
g	$cyclo$-C$_6$H$_{11}$-CO-	6
h	PhCO-	8
i	PhCH$_2$CO-	30
j	4-Me$_2$N-CH$_2$C$_6$H$_4$CO-	4
k	Ph-CH=CH-CO	15
l	MeOCH$_2$-CO-	3.5
m	EtO(Me$_2$)C-CO-	1.2
n	N-Me-pyrrole-2-CO-	2.9
o	t-BuCOCO-	10
p	n-BuSO$_2$-	1.8
q	Me$_2$NSO$_2$-	1.2
r	Me$_2$NCO-	1.6
s	(EtO)$_2$PO-	1.8

Source: From Woltersdorf, W. et al. (1989)
Journal of Medicinal Chemistry **32**,
2486–2492. With permission.

esterases from the ocular tissues, liberating the active inhibitor, the phenol **4.82** (R = OH), thereby acting as prodrugs (Woltersdorf et al. 1989). The pivalyl ester **4.83e** was shown to be a topically active ocular hypotensive CAI in rabbits (Woltersdorf et al. 1989). Not only were 6-substituted derivatives of type **4.83** reported in this study, but also the 5-substituted derivatives of type **4.84**, as well as the nitro- and amino-4,5-disubstituted benzothiazole-2-sulfonamides **4.85** were found to be as effective inhibitors as the esters **4.83** (Woltersdorf et al. 1989).

Compounds of types **4.83** to **4.85** could not be developed for clinical use (although they were very active CAIs) for two reasons: (1) they were rapidly metabolized, their sulfonamido group being nucleophilically displaced by reduced glutathione (G-SH **4.87**; see Scheme 4.11), forming a glutathione conjugate of type

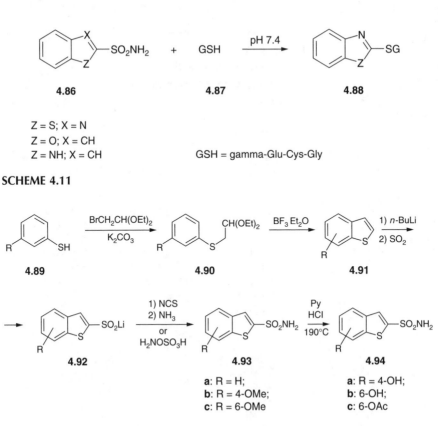

Z = S; X = N
Z = O; X = CH
Z = NH; X = CH

GSH = gamma-Glu-Cys-Gly

SCHEME 4.11

a: R = H;
b: R = 4-OMe;
c: R = 6-OMe

a: R = 4-OH;
b: 6-OH;
c: 6-OAc

SCHEME 4.12

4.88 (Conroy et al. 1984), and (2) clinical trials in rabbits proved **4.83e** to be a potent allergen, along with a number of related benzothiazole-, benzofuran- and indole-sulfonamides of type **4.86** (Woltersdorf et al. 1989).

To avoid these undesirable reactions, considered to be due to the highly electrophilic character of the benzothiazole ring, a large number of bycilic sulfonamides derived from benzo[*b*]thiophene-, benzofuran and indole were prepared and assayed for their inhibitory action by the Merck group (Graham et al. 1989, 1990). Scheme 4.12 presents some of the benzo[*b*]thiophenesulfonamides prepared by Graham et al. (1989) and the synthetic strategy used.

For the benzo[*b*]thiophene-2-sulfonamide series, this consisted of the alkylation of substituted thiophenols **4.89** with bromo-acetaldehyde diethyl acetal, followed by cyclization in the presence of boron trifluoride etherate, which formed (for R = OMe) a 1:10 mixture of the 4- and 6-methoxy[*b*]thiophenes **4.91** that was inseparable for the two components, so that the next steps were carried out on this mixture. They involved lithiation, followed by reaction with SO_2, with formation of the lithium sulfinates **4.92**, which were chlorinated with *N*-chlorosuccinimide (NCS) and amidated with aqueous ammonia or hydroxylamine-*O*-sulfonic acid, leading to the mixture of sulfonamides **4.93**. These two sulfonamides were readily separable by fractional crystallization from dichloroethane, and their demethylation in the

TABLE 4.12
CA Inhibition Data with Benzothiophene-2-Sulfonamide Derivatives 4.94

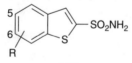

4.94

4.94	R	IC_{50} (nM)
a	H	5.2
b	6-OH	8.8
c	6-OAc	4.0
d	6-OCO-Bu-t	5.6
e	6-OSO$_3$Na	6.8
f	6-OPO$_3$Na$_2$	100
g	6-OMe	8.8
h	6-OCH$_2$COOH	14
i	6-NH$_2$	10
j	6-NHAc	4.8
k	6-Ac	6
l	4-OH	12
m	4-OMe	12
n	5-OH	6.2
o	5-OMe	23
p	5-CH$_2$NMe$_2$	12
q	5-CH$_2$SCH$_2$CH$_2$NMe	15
r	5-CH$_2$OH	9
s	5-CH$_2$OAc	10
t	7-OH	11
u	7-OMe	50

Source: From Graham, S.L. et al. (1989) *Journal of Medicinal Chemistry* **32**, 2548–2554. With permission.

presence of pyridine hydrochloride afforded the phenols **4.94a** and **b** (Graham et al. 1989). These compounds were used to prepare diverse esters and ethers (**4.94c–h**; Table 4.12), but derivatives bearing other groups and other substitution patterns were also prepared by variants of the synthesis shown in Scheme 4.12.

The inhibitory activity of these derivatives was very good, as seen from the data in Table 4.12. Furthermore, these compounds were devoid of allergenic properties, and some of them, such as the phenol **4.94b** or its acetate **4.94c**, were shown to act as potent ocular hypotensive agents in animal models (Graham et al. 1989).

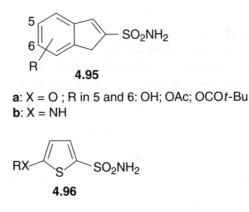

4.95

a: X = O ; R in 5 and 6: OH; OAc; OCO*t*-Bu
b: X = NH

4.96

Unfortunately, this was not the case for benzofuran and indole-2-sulfonamides of type **4.95a, b** prepared subsequently by Graham et al. (1990), using synthetic strategies similar to those for the benzo[*b*]thiophene derivatives. Compounds of this type were very effective inhibitors (IC$_{50}$ 8 to 30 n*M*) and showed topical IOP lowering effects in rabbits, but their development for clinical use was precluded because of their sensitization potential in guinea pigs, similarly to the ethoxzolamide congeners previously discussed.

Similar to the synthesis of hydroxyalkylthio- and hydroxyalkyl-sulfonyl-benzene-sulfonamide derivatives **4.12** and **4.13**, Shepard et al. (1991) also described the synthesis and CA inhibitory properties for derivatives of 5-hydroxyalkylthio- and 5-hydroxyalkyl-sulfonyl-thiophene-2-sulfonamides of type **4.96** (Table 4.13). They were prepared from 5-bromothiophene by the reactions showed in Scheme 4.13, as for the synthesis of the corresponding benzene derivatives **4.12** and **4.13** (Scheme 4.2). Some other systems related to **4.96** were also investigated to obtain topically active CAIs, such as the 4-substituted-thiophene/furan-2-sulfonamides (Scheme 4.13). The synthetic strategy for obtaining these compounds included lithiation of the 3-bromothiophenes/bromofurans **4.97**, conversion of the lithioderivatives **4.98** to the corresponding ketones **4.99** (by reaction with benzonitriles) or sulfoxides **4.101** (by reaction with aryldisulfides and oxidation with perbenzoic acid), followed by transformation into the corresponding sulfonamides **4.100** and **4.102**, respectively, by standard procedures (Shepard et al. 1991). Table 4.13 gives inhibition and pK_a data for some of these CAIs. As evident from the data, activity of these CAIs is influenced by the same structural elements as those for the related benzene derivatives. Thus, sulfoxides were more active than the corresponding sulfides and the enlargement of the hydroxyalkyl moiety was also beneficial for the inhibitory properties of these compounds. These inhibitors were active topically in lowering elevated IOP of experimental animals (Hartman et al. 1992).

Table 4.14 and Table 4.15 show CAIs of this type and their pK_a. As seen from the data, CAIs derived from these heterocyclic systems are highly effective, irrespective of the nature of groups present in position 4. They do not posses allergenic properties and are active as topical inhibitors in animal models (Hartman et al. 1992).

TABLE 4.13
IC$_{50}$ and pK_a Values for Hydroxyalkylthio- and Hydroxyalkylsulfonyl-Thiophene-2- Sulfonamide Derivatives 4.96 against hCA II

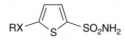

4.96

4.96	R	X	pK_a[a]	IC$_{50}$[b] (nM)
a	HO(CH$_2$)$_2$	S	9.46	30
b	HO(CH$_2$)$_2$	SO$_2$	8.65	6
c	AcO(CH$_2$)$_2$	SO$_2$	8.65	5.5
d	HO(CH$_2$)$_3$	S	9.32	10
e	HO(CH$_2$)$_3$	SO$_2$	8.32	6
f	AcO(CH$_2$)$_3$	SO$_2$	8.50	5
g	MeOCH$_2$COO(CH$_2$)$_3$	SO$_2$	8.10	4.5
h	i-BuOCOO(CH$_2$)$_3$	SO$_2$	9.50	c
i	HO(CH$_2$)$_4$	S	8.67	3
j	HO(CH$_2$)$_4$	SO$_2$	9.30	2.3
k	i-BuNH(CH$_2$)$_3$	SO$_2$	8.05	18

[a] In 30% EtOH/water.
[b] For CO$_2$ hydration.
[c] The compound was partly hydrolyzed and IC$_{50}$ could not be determined.

Source: From Shepard, K.L. et al. (1991) *Journal of Medicinal Chemistry* **34**, 3098–3105. With permission.

Many other annulated bicyclic systems were then investigated in which it was mandatory that one of the rings be thiophene. Prugh et al. (1991) reported the synthesis of a large series of CAIs derived from thieno[2,3-*b*]- and thieno[3,2-*b*]thiophene-2-sulfonamide, prepared as outlined in Scheme 4.14 and Scheme 4.15.

For annulation of the first heterocyclic ring, the starting material was the ketal of thiophene-3-carbaldehyde **4.103**, which was subjected to lithiation, treatment with elemental sulfur and alkylation with methylbromoacetate (Scheme 4.14). Deprotection of the aldehyde moiety of **4.104** was then done via a transketalization reaction with acetone/TsOH (tosic acid), followed by cyclization of **4.105** in the presence of 1,5-diaza-bicyclo[4.3.0]non-5-ene (DBN). Ester **4.106** was subsequently converted to aldehyde **4.107** by reduction with LiAlH$_4$, followed by oxidation with pyridinium chlorochromate (PCC). To introduce the SO$_2$NH$_2$ moiety, a new protection of the aldehyde group was needed (by using ethylene glycol), and the ketal **4.108** thus obtained was lithiated, treated with liquid SO$_2$ and the lithium chlorosulfinate obtained oxidized with *N*-chlorosuccinimide (NCS) to the corresponding sulfonyl

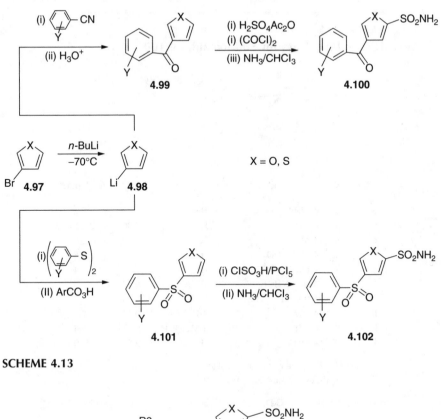

SCHEME 4.13

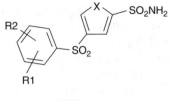

4.100

4.102

chloride **4.109**. Amidation of this last compound and deprotection of the aldehyde moiety afforded the key intermediate **4.110**. The aldehyde group was thereafter converted (via Schiff bases of types **4.111** and **4.113**) to secondary/tertiary amines to obtain a salt-like compound by reaction with strong acids. Compounds such as

TABLE 4.14

IC$_{50}$ and pK_a Data for Sulfonamides 4.100

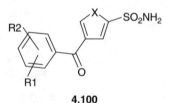

4.100

4.100	X	R1	R2	pK_a[a]	IC$_{50}$[b] (nM)
a	S	4-MeO	H	8.20	3.2
b	S	4-OH	H	7.90	2.5
c	S	4-Me	H	9.30	2.0
d	S	4-MeNHCH$_2$	H	9.57	4.5
e	S	4-PhCH$_2$NHCH$_2$	H	9.35	2.4
f	S	4-n-BuNHCH$_2$	H	9.32	12.0
g	S	4-[O(CH$_2$CH$_2$)$_2$NCH$_2$]	H	9.28	3.7
h	S	4-OH	3-Et$_2$NCH$_2$	9.22	7.0
i	S	4-OH	3-Me$_2$NCH$_2$	9.28	6.4
j	O	4-MeO	H	9.50	2.0

[a] For the SO$_2$NH$_2$ protons in 30% EtOH/water.
[b] For CO$_2$ hydration against hCA II.

Source: From Hartman, G.D. et al. (1992) *Journal of Medicinal Chemistry* **35**, 3027–3033. With permission.

4.112 or **4.114** possess for this reason an acceptable water solubility (Prugh et al. 1991).

To obtain sulfonamides derived from the isomeric system of thieno[3,2-*b*]thiophene-2-sulfonamide, the synthetic strategy used was that given in Scheme 4.15. Starting with 2,3-dibromothiophene **4.115**, the monoaldehyde **4.116** was obtained (via lithio derivatives), which after protection as the ethylene glycol ketal **4.117** was lithiated in position 3, and the methyloxycarbonylmethylthio moiety was introduced by reaction with elemental sulfur, followed by treatment with methyl bromoacetate, leading to **4.118**. Annulation of this intermediate was easier in this case (as compared with the isomeric system, **4.107**) in the presence of pyridinium acetate, followed by introduction of the sulfonamidic moiety as in the case of the thieno[2,3-*b*]thiophene derivatives, leading to the key intermediate **4.122**, which was then converted to the secondary/tertiary amines **4.123** and **4.124** via Schiff bases (Scheme 4.15). Table 4.16 and Table 4.17 give the CAI activities of compounds prepared by these procedures. From these data, one can see that both bicyclic systems led to strong CAIs. Both secondary and tertiary amines were equally active.

By an identical strategy, starting with the ethylene glycol ketal of furan-3-carbaldehyde, Hartman et al. (1992) prepared thieno[2,3-*b*]furan-2-sulfonamide

TABLE 4.15
IC_{50} and pK_a Data for Sulfonamides 4.102

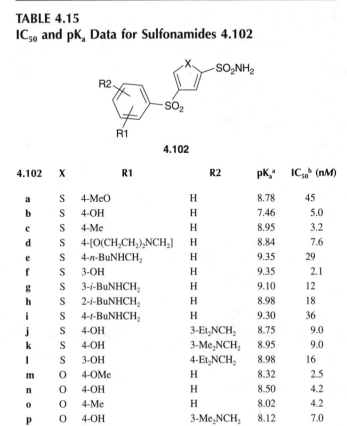

4.102

4.102	X	R1	R2	pK_a[a]	IC_{50}[b] (nM)
a	S	4-MeO	H	8.78	45
b	S	4-OH	H	7.46	5.0
c	S	4-Me	H	8.95	3.2
d	S	4-[O(CH$_2$CH$_2$)$_2$NCH$_2$]	H	8.84	7.6
e	S	4-n-BuNHCH$_2$	H	9.35	29
f	S	3-OH	H	9.35	2.1
g	S	3-i-BuNHCH$_2$	H	9.10	12
h	S	2-i-BuNHCH$_2$	H	8.98	18
i	S	4-t-BuNHCH$_2$	H	9.30	36
j	S	4-OH	3-Et$_2$NCH$_2$	8.75	9.0
k	S	4-OH	3-Me$_2$NCH$_2$	8.95	9.0
l	S	3-OH	4-Et$_2$NCH$_2$	8.98	16
m	O	4-OMe	H	8.32	2.5
n	O	4-OH	H	8.50	4.2
o	O	4-Me	H	8.02	4.2
p	O	4-OH	3-Me$_2$NCH$_2$	8.12	7.0
q	O	4-OH	3-Et$_2$NCH$_2$	8.72	10.0

[a] For the SO$_2$NH$_2$ protons in 30% EtOH/water.
[b] For CO$_2$ hydration against hCA II.
Source: From Hartman, G.D. et al. (1992) *Journal of Medicinal Chemistry* **35**, 3027–3033. With permission.

derivatives bearing 5-secondary/tertiary-amino groups, of type **4.125** (Table 4.18). Such compounds also possessed strong CA inhibitory properties. In contrast to the sulfonamides, compound **4.125** showed a strong binding to the ocular pigment and a moderate reactivity toward glutathione.

In addition to the Merck group, topically active CAIs were reported also by other researchers (Pierce et al. 1993; Sharir et al. 1994), this time incorporating the classical acetazolamide- or methazolamide-like ring systems in derivatives of types **4.126** to **4.131**.

By treating amine **4.43** with cyclic unsaturated anhydrides (e.g., maleic anhydride), the imide **4.126** was obtained, which on oxidation with t-BuOOH/OsO$_4$ led to the *trans*-diol **4.127**. This compound was reported to act as a topically active CAI in rabbits (Pierce et al. 1993). The same group condensed **4.43** with 3-oxoadipic

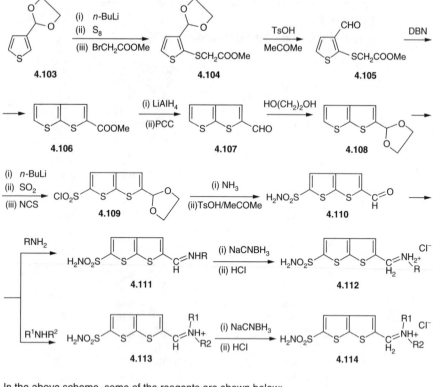

In the above scheme, some of the reagents are shown below;

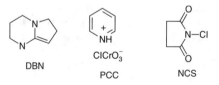

SCHEME 4.14

acid, obtaining a Schiff base, which on reduction with NaBH₃CN led to the secondary amine **4.128,** reported to act as a strong CAI (Sharir et al. 1994). Compounds of type **4.129** that possessed 5-ω-carboxypolymethylenecarboxamido moieties were reported by Antonaroli et al. (1992) to have IC$_{50}$ values of the order of 10^{-7} M and some topical activity in lowering IOP in rabbits, but detailed studies have not been performed with these derivatives.

The culmination of the ring approach was reached with the report of Ponticello et al. (1987) that thieno-thiopyran-2-sulfonamides of types **4.130** and **4.131** are water soluble, and this ring system led to the first clinically used topically acting sulfona-mide dorzolamide (Maren 1995; Supuran et al. 2003).

Two types of sulfonamides were prepared: the thieno[2,3-*b*]thiopyran- **4.130** and the thieno[3,2-*b*]thiopyran-2-sulfonamides **4.131**. Scheme 4.16 presents syntheses

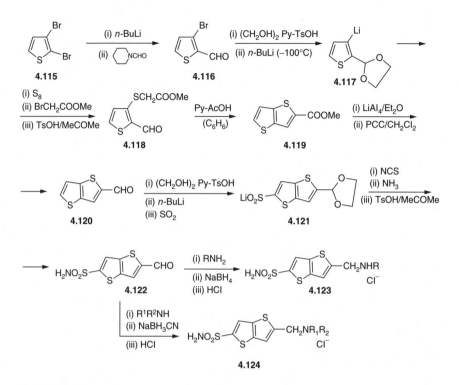

SCHEME 4.15

of derivatives from the first class, starting from the key intermediate ketone **4.132** reported earlier (Ponticello et al. 1987; Baldwin et al. 1989).

Treatment of **4.132** with sulfuric acid–acetic anhydride in methylene chloride provided in high yield the 2-sulfonic acid, which was converted to the sulfonyl chloride with PCl_5. Subsequent amidation let to sulfonamide **4.133** (Scheme 4.16). Its reduction with sodium borohydride afforded the alcohol **4.134,** which could be oxidized at the thiopyran moiety by different agents to the corresponding sulfone **4.135** and sulfoxide **4.136**, respectively. This last compound was then used to prepare the unsaturated derivative **4.137**, as well as the parent (unsubstituted) sulfonamide of the series **4.138**. By Ritter's reaction with nitriles in the presence of sulfuric acid, **4.136** was used for obtaining amides **4.139**, which were reduced by borane-dimethyl sulfide to amines **4.140**. Some compounds of these types showed a good water solubility (1 to 2%, which is excellent for a sulfonamide) and very good CA inhibitory activities (Table 4.19).

From the data from Table 4.19, it can be seen that generally compounds of type **4.141** are very strong CAIs, with important differences of activity seen for diverse groups substituted at position 4 as well as for different oxidation states of the sulfur moiety. The sulfones (**4.141**, $n = 2$) were the most active: very good activities were found for the ketone **4.141e**, the unsubstituted derivative **4.141f**, the primary amine **4.141g** and its alkyl derivatives **4.141h–n**. Compound **4.141m** (MK-927), originally

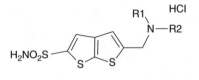

4.114

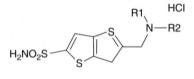

4.124

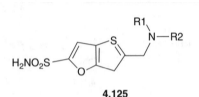

4.125

selected for clinical trials (Ponticello et al. 1987; Baldwin et al. 1989), needs special mention. Its racemate is a highly active inhibitor ($IC_{50} = 0.55 \times 10^{-8}\,M$), but by resolving the compound into its pure enantiomers, the (*R*)-optical antipode **4.141n** was found to be >10 times more active than its antipode. Moreover, x-ray crystallographic studies on enzyme inhibitor complexes by using **4.141n** and **4.141o** revealed differences in the relative orientation of these inhibitors within the active site of human CA II, as well as conformational change in the protein on binding, involving a side-chain movement of His 64 (Smith et al. 1994). In conclusion, binding of the (*R*)-enantiomer is favored over the (*S*)-enantiomer, which explains the differences in IC_{50} values shown in Table 4.19.

Shinkai (1992) also reported the enantioselective synthesis of the isomer (*R*)-MK-927, renamed MK-417 (**4.141n**). Scheme 4.17 outlines the most important steps of this synthesis developed for the industrial scale. Oxidation of ketone **4.132** with H_2O_2 in the presence of sodium wolframate afforded sulfone **4.133**, which was enantioselectively reduced to the key compound, alcohol **4.136**. Different reductions were tried: in the presence of (–)-B-chlorodiisopinocamphenylborane (when the (*R*):(*S*) ratio was 11:89), (*S*)-B-methyl-1,3,2-oxazaborolidine (*R*):(*S*) ratio of 9:1) or yeast. For large-scale applications, the second catalyst proved most suitable, being easily accessible from L-proline and affording good yields of the desired enantiomer (Shinkai 1992). Compound **4.136** was converted to the tosylate **4.142** possessing

TABLE 4.16
Inhibition and pK$_a$ Data for Thieno [2,3-b]-Thiophene-2-Sulfonamide Derivatives 4.114

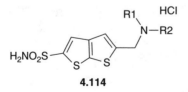

4.114

4.114	R1	R2	pK$_a$	IC$_{50}$ (nM)
a	H	Me	8.02	3
b	H	t-Bu	8.28	10
c	H	i-PrCH$_2$	7.28	2.5
d	H	(CH$_2$)$_4$OH	7.70	3
e	H	(CH$_2$)$_2$OH	7.18	6.3
f	H	CH$_2$CH$_2$OMe	7.00	5
g	H	(CH$_2$)$_3$OMe	7.52	4.9
h	H	CH$_2$CF$_3$	3.75	2.2
i	H	H	7.30	14
j	H	CH$_2$CH$_2$SMe	6.85	4
k	H	CH$_2$CH$_2$F	6.40	2.7
l	H	CH$_2$CH$_2$SOMe	5.58	8.4
m	H	CH$_2$CH$_2$SO$_2$Me	5.20	4.5
n		–CH$_2$CH$_2$– O – CH$_2$CH$_2$–	5.25	3
o		–CH$_2$CH$_2$– S – CH$_2$CH$_2$–	5.05	2.6
p		–CH$_2$CH$_2$– SO – CH$_2$CH$_2$–	4.23	4.0
q	CH$_2$CH$_2$OMe	CH$_2$CH$_2$OMe	5.34	4.2
r	(CH$_2$)$_2$O(CH$_2$)$_2$OMe	CH$_2$CH$_2$OMe	5.55	2.8
s	CH$_2$CH$_2$OH	CH$_2$CH$_2$OH	5.50	2.3

the same configuration (by treatment with TsCl and NaC≡CH), which by an SN$_2$ reaction with *iso*-butylamine (with inversion of configuration at C-4) led to **4.143**. The sulfonamido group was introduced by treatment with sulfonyl chloride and amidation, these reactions not affecting the chiral center at C-4, leading to MK-417 (compound **4.141n**). For reasons little understood Merck, Sharp & Dome chose for launching into the market a structurally related inhibitor, i.e., MK-507 (**4.144**), named dorzolamide.

The structure and the pharmacology of these CAIs are extremely similar to those of other thienothiopyransulfonamides discussed previously, but MK-507 contains two chiral centers because of the presence of a 6-methyl group. Its enantioselective synthesis has been reported and optimized for large-scale production (Scheme 4.18; Blacklock et al. 1993).

Starting from methyl-(*R*)-3-hydroxybutyrate tosylate **4.145**, which was treated with 2-lithiomercapto-thiophene **4.146** and in an SN$_2$ reaction (with inversion of

TABLE 4.17
Inhibition and pK_a Data for Thieno [3,2-*b*]-
Thiophene-2-Sulfonamide Derivatives 4.124

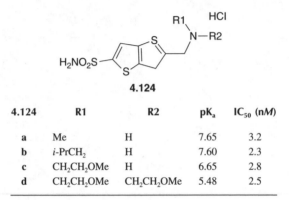

4.124

4.124	R1	R2	pK_a	IC$_{50}$ (n*M*)
a	Me	H	7.65	3.2
b	*i*-PrCH$_2$	H	7.60	2.3
c	CH$_2$CH$_2$OMe	H	6.65	2.8
d	CH$_2$CH$_2$OMe	CH$_2$CH$_2$OMe	5.48	2.5

TABLE 4.18
Inhibition and pK_a Data for Thieno[2,3-*b*]Furan-2-
Sulfonamide Derivatives 4.125

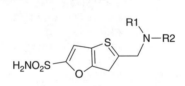

4.125

4.125	R1	R2	pK_a	IC$_{50}$ (n*M*)
a	H	Me	7.85	9.0
b	H	Et	8.05	11.0
c	H	*n*-Pr	7.85	8.0
d	H	*i*-Bu	7.55	6.0
e	H	(CH$_2$)$_2$OMe	7.00	11.0
f	CH$_2$CH$_2$OMe	(CH$_2$)$_2$OMe	5.57	5.1
g	H	(CH$_2$)$_2$OEt	6.85	8.0
h	H	(CH$_2$)$_4$OH	7.65	11.0
i	H	(CH$_2$)$_3$OH	7.25	12.0
j	H	CH$_2$CH$_2$F	6.55	12.0
k	H	–CH$_2$–2-pyridyl	9.08	8.2
l		–CH$_2$CH$_2$–O–CH$_2$CH$_2$–	9.20	5.6

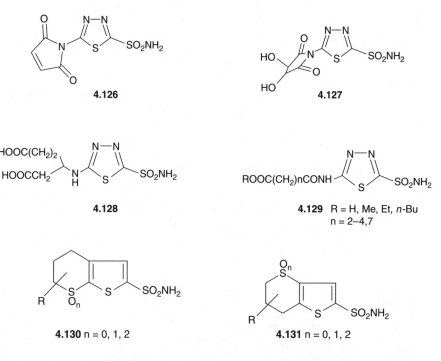

4.126

4.127

4.128

4.129 R = H, Me, Et, *n*-Bu
n = 2–4,7

4.130 n = 0, 1, 2

4.131 n = 0, 1, 2

configuration), the alkylation product with the desired stereochemistry, **4.147**, was obtained. Cyclization of **4.147** (after hydrolysis of the ester moiety) was achieved in the presence of trifluoroacetic anhydride in toluene, when ketone **4.148** was obtained in good yields. Its reduction with LiAlH₄ in 95% yield gave the *cis*-alcohol **4.149a**, which was epimerized to **4.149b** in the presence of cold 1 *N* H₂SO₄. A mixture of 76:24 *trans/cis* epimeric alcohols resulted, which was oxidized with H₂O₂/Na₂WO₄ to the sulfone **4.150**. By a Ritter reaction with acetonitrile, **4.150** was transformed into the acetamido derivative **4.151**. An interesting discovery in these syntheses was that the Ritter reaction occured with retention of configuration for the *trans* alcohol and with considerable inversion for the *cis* alcohol. However, finally the desired compound **4.151** was formed preferentially. This was converted to the corresponding sulfonyl chloride **4.152** and sulfonamide **4.153**, followed by reduction of the acetamide to the ethylamino moiety with borane-dimethylsulfide, to obtain **4.144**. The hydrochloride salt of **4.144** is dorzolamide (MK-507).

The second clinically used topical CAI brinzolamide (**4.154**) was developed by Alcon Laboratories, probably by using dorzolamide (**4.144**) as lead (Dean et al. 1993). The two compounds are structurally very similar, with brinzolamide possessing a slightly modified ring, i.e., the 2-substituted-2*H*-thieno[3,2-*e*]-1,2-thiazine-6-sulfonamide class. The main difference between the Merck and Alcon work was that Alcon researchers did not generally publish their research in scientific journals but only patented these compounds (Dean et al. 1993). Only recently have some brinzolamide congeners been described in some detail in a published paper (Chen et al. 2000).

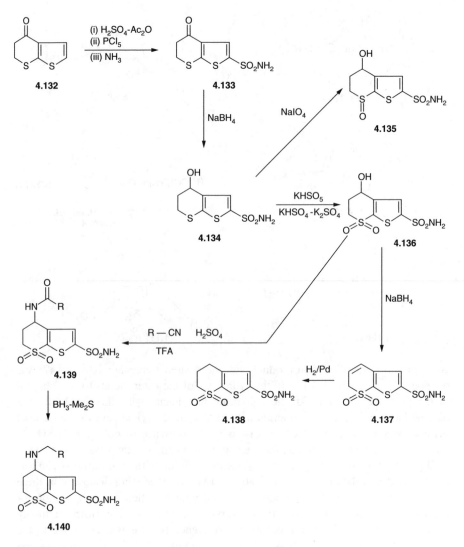

SCHEME 4.16

Similar to the thienothiopyran sulfonamides developed by Merck, the 2-substituted-2*H*-thieno[3,2-*e*]-1,2-thiazine-6-sulfonamides to which brinzolamide (**4.154**) belongs act as a low nanomolar inhibitor for hCA II, being slightly less effective as hCA IV inhibitors. However, the brinzolamide type compounds are less water soluble as compared to dorzolamide, and thus are formulated as suspensions for topical administration. Another important characteristic is that they are more effective than dorzolamide and produce less eye stinging and burning. A defect is that they provoke much more blurred vision after administration, obviously because of the suspension nature of the formulation (Supuran et al. 2003).

TABLE 4.19
Inhibition Data for Thieno-Thiopyran-2-Sulfonamide Derivatives 4.141

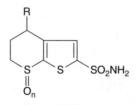

4.141

4.141	n	R	$IC_{50} \times 10^8$ (M)
a	0	=O	0.85
b	0	OH	3.0
c	1	OH	10.0
d	2	OH	1.3
e	2	=O	0.5
f	2	H	0.45
g	2	NH_2	0.92
h	2	NHMe	1.4
i	2	NHEt	0.73
j	2	NEt_2	7.0
k	2	n-PrNH–	0.92
l	2	n-BuNH–	1.24
m	2	$Me_2CH_2CH_2NH$–	0.59
n	2	(R)-**4.141m**	0.40
o	2	(S)-**4.141m**	5.31

Source: From Ponticello, G.S. et al. (1987) *Journal of Medicinal Chemistry* **30**, 591–597. With permission.

A common problem of both dorzolamide and brinzolamide is that they contain chiral centers, and the preparation of the pure enantiomer (4S,6S) in the case of dorzolamide and (4R) in the case of brinzolamide is rather expensive. Thus, the Alcon group has recently reported some brinzolamide-like compounds of type **4.155** (Chen et al. 2000) that do not contain chiral centers. These derivatives principally differ from brinzolamide by the absence of the 4-substituent (that induced the chirality) and a rather large number of substituents in position 2 of the heterocyclic ring (the R group of formula **4.155**), as well as by the presence of an additional double bond in the six-membered heterocycle annulated to the thiophene nucleus, which represents an innovative feature over the previously topically active sulfonamides prepared by the ring approach. These compounds were effective nanomolar inhibitors against hCA II and hCA IV, and showed good IOP lowering (20 to 30%) properties in naturally hypertensive Dutch-belted rabbits after administration as

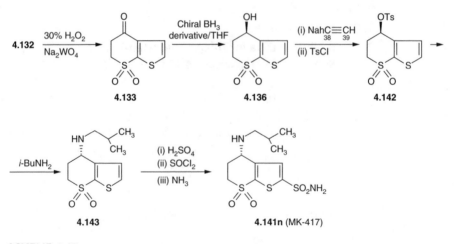

SCHEME 4.17

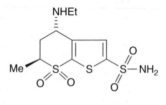

4.144

suspensions (except for two derivatives that were soluble enough to be administered in solution at pH 5.5; Chen et al. 2000). It is not clear at present whether such compounds can substitute brinzolamide as second-generation topically acting sulfonamides in the near future.

4.3.2 THE TAIL APPROACH

This approach has been developed in our laboratory (Scozzafava et al. 1999a, 1999b; Supuran and Scozzafava 2001) and consists in using well-known aromatic/heterocyclic sulfonamide scaffolds (of types **A–Y**) to which tails that will induce water solubility (or other desired physicochemical properties, see later) are attached at the amino, hydroxy, imino or hydrazino moieties contained in the precursor sulfonamides **A–Y** (Figure 4.2).

The parent sulfonamides derivatized by this simple approach included 2-, 3- or 4-amino-benzenesulfonamides, 4-(ω-aminoalkyl)-benzenesulfonamides/thiadiazole-sulfonamides, 3-halogeno-substituted-sulfanilamides, 1,3-benzene-disulfonamides, 1,3,4-thiadiazole-2-sulfonamides, benzothiazole-2-sulfonamides, as well as sulfanilyl-substituted aromatic/heterocyclic sulfonamides (structures **U–X**), and were chosen so as to prove that the tail approach is a general approach (Supuran et al. 1999b, 1999c, 2000a; Borras et al. 1999). Tails introduced in the molecules of such

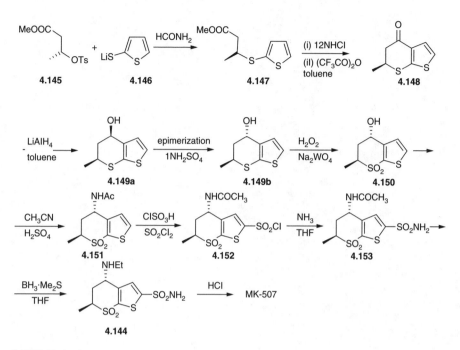

SCHEME 4.18

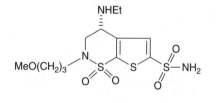

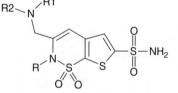

4.154

4.155: R = Me, Et, Pr, allyl, MeO-(CH$_2$)$_3$, etc.
NR1R2= 4-morpholinyl; N(CH$_2$CH$_2$OMe)$_2$, etc.

novel CAIs contained either moieties protonable at endocyclic nitrogen atoms (such as pyridine- or quinoline rings **4.156** to **4.161**) or at amino groups belonging to amino acids and some of their derivatives (such as in tails derived from glycine, β-alanine, GABA, sarcosine, creatine and glycyl-glycine moieties of types **4.162** to **4.167**), polyaminopolycarboxylic acid tails (of types **4.168** to **4.174**), protonated by acids and ionized by bases as well as perfluoroalkyl/aryl moieties (which are not protonable at pH values in the neutral range, of types **4.175** to **4.180**). The water solubility of the first type of such derivatives is assured by formation of salts with hydrochloric, trifluoroacetic or triflic acid, or by formation of sodium salts for derivatives possessing carboxylic acid moieties (tails of types **4.160** and **4.168** to **4.174**). The perfluoroderivatives (incorporating tails **4.175** to **4.180**) constitute a special case that is discussed later. The tail was generally attached to the sulfonamide scaffold as an amide (CONH), ester (COOR), sulfonamide (SO$_2$NH) or imide moiety

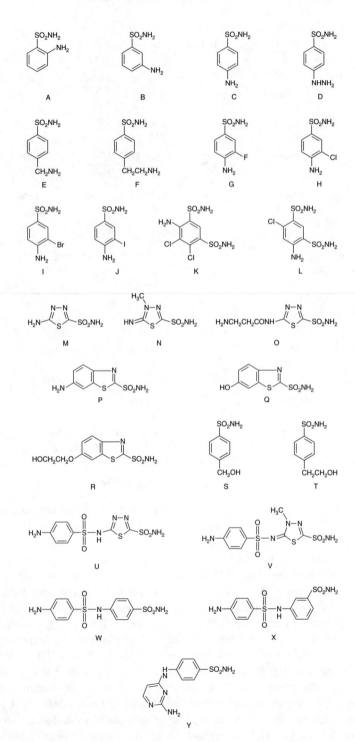

FIGURE 4.2 The tail approach.

(in one case, the tail **4.161** was obtained by reaction of pyridine-2,3-dicarboxylic acid anhydride with amino-sulfonamides), leading to phthalimide-like compounds (Scozzafava et al. 1999b). Thus, a sulfonamide CAI obtained by the tail approach is generally conveniently described by a figure (corresponding to the tail moiety) and a letter (corresponding to the sulfonamide scaffold to which the tail has been attached, through a carboxamide, ester or secondary sulfonamide bond). For instance, **4.158C** should be the isonicotinoylamido derivative of sulfanilamide or **4.162M** the glycinamido derivative of 5-amino-1,3,4-thiadiazole-2-sulfonamide (Scozzafava et al. 1999a, 1999b, 2000c, 2001, 2002; Supuran et al. 1999b, 1999c, 2000a; Menabuoni et al. 1999; Mincione et al. 1999; Barboiu et al. 1999; Casini et al. 2001).

The main advantages of this approach over the ring approach are that it is much simpler and allows a parallel type synthesis easily, so that a large number of different derivatives can be prepared and their physicochemical properties modulated in such a way as to assure possession of the desired pharmacological properties. By choosing different tails, it is possible to attain the much desired water solubility of these CAIs (as salts of acids or bases) at pH values around neutrality, thus avoiding the irritation problems observed with the strongly acidic dorzolamide solutions. Many of the protonatable moieties of tails **4.156** to **4.174** possess pK_a values of 6 to 7, which is quite advantageous for the solubility of these derivatives at pH values in the almost neutral range. Furthermore, some of these derivatives also contain carboxylic acid moieties in addition to the protonatable amino moieties, which provide aqueous solutions as their sodium salts with pH values of 7.0 to 7.5. Such solutions never irritate the eyes of experimental animals (Scozzafava et al. 1999a, 1999b, 2000c, 2001, 2002; Supuran et al. 1999b, 1999c, 2000a; Menabuoni et al. 1999; Mincione et al. 1999; Barboiu et al. 1999; Casini et al. 2001).

Sulfonamides obtained by incorporating pyridine-carboxamido- or quinoline-sulfonamido tails (of types **4.156** to **4.161**) into the scaffolds **A–Y** were in many cases nanomolar inhibitors of isozymes hCA II and bCA IV (Scozzafava et al. 1999a, 1999b; Borras et al. 1999; Supuran et al. 1999b, 1999c), possessed a good water solubility of 1.5 to 2%, and the pH of their solutions used for *in vivo* experiments was 6.5 to 7.5. Many of these compounds were very effective IOP-lowering agents in normotensive and glaucomatous albino rabbits, with potencies two or three times greater than dorzolamide.

New derivatives prepared by the tail approach, incorporating amino acyl-, oligo-peptidyl- or polyaminopolycarboxyl- moieties (of types **4.162** to **4.174**) were again potent inhibitors of isozymes CA I, CA II and CA IV, and could be formulated in water solutions in at concentrations of 2 to 2.5% at almost neutral pH values (pH 6 to 7). They produced powerful and prolonged IOP lowering after topical adminis-tration in normotensive and glaucomatous rabbits (Scozzafava et al. 1999a; Supuran et al. 2000a; Mincione et al. 1999; Barboiu et al. 1999).

A special case of this approach was constituted by the perfluoroalkyl-/aryl-containing derivatives, which incorporated tails **4.175** to **4.180** (Scozzafava et al. 2000d). Such derivatives cannot form water-soluble salts with acids or bases at neutral pH. Unexpectedly, this class of derivatives possessed a high water solubility (1.5 to 2%), balanced by a significant lipid solubility (because of the presence of

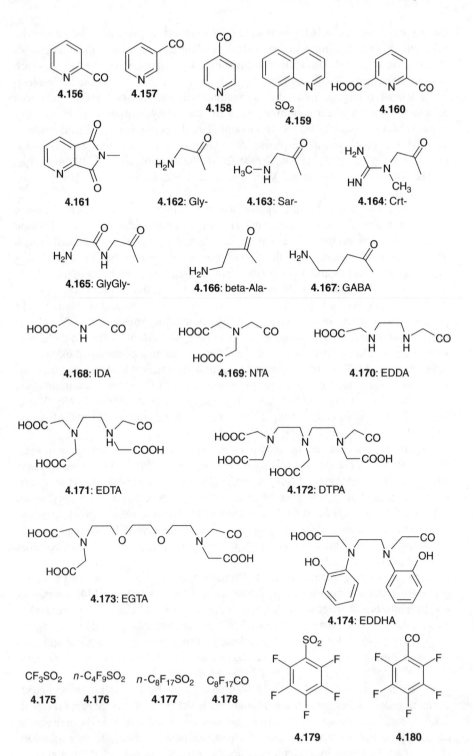

4.156

4.157

4.158

4.159

4.160

4.161

4.162: Gly-

4.163: Sar-

4.164: Crt-

4.165: GlyGly-

4.166: beta-Ala-

4.167: GABA

4.168: IDA

4.169: NTA

4.170: EDDA

4.171: EDTA

4.172: DTPA

4.173: EGTA

4.174: EDDHA

CF_3SO_2 n-$C_4F_9SO_2$ n-$C_8F_{17}SO_2$ $C_8F_{17}CO$

4.175   **4.176**   **4.177**   **4.178**

4.179

4.180

the perfluoroalkyl-/aryl moieties; Scozzafava et al. 2000d). Correlated with their very good CA inhibitory properties against isozymes CA I, CA II and CA IV (in the low nanomolar range), such derivatives possessed optimal properties to act as efficient IOP-lowering agents: they were formulated as aqueous solutions at neutral pH; they showed very good penetrability through the cornea (because of their good lipophilic character), arriving thus at the ciliary processes enzyme; and their duration of action was much more prolonged than that of dorzolamide. Compounds such as **4.179M** or **4.178N**, among others, showed a very strong IOP-lowering effect in both normo-tensive and glaucomatous rabbits, and this IOP-lowering effect lasted for 5 to 6 h. (In the case of dorzolamide, pressure returns to basal levels after a much shorter period; Scozzafava et al. 2000d.)

It was obvious from these studies that the tail incorporated into the molecules of such CAIs is important for at least three critical properties of the topical anti-glaucoma agent: (1) to assure the water solubility of the drug in order to formulate it as a solution for ophthalmologic use. By the tail approach it is possible to formulate topically acting antiglaucoma sulfonamides as 1.5 to 2% solutions at pH values in the neutral range (pH 6.5 to 7.5, salts of either a strong acid or a strong base); (2) to assure the optimal penetration of the drug through the cornea to inhibit the ciliary body enzymes (CA II and CA IV), this being mainly possible if the inhibitors possess a modest, but not insignificant, lipophilicity. Highly lipophilic sulfonamides are readily washed away from the eye by blood circulation (where high amounts of CA I and CA II are present), whereas compounds that are too hydrophilic do not have the chance to penetrate through the membranes to reach the ciliary process enzyme; and (3) to assure high affinity for the enzyme (mainly the isozyme II), with nano-molar inhibitors as the best candidates for clinical development as antiglaucoma drugs (Scozzafava et al. 1999a, 1999b, 2000c, 2001, 2002; Supuran et al. 1999b, 1999c, 2000a; Menabuoni et al. 1999; Mincione et al. 1999; Barboiu et al. 1999; Casini et al. 2001).

The tail approach is a general one, as shown by the multitude of highly active topically effective antiglaucoma sulfonamides reported in the last few years. Several variants of the main approach summarized here have also been designed and exam-ples are provided. Thus, sulfonamides of type **A–Y** possessing free amino, imino or hydrazino moieties were reacted with 7-chloro-4-chloromethylcoumarin **4.181**, affording a series of secondary amines possessing *N*-[(7-chloro-4-coumarinyl)-methyl]-moieties in their molecules. These showed effective inhibition of three CA isozymes (CAs I, II and IV) and were water soluble as hydrochloride salts (Renzi et al. 2000). Some of these derivatives also showed IOP-lowering properties in the normotensive rabbits after topical application as 2% aqueous solutions, but their

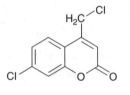

4.181

A: X = O
4.182 B: X = NH
C: X = S

efficiency was not as good as that of the previously mentioned derivatives. In another paper, some of the sulfonamides (**A–Y**) were derivatized by furan-, pyrrole- and thiophene-carboxamido moieties of type **4.182** (Ilies et al. 2000). The new derivatives obtained this way were not water soluble and were formulated as suspensions for *in vivo* experiments, similarly to that for brinzolamide. The compounds incorporating furan- and pyrrole-carboxamido moieties (but not the corresponding thiophene-substituted derivatives), showed effective and long-lasting IOP lowering both in normotensive and glaucomatous animals, with potencies superior to those of dorzolamide and brinzolamide, and this was explained by the insufficient lipophilicity of the later derivatives as compared with those of the structurally related furan- and pyrrole-carboxamido-containing compounds, because all of them showed nanomolar affinity for hCA II and bCA IV (Ilies et al. 2000).

4.4 ISOZYME-SPECIFIC INHIBITORS

Although many sulfonamide CAIs possess high affinity for the major isozymes (such as CA II, CA IV and CA V) considered to play important physiological functions (Supuran et al. 2003), the critical challenge for designing novel pharmacological agents from this class is their lack of specificity toward different isozymes. Among the 14 isozymes described, several, such as CA II, CA VII and CA IX, have very similar affinities for sulfonamide inhibitors, although small differences exist between them (Supuran et al. 2003). This fact, as well as the physiological importance of these different isozymes, prompted much research in many laboratories to find compounds that might discriminate between them. Some progress has been made recently in designing compounds with some selectivity toward CA I, CA IV and CA III, and these data are presented later.

4.4.1 ISOZYME I

The main difference in the active-site architecture of isozymes CA I and CA II is the presence of more histidine residues in CA I. Thus, in addition to the zinc ligands (His 94, His 96 and His 119), His 64 plays an important role in catalysis. This is the only other histidine residue present in the active site of CA II, whereas in CA I there are three such additional residues, His 67, His 200 and His 243 (Ferraroni et al. 2002a, 2002b). Another important difference between the two isozymes is that CA II contains a histidine cluster, consisting of residues His 64, His 4, His 3, His 10, His 15 and His 17 (streching from the middle of the active site to the rim of the cavity and protruding on the surface of the protein), which is absent in CA I (Supuran et al. 2003). These two isozymes also possess a different affinity for the two main classes of inhibitors: CA I has higher affinity than CA II for anions (such as cyanide, thiocyanate, cyanate and halides), whereas CA II has generally a higher affinity than CA I for sulfonamides (Supuran et al. 2003). As a consequence, it is relatively difficult to obtain sulfonamide inhibitors with higher affinity for CA I than for CA II, although the two isozymes possess significant differences in the active-site architecture. The first of such compounds was only recently reported by this group (Supuran et al. 1998c; Scozzafava and Supuran 1998a, 1999; Scozzafava et al.

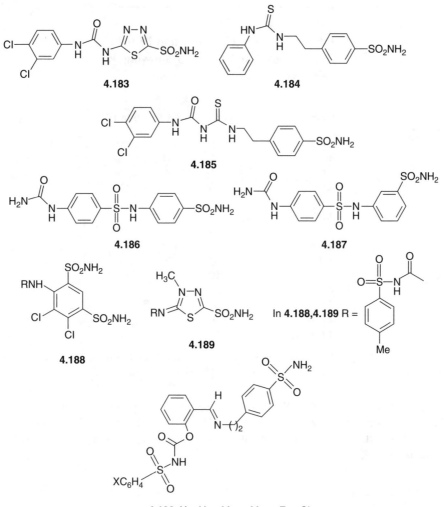

4.190: X = H, p-Me; o-Me; p-F; p-Cl

2000a) and were discovered serendipitously by screening a large number of sulfonamides possessing different structural motifs in their molecules. Remarkably, all the
compounds possessing higher affinity for CA I than for CA II (and CA IV), of types
4.183 to **4.190**, contain ureido or thioureido moieties in their molecules. Table 4.20
gives their inhibition data against the three isozymes.

Such isozyme I avid inhibitors belong to both the aromatic sulfonamide class
and the heterocyclic sulfonamide class, whereas the ureido/thioureido moieties
present in their molecules might be unsubstituted or substituted with bulkier groups
(3,4-dichlorophenyl, phenyl, substituted-phenylsulfonyl, etc.). Compounds **4.190**
containing arylsulfonylcarbamate instead of arylsulfonylureido moieties were investigated in more detail (Supuran et al. 1998c; Scozzafava and Supuran 1998a, 1999;
Scozzafava et al. 2000a). These compounds also significantly inhibit isozymes II

TABLE 4.20
Inhibition of Isozymes I, II and IV with
Compounds 4.183 to 4.195 Showing
Selectivity Toward One of Them

Inhibitor	K_I (nM)		
	hCA I	hCA II	bCA IV
4.183	3	6	8
4.184	50	53	70
4.185	7	10	24
4.186	3	8	20
4.187	4	10	25
4.188	8	12	14
4.189	4	5	11
4.190a (X = 4-Me)	40	110	120
4.190b (X = 4-Cl)	60	100	160
4.191	1100	150	140
4.192	200	20	10
4.193	200	10	8
4.194	620	12	10
4.195	180	15	12

Source: From Supuran, C.T. et al. (2003) *Medicinal Research Reviews* **23**, 146–189. With permission.

and IV and are thus not really isozyme-I specific, but represent an important step toward generating isozyme-specific CAIs. Dorzolamide has a very low affinity for hCA I, but its deethylated metabolite is a very potent inhibitor of this isozyme (Supuran et al. 2003).

4.4.2 ISOZYME IV

Isozyme CA IV contains only one histidine residue, besides the zinc ligands, within its active site His 64, which, as in hCA II, plays a critical role in catalysis as proton shuttle residue between the active site and the environment (Supuran et al. 2003). The most characteristic feature of the active site of this isozyme is related to the presence of four cysteine residues, which form two disulfide bonds situated at the entrance within the cavity (Cys 6 – Cys 11G, and Cys 23 – Cys 203; Supuran et al. 2003). These residues occupy practically the same region of the active site as the histidine cluster in hCA II does, and it was hypothesized that this might be the most relevant aspect explaining the difference in affinity for sulfonamide inhibitors of these two isozymes (Supuran et al. 2003). Even so, similar to those for CA I, the first compounds with some specificity for CA IV, of type **4.191** to **4.195**, were again discovered serendipitously, and they all belong to the same class of Schiff bases of aromatic/heterocyclic sulfonamides (Supuran et al. 1996b, 1996c, 1997b; Popescu et al. 1999).

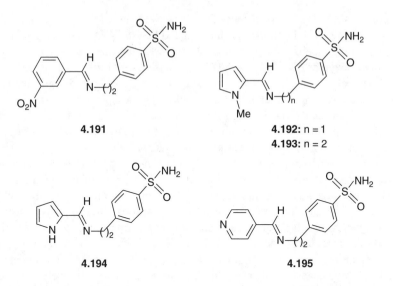

4.191

4.192: n = 1
4.193: n = 2

4.194

4.195

Only Schiff bases of aromatic sulfonamides were investigated in some detail, and it was shown that the best CA IV inhibition patterns are connected to the presence of heterocyclic moieties (in the original aldehyde used for the preparation of the Schiff base) or aromatic moieties substituted with electron-attracting groups, such as the nitro group. Such compounds also appreciably inhibited CA II, and, to a smaller extent, CA I (Table 4.20; Supuran et al. 1996b, 1996c, 1997b; Popescu et al. 1999).

4.4.3 ISOZYME III

Although the structure of this isozyme is relatively similar to that of hCA II, CA III's CO_2 hydration activity is ca. 0.3% that of hCA II, because it does not possess a His but a Lys residue in position 64, which is much less effective as a proton shuttle (Erikson and Liljas 1993). Furthermore, position 198 in CA III is occupied by a Phe, which possesses a very bulky side chain, whereas the water bound to Zn(II) has a pK_a ca. 5.5. All these characteristics might explain the low catalytic activity of CA III as well as its insensitivity to sulfonamide inhibitors that do not have sufficient space to bind in the neighborhood of the Zn(II) ion, principally because of steric impairment of Phe 198 (Erikson and Liljas 1993). Only the very small sulfonamide $CF_3SO_2NH_2$ acts as an efficient CA III inhibitor, having an inhibition constant of 0.9 μM (but this compound is a nanomolar inhibitor of CAs I, II and IV; Maren and Conroy 1993). Other sulfonamides (such as acetazolamide or methazolamide) inhibit CA III, with inhibition constants in the millimolar range (Maren and Conroy 1993).

4.5 SELECTIVE INHIBITORS FOR MEMBRANE-ASSOCIATED CAs

At least four CA isozymes (CA IV, CA IX, CA XII and CA XIV) are associated with cell membranes, with the enzyme active site generally oriented extracellularly

(Supuran et al. 2003). Some of these isozymes were shown to play pivotal physiological roles (e.g., CA IV in the eye, lungs and kidneys, CA IX in the gastric mucosa and many tumor cells; Supuran et al. 2003), whereas the function of the other membrane isozymes (CA XII, CA XIV) is at present less well understood. Because of the extracellular location of these isozymes, it is possible to design membrane-impermeant CAIs, which in this way would become specific inhibitors for the membrane-associated CAs. This possibility has been fully explored in our laboratory by designing positively charged sulfonamides (Supuran et al. 2003), whereas an alternative approach consisted of designing polymeric (high-molecular-weight) inhibitors, but such compounds were not very useful *in vivo* because of the usual problems connected with polymers (i.e., allergic reactions or problems of bioavailability).

Thus, the first historical approach for inducing membrane impermeability to CAIs was that of attaching aromatic/heterocyclic sulfonamides to polymers such as polyethyleneglycol (Maren et al. 1997), aminoethyldextran (Lucci et al. 1983; Tinker et al. 1981) or dextran (Heming et al. 1986). Compounds such as **4.196** to **4.198**, possessing molecular weights of 3.5 to 99 kDa, prepared this way showed membrane impermeability because of their high molecular weights and selectively inhibited *in vivo* CA IV only and not the cytosolic isozymes (primarily CA II), being used in several renal and pulmonary physiological studies. Because of their macromolecular nature, such inhibitors could not be developed as drugs or diagnostic tools, because *in vivo* they induced potent allergic reactions. A second approach for achieving membrane impermeability was that of using highly polar salt-like compounds. Only one such sulfonamide, quaternary ammonium sulfanilamide (QAS) **4.199**, has been used in physiological studies, which has been reported by Henry (1996) to inhibit only extracellular CAs in a variety of arthropods (such as the crab *Callinectes sapidus*) and fish. The main drawback of QAS is its high toxicity in higher vertebrates (Maren 1967).

Thus, a program of developing cationic sulfonamides was initiated in our laboratory (Supuran et al. 1992; Supuran and Clare 1995) by using QAS (**4.199**) as lead molecule, which is also a relatively weak CAI, with micromolar affinity for hCA II (Supuran and Clare 1995). The first such compounds of types **4.200** to **4.203** were prepared by reacting aromatic/heterocyclic sulfonamides containing free NH_2 groups with pyrylium salts, affording pyridinium derivatives (Supuran et al. 1992, 1998b; Supuran and Clare 1995). These compounds were moderately active CA II and CA IV inhibitors, with affinities of 10^{-6} to 10^{-7} M. By using QSAR data from our laboratory (Supuran and Clare 1995), it was shown that increased CA II and CA IV inhibitory properties of aromatic/heterocyclic sulfonamides were connected with the presence of an elongated inhibitor molecule [on the axis passing through the Zn(II) of the enzyme, the sulfonamide nitrogen atom and the long axis of the inhibitor molecule itself]. Thereby, such elongated molecules have been designed (Supuran et al. 2000b; Scozzafava et al. 2000c) by reacting pyrylium salts with amino acids (such as glycine or β-alanine) and coupling the pyridinium derivatives thus obtained of types **4.204** to **4.207** with the aromatic/heterocyclic sulfonamides possessing free amino, hydroxy, imino or hydroxyl moieties of types A-Y (a variant of the tail approach, not intended in this case for obtaining antiglaucoma sulfonamides). The inhibitors obtained this way, such as **4.208** to **4.211**, showed nanomolar affinities

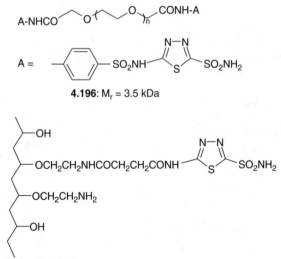

4.196: M_r = 3.5 kDa

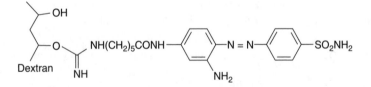

Aminoethyldextran

4.197: M_r = 6.7- 99 kDa

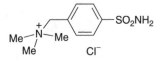

4.198: M_r = 5, 100 and 1000 kDa

4.199: QAS

for both CA II and CA IV, and, more importantly, were unable to cross the plasma membranes *in vivo* (Supuran et al. 2000b; Scozzafava et al. 2000c). In two model systems (human red cells and perfusion experiments in rats), this new class of potent, positively charged CAIs was able to discriminate the membrane-bound from the cytosolic isozymes, selectively inhibiting CA IV only (Supuran et al. 2000b; Scozzafava et al. 2000c). Such data are important both for specific *in vivo* inhibition of membrane-associated isozymes and for the eventual development of some novel anticancer therapies, because it has been shown that some tumor cells predominantly express only some membrane-associated CA isozymes, such as CA IX and CA XII.

This type of selective CAI might be of great relevance to different physiological studies. For example, Sterling et al. (2002) investigated the functional and physical

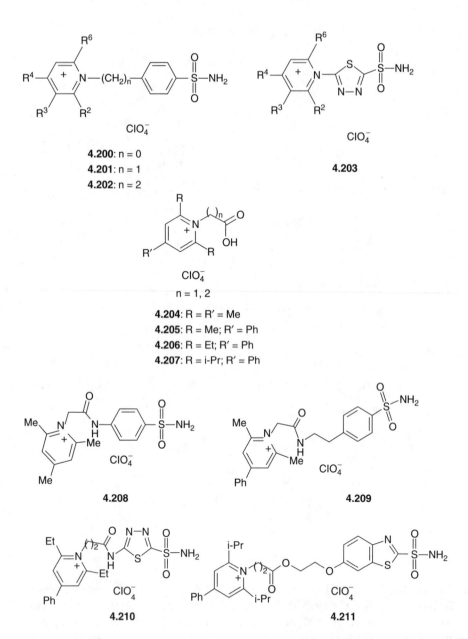

4.200: n = 0
4.201: n = 1
4.202: n = 2

4.203

n = 1, 2

4.204: R = R′ = Me
4.205: R = Me; R′ = Ph
4.206: R = Et; R′ = Ph
4.207: R = i-Pr; R′ = Ph

4.208

4.209

4.210

4.211

relationship between the downregulated, in adenoma, bicarbonate transporter and CA II, by using membrane-impermeant inhibitors of type **4.210** (in addition to classical inhibitors such as acetazolamide), which could clearly discriminate between the contribution of the cytosolic and membrane-associated isozymes in these physiological processes.

4.6 ANTITUMOR SULFONAMIDES

There are many connections between CA and cancer; for example, some CA isozymes (CA IX and XII) are predominantly found in cancer cells and not in the normal counterparts (Pastorek et al. 1994; Pastorekova et al. 1997; Chegwidden et al. 2001). Teicher et al. (1993) reported that acetazolamide (**4.44**) functions as a modulator in anticancer therapies in combination with different cytotoxic agents, such as alkylating agents, nucleoside analogs or platinum derivatives. It was hypothesized that the anticancer effects of acetazolamide (alone or in combination with such drugs) might be due to acidification of the intratumoral environment occurring after CA inhibition, although other mechanisms of action of this drug were not excluded (Treicher et al. 1993). Chegwidden et al. (2001) hypothesized that *in vitro* inhibition of growth in cell cultures of human lymphoma cells with two other potent, clinically used sulfonamide CAIs, methazolamide (**4.48**) and ethoxzolamide (**4.81**), is probably due to the reduced provision of bicarbonate for nucleotide synthesis (HCO_3^- is the substrate of carbamoyl phosphate synthetase II) as a consequence of CA inhibition.

The development of CAIs possessing potent tumor cell growth inhibitory properties was reported by this group (Supuran and Scozzafava 2000b, 2000c; Scozzafava and Supuran 2000a; Supuran et al. 2001). Such compounds were discovered in a large screening program [in collaboration with the National Institutes of Health (NIH)] of sulfonamide CAIs. Several hundred aromatic/heterocyclic sulfonamides were assessed *in vitro* as potential inhibitors of growth of a multitude of tumor cell lines, such as leukemia, nonsmall cell lung cancer, ovarian, melanoma, colon, CNS, renal, prostate and breast cancers. The active compounds (most of them nanomolar inhibitors of CA II and CA IV), of types **4.212** to **4.223**, belong to both the aromatic and the heterocyclic sulfonamide classes and showed GI_{50} values (molarity of inhibitor producing a 50% inhibition of tumor cell growth after a 48-h exposure to the drug) in the micromolar range (Supuran and Scozzafava 2000b, 2000c). Better antitumor compounds were then developed by an original strategy, and they incorporated in their molecules *N*,*N*-dialkylthiocarbonylsulfenylamino moieties (Scozzafava and Supuran 2000a; Supuran et al. 2001). Thus, aromatic/heterocyclic sulfonamides possessing free amino, imino or hydrazino groups of types A–Y (see the tail approach) were transformed to the corresponding *N*-morpholyl-thiocarbonyl-sulfenyl or *N*,*N*-dimethyl/diethyl-thiocarbonylsulfenylamino derivatives **4.224** to **4.226** by reaction with dithiocarbamates in the presence of oxidizing agents (NaClO or iodine).

Sulfonamides of the types **4.224** to **4.226** showed nanomolar affinity for CA II and CA IV, but, more importantly, some of them inhibited the growth of several tumor cell lines at concentrations as low as 10 n*M* (Scozzafava and Supuran 2000a; Supuran et al. 2001), thus showing a highly increased antitumor efficacy as compared with classical CAIs (acetazolamide, methazolamide) or compounds **4.212** to **4.218**.

The antitumor sulfonamide indisulam, E7070 (**4.227**), is in Phase II clinical trials in Europe and the U.S. as a novel anticancer agent to treat solid tumors (Supuran 2003). This compound is also a very potent inhibitor of many CA isozymes, including CA I, II and IX (unpublished results from our laboratory).

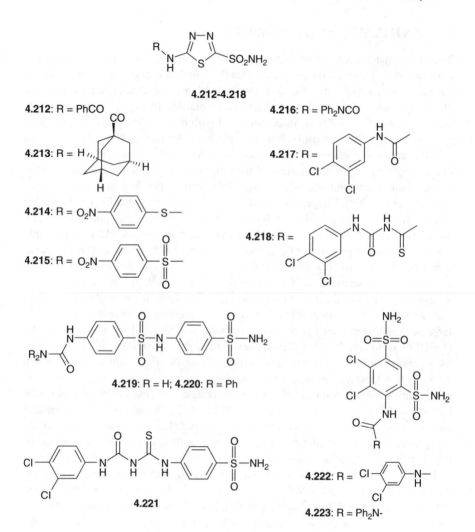

4.212-4.218

4.212: R = PhCO

4.213: R = H

4.214: R = O$_2$N—⟨⟩—S—

4.215: R = O$_2$N—⟨⟩—S—

4.216: R = Ph$_2$NCO

4.217: R =

4.218: R =

4.219: R = H; **4.220**: R = Ph

4.221

4.222: R =

4.223: R = Ph$_2$N-

4.7 SULFONAMIDES WITH MODIFIED MOIETIES AND OTHER ZINC-BINDING GROUPS

Krebs (1948) reported that substitution of the sulfonamido moiety to give compounds of type ArSO$_2$NHR drastically reduced the CA inhibitory properties compared with those of the corresponding derivatives possessing primary sulfonamido groups, ArSO$_2$NH$_2$. As a consequence, other zinc-binding functions except for the SO$_2$NH$_2$ group have rarely been considered in the design of CAIs, although many other zinc enzymes are inhibited by a multitude of derivatives possessing an entire range of zinc-binding functions, such as thiols, phosphonates, carboxylates and hydroxamates (Supuran and Scozzafava 2002a, 2002b). Only recently, several detailed studies on the possible modifications of the sulfonamido moiety, compatible with the retention of strong binding to the enzyme, have been reported (Briganti et al. 1996; Mincione

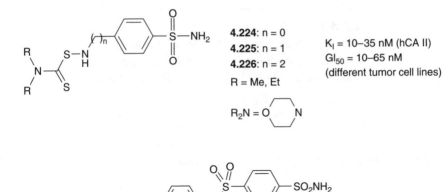

4.224: n = 0
4.225: n = 1
4.226: n = 2

R = Me, Et

K_I = 10–35 nM (hCA II)
GI_{50} = 10–65 nM
(different tumor cell lines)

$R_2N = O$ ⟨N⟩

4.227: E7070 (indisulam)

et al. 1998; Supuran et al. 1999a; Scozzafava and Supuran 2000b; Scozzafava et al. 2000b; Fenesan et al. 2000). Compounds of types **4.228** to **4.230** were studied kinetically for inhibition of reactions catalyzed by CA I and II (CO_2 hydration and ester hydrolysis), but their binding to the enzyme has also been monitored spectroscopically by studying the electronic and ^{1}H-NMR spectra of adducts of such inhibitors with Co(II)–CA II (Briganti et al. 1996).

Thus, for the series of derivatives with modified sulfonamido moieties **4.228** to **4.230** (Table 4.21), it has been observed that the presence of bulky substituents at the sulfonamido moiety (such as phenylhydrazino, ureido, thioureido or guanidino) led to compounds with weak inhibitory properties, whereas moieties present in inorganic anion CAIs (such as NO, NCS or N_3) or compact moieties substituting the sulfonamide nitrogen (such as OH, NH_2, CN or halogeno) led to compounds with appreciable inhibitory properties. Thus, the N-hydroxy sulfonamide **4.228b**, the N-chloro-substituted derivatives **4.228j, k** and the nitroso- and thiocyanato derivatives **4.228d, e** possessed the same affinity for the two investigated isozymes as did the unsubstituted sulfonamide **4.228a**. Interestingly, the thiosulfonic acid (as sodium salt) **4.228q** was one of the best inhibitors in this series, in contrast to the sulfonate (as sodium salt) **4.228p**, which behaved as a very weak inhibitor. Sulfamide **4.229** (the most simple compound containing a sulfonamido moiety) behaves as a weak inhibitor, but it binds to the Zn(II) ion, as shown by electronic spectroscopic studies on the Co(II)-substituted enzyme, whereas sulfamic acid **4.230** is a much stronger inhibitor. Furthermore, this compound has a much higher affinity for CA I than for CA II, and this might be exploited for designing CA I-specific inhibitors based on this type of zinc-binding function. Recently, the binding of sulfamide and sulfamic acid to hCA II has been investigated in detail by x-ray crystallography (Abbate et al. 2002). The binegatively charged $(NH)SO_3^{2-}$ sulfamate ion and the monoanion of sulfamide $NHSO_2NH_2^-$ bound to the Zn(II) ion within the enzyme

TABLE 4.21

Inhibition of hCA I and hCA II by Compounds Incorporating Modified Sulfonamide Moieties 4.228a–t, Sulfamide 4.229 and Sulfamic Acid 4.230

		4-Me-C$_6$H$_4$SO$_2$-X 4.228a–t	H$_2$NSO$_2$NH$_2$ 4.229	HOSO$_2$NH$_2$ 4.230
			K_I (μM)	
Inhibitor	X		hCA I	hCA II
4.228a	NH$_2$		50	11
4.228b	NHOH		41	9
4.228c	NHOMe		220	173
4.228d	NO		35	24
4.228e	NCS		30	18
4.228f	N$_3$		27	45
4.228g	Imidazol-1-yl		160	34
4.228h	NHNH$_2$		70	53
4.228i	NHNHPh		>1000	120
4.228j	NHCl		19	2.1
4.228k	NCl$_2$		12	3.6
4.228m	NHCN		210	125
4.228n	NHOCH$_2$COOH		150	85
4.228p	OH		130	460
4.228q	SH		5	10
4.228r	NHCONH$_2$		>1000	460
4.228s	NHCSNH$_2$		>1000	410
4.228t	NHC(NH)NH$_2$		>1000	540
4.229			310	1130
4.230			21	390

Source: From Briganti, F. et al. (1996) *European Journal of Medicinal Chemistry* **31**, 1001–1010. With permission.

active site (Abbate et al. 2002). These two structures provided some close insights into why the sulfonamide functional group appears to have unique properties for CA inhibition: (1) it exhibits a negatively charged, most likely monoprotonated nitrogen coordinated to the Zn(II) ion; (2) simultaneously this group forms a hydrogen bond as donor to the oxygen Oγ of the adjacent Thr 199; and (3) a hydrogen bond is formed between one of the SO$_2$ oxygens to the backbone NH of Thr 199. Thus, the basic structural elements explaining the strong affinity of the sulfonamide moiety for the Zn(II) ion of CAs were demonstrated in detail by using these simple compounds as prototypical CAIs, without the need to analyze the interactions of the organic scaffold usually present in other inhibitors (generally belonging to the aromatic/heterocyclic sulfonamide class; Abbate et al. 2002). Despite important similarities for the binding of these two inhibitors to the enzyme with that of aromatic/heterocyclic sulfonamides of the type RSO_2NH_2 previously investigated, the absence of a $C–SO_2NH_2$ bond in sulfamide/sulfamic acid leads to a different hydrogen bond network in the neighborhood of the catalytical Zn(II) ion, which was shown to be useful for the drug design of more potent CA inhibitors possessing zinc-binding functions different from those of the classical sulfonamide group (Abbate et al. 2002). By using such compounds as leads, several series of much stronger inhibitors were subsequently reported, possessing modified sulfonamido moieties as zinc-binding functions of the type SO_2NHOH, SO_2NHCN, $SO_2NHPO_3H_2$, $SO_2NHSO_2NH_2$, SO_2NHSO_3H or $SO_2NHCH_2CONHOH$, among others (Mincione et al. 1998; Supuran et al. 1999a; Scozzafava et al. 2000b; Fenesan et al. 2000).

Thus, compounds such as **4.231** to **4.240** possessing *N*-cyano, *N*-hydroxy or *N*-phosphoryl-sulfonamido moieties, or the related modified sulfamide/sulfamic acid zinc-binding functions, and diverse alkyl, aryl or heterocyclic moieties in their molecules showed affinities in the low nanomolar range for hCA II (except **4.237**), being equipotent or better inhibitors than the corresponding unsubstituted sulfonamides (Mincione et al. 1998; Supuran et al. 1999a; Scozzafava et al. 2000b; Fenesan et al. 2000). Compound **4.237** is a weak inhibitor of hCA II (affinity constant of 1.2 μM), but has a much higher affinity (50 n*M*) for hCA I, thus being one of the most selective hCA I inhibitors reported to date (Scozzafava et al. 2000b).

$n\text{-}C_8F_{17}SO_2NHCN$
4.231: K_i = 5 nM (hCA II)

4.232: K_i = 25 nM (hCA II)

4.233: K_i = 4 nM (hCA II)

4.234: K_i = 0.8 nM (hCA II)

4.235: K_i = 2 nM (hCA II)

4.236: K_i = 10 nM (hCA II)

4.237: K_i = 1200 nM (hCA II)
K_i = 50 nM (hCA I)

4.238: K_i = 20 nM (hCA II)

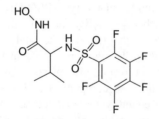

4.239
(K$_i$ hCA I= 8 nM; hCA II = 11 nM; bCA IV = 13 nM)

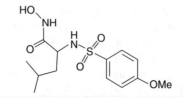

4.240
(K$_i$ hCA I= 50 nM; hCA II = 5 nM; bCA IV = 39 nM)

Sulfonylated amino acid hydroxamates were also shown to possess strong CA inhibitory properties (Scozzafava and Supuran 2000b). Such hydroxamates generally act as potent inhibitors of metalloproteases containing catalytic zinc ions, such as matrix metalloproteinases (MMPs) or bacterial collagenases (Scozzafava and Supuran 2002a). They bind to the Zn(II) ions present in these enzymes bidentately, coordinating through the hydroxamate (ionized) moiety (Supuran and Scozzafava 2002b). Scolnick et al. (1997) showed that two simple hydroxamates of the type RCONHOH (R = Me, CF$_3$) act as micromolar inhibitors of hCA II and bind to the Zn(II) ion of this enzyme, as demonstrated by x-ray crystallography. By using these two derivatives as lead molecules, Scozzafava and Supuran (2000b) designed a series of sulfonylated amino acid hydroxamate derivatives possessing the general formula RSO$_2$NHCH(R')CONHOH and showed by electron spectroscopic studies on the Co(II)-substituted CA that they bind to the Zn(II) ion of CA. Some of these compounds, such as **4.239** and **4.240**, showed affinity in the low nanomolar range for the major CA isozymes (CA I, II and IV), but substitution of the sulfonamide nitrogen by a benzyl or a substituted-benzyl moiety led to a drastic reduction of the CA inhibitory properties and to an enhancement of the MMP inhibitory properties. Thus, between the two types of zinc enzymes, the zinc proteases and the CAs, there exist some cross-reactivity as regards hydroxamate inhibitors, but generally strong MMP inhibitors are weak CAIs and vice versa.

All these data demonstrate that in addition to the classical CAIs of the aromatic/heterocyclic sulfonamide type, other compounds can be designed with very strong affinity for the active site of different isozymes, a fact that might be relevant for obtaining diverse pharmacological agents that modulate the activity of these widespread enzymes.

4.8 ANTIEPILEPTIC SULFONAMIDES AND OTHER MISCELLANEOUS INHIBITORS

Several sulfonamide CA inhibitors such as acetazolamide **4.44**, methazolamide **4.48**, topiramate **4.241** or zonisamide **4.242** were and are still used as antiepileptic drugs (Masereel et al. 2002). The anticonvulsant effects of these or related sulfonamides are probably due to CO_2 retention following inhibition of the red cell and brain enzymes, but other mechanisms of action such as blockade of sodium channels and kainate/AMPA receptors as well as enhancement of GABA-ergic transmission were also hypothesized or proved for some of these drugs (Masereel et al. 2002). Acetazolamide and methazolamide are still clinically used at present in some forms of epilepsy, but are considered to belong to a minor class of antiepileptic agents, whereas the more recently developed drug topiramate **4.241**, a very effective antiepileptic, was also shown to act as a strong CA inhibitor, with a potency similar to that of acetazolamide against the physiologically important isozyme CA II (Masereel et al. 2002). Furthermore, its x-ray crystal structure in complex with hCA II has recently been reported by Casini et al. (2003) and reveals a very tight association of the inhibitor with a network of seven strong hydrogen bonds fixing topiramate within the active site, in addition to Zn(II) coordination through the ionized sulfamate moiety. (See also Chapter 3 of this book.)

Recently, a series of aromatic/heterocyclic sulfonamides incorporating valproyl moieties was prepared to design antiepileptic compounds possessing in their structure two moieties known to induce such a pharmacological activity: valproic acid (**4.243**), one of the most widely used antiepileptic drugs, and the sulfonamide residue included in acetazolamide and topiramate, two CAIs with antiepileptic properties (Masereel et al. 2002). The valproyl derivative of acetazolamide (5-valproylamido-1,3,4-thiadiazole-2-sulfonamide, **4.244**) was one of the best hCA I and hCA II inhibitors in the series and exhibited very strong anticonvulsant properties in the MES test in mice (Masereel et al. 2002). Consequently, other 1,3,4-thiadiazole-sulfonamide derivatives possessing potent CA inhibitory properties (of types **4.245**

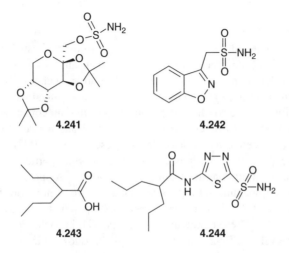

4.241   4.242

4.243   4.244

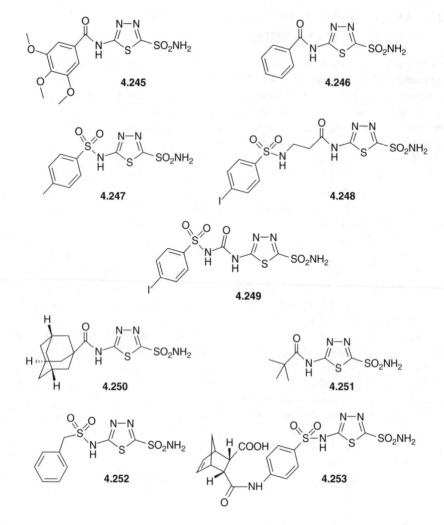

to **4.253**) and substituted with different alkyl/arylcarboxamido/sulfonamido/ureido moieties in position 5 have been investigated for their anticonvulsant effects in the same animal model. Some structurally related derivatives such as 5-benzoylamido-, 5-toluenesulfonylamido-, 5-adamantylcarboxamido- and 5-pivaloylamido-1,3,4-thiadiazole-2-sulfonamide also showed promising *in vivo* anticonvulsant properties, and these compounds can be considered interesting leads for developing anticonvulsant or selective cerebrovasodilator drugs (Masereel et al. 2002).

Other miscellaneous CAIs possessing interesting properties have also been reported. Thus, in an attempt to obtain gastric mucosa CA-specific CAIs, a group of sulfenamido sulfonamides of type **4.254 – 4.257** have been reported that possess powerful CA II and CA IV inhibitory properties (Supuran et al. 1997a; Scozzafava and Supuran 1998b). These compounds were designed in such a way as to liberate aromatic/heterocyclic sulfonamides and sulfenyl chlorides in the presence of gastric

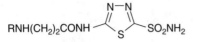

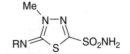

4.254: R = 2-O$_2$N-C$_6$H$_4$S
4.255: R = 4-O$_2$N-C$_6$H$_4$S

4.256: R = 2-O$_2$N-C$_6$H$_4$S
4.257: R = 4-O$_2$N-C$_6$H$_4$S

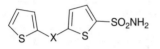

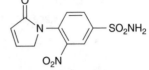

4.258: X = S (K$_i$ = 0.9 nM, hCA II)
4.259: X = SO$_2$ (K$_i$ = 0.8 nM, hCA II)

4.260: (K$_i$ = 0.6 nM, hCA II)

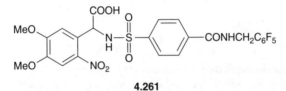

4.261

hydrochloric acid. The liberated compounds would then inhibit the enzymes involved in gastric acid production: the sulfonamides would bind within the CA active site, whereas the sulfenyl chlorides would inactivate (in an omeprazole-like manner) the gastric H$^+$/K$^+$-ATPase by alkylating the critical cysteine residues of this enzyme (Supuran et al. 1997a; Scozzafava and Supuran 1998b).

Klebe's group detected several potent hCA II inhibitors by computer-aided drug design (Gruneberg et al. 2001). Three of these compounds, **4.258** to **4.260** showed nanomolar affinity to hCA II, and the x-ray structure has also been resolved for two of them. Kehayova et al. (1999) reported a hydrophobic CAI of type **4.261** that possessed a photolabile group derived from *o*-nitrophenylglycine. It is stated that such compounds might be useful for the site-specific delivery of prodrugs, for instance to the eye, but the high-energy UV radiation needed to liberate the active inhibitor (*p*-H$_2$NO$_2$S-C$_6$H$_4$CONHCH$_2$C$_6$F$_5$) would probably do more harm than good to the eye tissues.

Reaction of *tert*-butyl-dimethylsilyl (TBDMS)-protected bile acids (cholic, chenodeoxycholic, deoxycholic, lithocholic, ursodeoxycholic acids) or dehydro-cholic acid with aromatic/heterocyclic sulfonamides possessing free amino/hydroxy moieties, of types **A–T**, in the presence of carbodiimides afforded after deprotection of the OTBDMS ethers a series of sulfonamides incorporating bile acid moieties in their molecules (Scozzafava and Supuran 2002). Many such derivatives, of which **4.262** are illustrative representatives, showed strong inhibitory properties against three isozymes CA I, CA II and CA IV. Some of the most active derivatives,

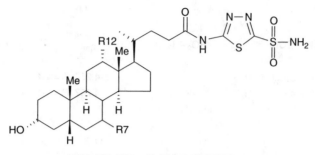

4.262 (R7, R12 = H, OH, O-TBDMSi)

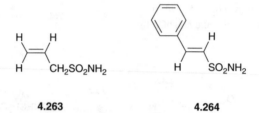

4.263   **4.264**

incorporating 1,3,4-thiadiazole-2-sulfonamide or benzothiazole-2-sulfonamide func-
tionalities in their molecules, showed low nanomolar affinity for CA II and CA IV.
Furthermore, the bioavailability of these derivatives in rabbits was comparable to
that of acetazolamide, being ca. 85 to 90%, making them promising candidates for
systemically acting CA inhibitors (Scozzafava and Supuran 2002).

Unsaturated primary sulfonamides have only recently been investigated for their
interaction with CA (Chazalette et al. 2001). Such compounds, and more precisely
allyl-sulfonamide **4.263** and *trans*-styrene sulfonamide **4.264**, behave as nanomolar
inhibitors of the physiologically relevant isozymes CA I and CA II (Chazalette et al.
2001).

4.9 ALIPHATIC SULFONAMIDES

Aliphatic sulfonamides of the type $R-SO_2NH_2$ (R = Me, $PhCH_2$) were investigated
as CAIs by Maren (1967), who showed that in contrast to aromatic/heterocyclic
sulfonamides, such compounds are extremely weak inhibitors ($IC_{50} > 10^{-4}$ M). This
lack of activity was correlated with the absence of the cyclic moiety that would
influence the stability of the enzyme–inhibitor (EI) complexes in two ways: (1) by
an acidifying effect on the SO_2NH_2 protons (because of the electron-attracting properties of the aryl/hetaryl moieties), and (2) by the cyclic moiety in the molecule of
CAIs participating in hydrophobic and van der Waals interactions, as well as hydrogen bond networks with amino acid side chains within the active site.

In reality, the situation was much more complex, and the simple assertions made
were shown to be basically incorrect by the report of Maren and Conroy (1993) that

TABLE 4.22
Dog CA II Inhibition with
Aliphatic Sulfonamides 4.265

RSO_2NH_2
4.265

4.265b	R	pK_a	K_i (M)
a	Me	10.5	10^{-4}
b	$ClCH_2$	9.2	7×10^{-6}
c	F_3CCH_2	9.1	2×10^{-6}
d	FCH_2	9.0	10^{-6}
e	F_2CH	7.5	8×10^{-8}
f	Cl_3C	6.5	2×10^{-8}
g	C_4F_9	6.2	5×10^{-9}
h	C_4F_9	6.0	10^{-8}
i	F_3C	5.8	2×10^{-9}

Source: From Maren T.H., and Conroy, C.W. (1993) *Journal of Biological Chemistry* **268**, 26233–26239. With permission.

some types of aliphatic sulfonamides can act as very strong CAIs. In Table 4.22, data are presented for some halogeno/polyhalogeno-alkylsulfonamides (**4.265b–i**), as compared to the simplest such sulfonamide methanesulfonamide **4.265a**. The data show that an increasing number of halogen atoms in position 1 or 2, or both, from the SO_2NH_2 group has the effect of strengthening the acidity of the SO_2NH_2 protons and a concomitant dramatic increase in CA inhibitory properties. In practice, trichloromethyl and pentafluoroethyl derivatives **4.265f, h** are as strong CAIs as acetazolamide, whereas the perfluoroderivatives **4.265g, i** are even more potent against dog CA II (Maren and Conroy 1993).

Even more interesting inhibitors of this class, of types **4.266** to **4.270**, have been reported by Scholz et al. (1993) in their search for antiglaucoma agents with a topical action. Commencing from the modestly active lead **4.266** (a micromolar inhibitor), it has been observed that (1) the aromatic ring was not necessary for inducing good CA inhibitory properties in this type of compounds; (2) lengthening the alkyl chain connected to the sulfamoylmethyl-sulfoxide moiety of the molecule was highly beneficial for achieving very good affinity for the enzyme (the study conducted on hCA II; Scholz et al. 1993). Indeed, the phenethyl **4.267** and especially n-octyl **4.268** derivatives show an increased affinity for CA II as compared to the lead **4.266**. Because the purely aliphatic compound **4.268** was the best inhibitor (IC_{50} 22 nM), the hydrogen atoms in position α to the sulfamoyl moiety were substituted by one or two halogens (fluorine and chlorine) or by methyl groups. Only the monohalogeno derivatives **4.269** and **4.270** possessed good CA inhibitory activities; the dihalogeno-substituted or the methyl derivatives were only micromolar inhibitors.

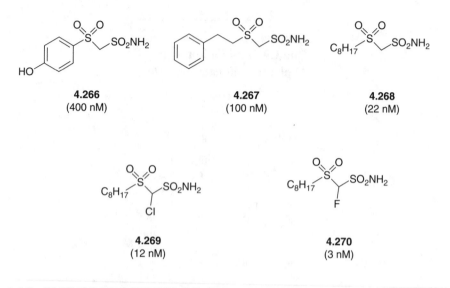

4.266
(400 nM)

4.267
(100 nM)

4.268
(22 nM)

4.269
(12 nM)

4.270
(3 nM)

4.10 FUTURE PROSPECTS OF CAIs

CAIs are widely used as therapeutic agents to manage or prevent many diseases. This is mainly because of the wide distribution of the 14 vertebrate CAs in many cells, tissues and organs, wherein they play crucial physiological functions. The available pharmacological agents are still far from perfect. They possess many undesired side effects, mainly because they lack selectivity for the different CA isozymes. Thus, developing isozyme-specific or at least organ-selective sulfonamide inhibitors will be highly beneficial both for obtaining novel types of drugs, devoid of major side effects, as well as for many physiological studies in which specific/selective inhibitors will be valuable tools for understanding the physiology of these enzymes. Prospects for achieving such a goal are not very optimistic at present, because of the high similarity between several isozymes in their interaction with sulfonamide inhibitors. Some progress in this field has been recently recorded, as shown in this chapter, by developing low-molecular-weight membrane-impermeant inhibitors, which being excluded from the intracellular space inhibit selectively only membrane-associated and not cytosolic CA isozymes.

Important advances have been made in the past few years in developing topically effective CAIs for treating glaucoma, with two drugs being available: dorzolamide and brinzolamide. These drugs have reduced many of the undesired side effects observed with the systemically used CAIs for treating glaucoma. Both dorzolamide and brinzolamide are effective antiglaucoma agents, but because of local side effects they tend to pose tolerability problems in many patients. This is probably because these two compounds are salts of weak bases with a very strong acid, and thus the pH of the administered drug is relatively acidic. Thus, the search for novel types of topically acting antiglaucoma sulfonamides continues. Recently, an approach different (and more general) from the ring approach, which led to the drugs mentioned previously, has been reported for preparing topical antiglaucoma CAIs. This consists of introducing water-solubilizing tails to the molecules of aromatic/heterocyclic sulfonamides. Several of the compounds obtained this way, which can form salts with pH values in both the slightly acidic and the slightly basic region (pH 6.5 to 7.5), showed very potent IOP-lowering properties in several animal models of glaucoma, and some of them are under clinical evaluation. Furthermore, such agents showed a longer IOP-lowering effect than the first-generation topically acting sulfonamides did, which might be a valuable feature in a new drug from this family of pharmacological agents. In our view, another important aspect for the future applications of CAIs in ophthalmology regards their possible use in treating macular edema and related degenerative pathologies, for which no effective treatment is known at present. (See Chapter 8 of this book.)

The potential of CAIs to treat epilepsy and other neurological/neuromuscular disorders is far from having been fully exploited. Thus, no specific brain enzyme CAI has been reported yet, but a specific cerebrovasodilator from this class of pharmacological agents will be a valuable drug and also an interesting diagnostic tool, because acetazolamide is widely employed (the so-called acetazolamide-test) for assessing cerebrovascular reactivity in normal and pathological states (Supuran and Scozzafava 2000a). Even so, the "imperfect" drug acetazolamide represents the only available therapy of choice in many minor neurological disorders, such as familial hemiplegic migraine and ataxia, tardive dyskinesia and hypo- and hyperkalemic periodic paralysis (Supuran and Scozzafava 2000a). The development of more selective CAIs of this type will be an interesting challenge for medicinal chemists and pharmacologists.

Sulfonamide CAIs played a crucial role in the understanding of renal physiology and pharmacology and led to the development of widely used diuretic drugs, such as the benzothiadiazine and high-ceiling diuretics. Still, few advances have been ultimately recorded in this field, probably because a large number of clinically approved diuretics are available. An orphan drug of this class, benzolamide, still awaits a wider use. Few studies are also available on the therapeutic potential of CAIs in osteoporosis. The development of bone-targeted sulfonamides will lead to drugs devoid of severe systemic effects, but little progress has been registered in this field (Supuran and Scozzafava 2000a).

A recent and new field in CAI research has been opened by the report of the potent antitumor properties of an entire class of sulfonamide CAIs, as well as by the isolation of some CA isozymes predominantly present in tumor cells. The mechanisms by which sulfonamides inhibit tumor cell growth are only beginning to be understood, and we predict important advances in this direction because several laboratories are involved in the synthesis, evaluation and *in vitro/in vivo* antitumor testing of novel classes of sulfonamides with potential applications as anticancer therapeutic agents. Indeed, a compound of this type, indisulam, has progressed to Phase II clinical trials for use in treating solid tumors.

No pharmacological agents from this class of compounds have so far been developed for inhibiting the liver enzyme (predominantly CA V), and because this is involved in biosynthetic reactions (urea synthesis, glucogenesis, etc.), such agents might be useful in some metabolic dysfunctions. In fact, CA V remains one of the least studied major CA isozymes, and even its catalytic mechanism is not fully understood.

Very few studies are available on the inhibition of nonvertebrate CAs. Because CAs were recently shown to be present in a multitude of parasites (e.g., *Plasmodium* spp.), bacteria and archaea, it is possible to develop CAI-based antibiotics or antimalarial therapies.

CAs and their inhibitors are indeed remarkable — after many years of intense research in this field, they continue to offer interesting opportunities for developing novel drugs and new diagnostic tools and for understanding in greater depth the fundamental processes of the life sciences.

REFERENCES

Abbate, F., Supuran, C.T., Scozzafava, A., Orioli, P., Stubbs, M., and Klebe, G. (2002) Nonaromatic sulfonamide group as an ideal anchor for potent human carbonic anhydrase inhibitors: Role of hydrogen-bonding networks in ligand binding and drug design. *Journal of Medicinal Chemistry* **45**, 3583–3587.

Antonaroli, S., Bianco, A., Brufani, M., Cellai, L., Lo Baido, G., Potier, E., Bonomi, L., Perfetti, S., Fiaschi, A.I., and Segre, G. (1992) Acetazolamide-like carbonic anhydrase inhibitors with topical ocular hypotensive activity. *Journal of Medicinal Chemistry* **35**, 2697–2703.

Arslan, O., Cakir, U., and Ugras, H.I. (2002) Synthesis of new sulfonamide inhibitors of carbonic anhydrase. *Biochemistry (Moscow)* **67**, 1273–1276.

Avila, L.Z., Chu, Y.H., Blossey, E.C., and Whitesides, G.M. (1993) Use of affinity capillary electrophoresis to determine kinetic and equilibrium constants for binding of arylsulfonamides to bovine carbonic anhydrase. *Journal of Medicinal Chemistry* **36**, 126–133.

Baldwin, J.J., Ponticello, G.S., Anderson, G.S., Christy, M.E., Murcko, M.A., Randall, W.C., Schwam, H., Sugrue, M.F., Springer, J.B., Gautheron, P., Grove, J., Mallorga, P., Viader, M.P., McKeever, B.M., and Navia, M.A. (1989) Thienothiopyran-2-sulfonamides: Novel topically active carbonic anhydrase inhibitors for the treatment of glaucoma. *Journal of Medicinal Chemistry* **32**, 2510–2513.

Barboiu, M., Supuran, C.T., Menabuoni, L., Scozzafava, A., Mincione, F., Briganti, F., and Mincione, G. (1999) Carbonic anhydrase inhibitors: Synthesis of topically effective intraocular pressure lowering agents derived from 5-(ω-aminoalkylcarboxamido)-1,3,4-thiadiazole-2-sulfonamide. *Journal of Enzyme Inhibition* **15**, 23–46.

Barnish, I.T., Cross, P.E., Dickinson, R.P., Gadsby, B., Parry, M.J., Randall, M.J., and Sinclair, I.W. (1980) Cerebrovasodilation through selective inhibition of the enzyme carbonic anhydrase. 2. Imidazo[2,1-b]thiadiazole and imidazo[2,1-b]thiazolesulfonamides. *Journal of Medicinal Chemistry* **23**, 117–121.

Barnish, I.T., Cross, P.E., Dickinson, R.P., Parry, M.J., and Randall, M.J. (1981) Cerebrovasodilation through selective inhibition of the enzyme carbonic anhydrase. 3. 5-(Arylthio)-, 5-(arylsulfinyl)- and 5-(arylsulfonyl)thiophene-s-sulfonamides. *Journal of Medicinal Chemistry* **24**, 959–964.

Beasley, Y.M., Overell, B.G., Petrow, V., and Stephenson, O. (1958) Some *in vitro* inhibitors of carbonic anhydrase. *Journal of Pharmacy and Pharmacology* **10**, 696–705.

Blacklock, T.J., Sohar, P., Butcher, J.W., Lamanec, T., and Grabowski, E.J.J. (1993) An enantioselective synthesis of the topically-active carbonic anhydrase inhibitor MK-0507: 5,6-dihydro-(*S*)-4-(ethylamino)-(*S*)-6-methyl-4H-thieno[2,3-b]thiopyran-2-sulfonamide-7,7-dioxide hydrochloride. *Journal of Organic Chemistry* **58**, 1672–1679.

Boddy, A., Edwards, P., and Rowland, M. (1989) Binding of sulfonamides to carbonic anhydrase: Influence on distribution within blood and on pharmacokinetics. *Pharmaceutical Research* **6**, 203–209.

Boriack, P.A., Christianson, D.W., Kingery-Wood, J., and Whitesides, G.M. (1995) Secondary interactions significantly removed from the sulfonamide binding pocket of carbonic anhydrase II influence inhibitor binding constants. *Journal of Medicinal Chemistry* **38**, 2286–2291.

Borras, J., Scozzafava, A., Menabuoni, L., Mincione, F., Briganti, F., Mincione, G., and Supuran, C.T. (1999) Carbonic anhydrase inhibitors: Synthesis of water-soluble, topically effective intraocular pressure lowering aromatic/heterocyclic sulfonamides containing 8-quinoline-sulfonyl moieties — Is the tail more important than the ring? *Bioorganic and Medicinal Chemistry* **7**, 2397–2406.

Briganti, F., Pierattelli, A., Scozzafava, A., and Supuran, C.T. (1996) Carbonic anhydrase inhibitors: Part 37. Novel classes of carbonic anhydrase inhibitors and their interaction with the native and cobalt-substituted enzyme: kinetic and spectroscopic investigations. *European Journal of Medicinal Chemistry* **31**, 1001–1010.

Burbaum, N.J., Ohlmeyer, M.H.J., Reader, J.C., Henderson, I., Cillard, L.W., Li, G., Randle, T.L., Sigal, N.H., Chelsky, D., and Baldwin, J.J. (1995) A paradigm for drug discovery employing encoded combinatorial libraries. *Proceedings of the National Academy of Sciences of the United States of America* **92**, 6027–6031.

Buzas, A., Frossard, J., and Teste, J. (1961) Etude de l'activité diuretique de quelques dérivés du thiophene. *Annales Pharmaceutique Française* **19**, 449–458.

Casini, A., Antel, J., Abbate, F., Scozzafava, A., David, S., Waldeck, H., Schäfer, S., and Supuran, C.T. (2003) Carbonic anhydrase inhibitors: SAR and x-ray crystallographic study for the interaction of sugar sulfamates/sulfamides with isozymes I, II and IV. *Bioorganic and Medicinal Chemistry Letters* **13**, 841–845.

Casini, A., Mincione, F., Ilies, M.A., Menabuoni, L., Scozzafava, A., and Supuran, C.T. (2001) Carbonic anhydrase inhibitors: Synthesis and inhibition against isozymes I, II and IV of topically acting antiglaucoma sulfonamides incorporating *cis*-5-norbornene-*endo*-3-carboxy-2-carboxamido moieties. *Journal of Enzyme Inhibition* **16**, 113–123.

Casini, A., Scozzafava, A., Mincione, F., Menabuoni, L., Ilies, M.A., and Supuran, C.T. (2000) Carbonic anhydrase inhibitors: Water soluble 4-sulfamoylphenylthioureas as topical intraocular pressure lowering agents with long lasting effects. *Journal of Medicinal Chemistry* **43**, 4884–4892.

Chazalette, C., Rivière-Baudet, M., Supuran, C.T., and Scozzafava, A. (2001) Carbonic anhydrase inhibitors: Allylsulfonamide, styrene sulfonamide, *N*-allyl sulfonamides and some of their metal (Si, Ge, B) derivatives. *Journal of Enzyme Inhibition* **16**, 475–489.

Chegwidden, W.R., Spencer, I.M., and Supuran, C.T. (2001) The roles of carbonic anhydrase isozymes in cancer. In *Gene Families: Studies of DNA, RNA, Enzymes and Proteins*, Xue, G., Xue, Y., Xu, Z., Hammond, G.L., Lim, H.A., Eds., World Scientific, Singapore, pp. 157–169.

Chen, H.H., Gross, S., Liao, J., McLaughlin, M., Dean, T.R., Sly, W.S., and May, J.A. (2000) 2H-Thieno[3,2-e]- and [2,3-e]-1,2-thiazine-6-sulfonamide 1,1-dioxides as ocular hypotensive agents: synthesis, carbonic anhydrase inhibition and evaluation in the rabbit. *Bioorganic and Medicinal Chemistry* **8**, 957–975.

Chow, K., Lai, R., Holmes, J.M., Wijono, M., Wheeler, L.A., and Garst, M.E. (1996) 5-Substituted-3-thiophenesulfonamides as carbonic anhydrase inhibitors. *European Journal of Medicinal Chemistry* **31**, 175–186.

Conroy, C.W., Schwam, H., and Maren, T.H. (1984) The nonenzymatic displacement of the sulfamoyl group from different classes of aromatic compounds by glutathione and cysteine. *Drug Metabolism and Disposition* **12**, 614–618.

Cross, P.E., Gadsby, B., Holland, G.F., and McLamore, W.M. (1978) Cerebrovasodilatation through selective inhibition of the enzyme carbonic anhydrase. 1. Substituted benzene-disulfonamides. *Journal of Medicinal Chemistry* **21**, 845–850.

Davenport, H.W. (1945) The inhibition of carbonic anhydrase by thiophene-2-sulfonamide and sulfanilamide. *Journal of Biological Chemistry* **158**, 567–571.

Dean, T.R., Chen, H.H., and May, J.A. (1993) Thiophenesulfonamides useful as inhibitors of carbonic anhydrase. United States Patent 5,240,923, 1993.

Doyon, J.B., and Jain, A. (1999) The pattern of fluorine substitution affects binding affinity in a small library of fluoroaromatic inhibitors for carbonic anhydrase. *Organic Letters* **1**, 183–185.

Eller, M.G., Schoenwald, R.D., Dixson, J.A., Segarra, T., and Barfknecht, C.F. (1985) Topical carbonic anhydrase inhibitors. III. Optimization model for corneal penetration of ethoxzolamide analogues. *Journal of Pharmaceutical Sciences* **74**, 155–160.

Eriksson, A.E., and Liljas, A. (1993) Refined structure of bovine carbonic anhydrase III at 2.0 Å resolution. *Proteins: Structure, Function, and Genetics* **16**, 29–42.

Fenesan, I., Popescu, R., Scozzafava, A., Crucin, V., Mateiciuc, E., Bauer, R., Ilies, M.A., and Supuran, C.T. (2000) Carbonic anhydrase inhibitors: Phosphoryl-sulfonamides: A new class of high affinity inhibitors of isozymes I and II. *Journal of Enzyme Inhibition* **15**, 297–310.

Gao, J.M., Cheng, X.H., Chen, R.D., Sigal, G.B., Bruce, J.E., Schwartz, B.L., Hofstadler, S.A., Anderson, G.A., Smith, R.D., and Whitesides, G.M. (1996) Screening derivatized peptide libraries for tight binding inhibitors to carbonic anhydrase II by electrospray ionization: Mass spectroscopy. *Journal of Medicinal Chemistry* **39**, 1949–1955.

Gao, J.M., Qiao, S., and Whitesides, G.M. (1995) Increasing binding constants of ligands to carbonic anhydrase by using "greasy tails." *Journal of Medicinal Chemistry* **38**, 2292–2301.

Graham, S.L., Hoffman, J.M., Gautheron, P., Michelson, S.R., Scholz, T.H., Schwam, H., Shepard, K.L., Smith, A.M., Sondey, J.M., Sugrue, M.F., and Smith, R.L. (1990) Topically active carbonic anhydrase inhibitors. 3. Benzofuran- and indole-2-sulfonamides. *Journal of Medicinal Chemistry* **33**, 749–754.

Graham, S.L., Shepard, K.L., Anderson, P.S., Baldwin, J.J., Best, D.B., Christy, M.E., Freedman, M.B., Gautheron, P., Habecker, C.N., Hoffman, J.M., Lyle, P.A., Michelson, S.R., Ponticello, G.S., Robb, C.M., Schwam, H., Smith, A.M., Smith, R.L., Sondey, J.M., Strohmaier, K.M., Sugrue, M.F., and Varga, S.L. (1989) Topically active carbonic anhydrase inhibitors. 2. Benzo[b]thiophenesulfonamide derivatives with ocular hypotensive activity. *Journal of Medicinal Chemistry* **32**, 2548–2554.

Gruneberg, S., Wendt, B., and Klebe, G. (2001) Subnanomolar inhibitors from computer screening: A model study using human carbonic anhydrase II. *Angewandte Chemie International Edition* **40**, 389–393.

Hartman, G.D., and Halczenko, W. (1989) The synthesis of 5-alkylaminomethylthieno[2,3-b]pyrrole-5-sulfonamides. *Heterocycles* **29**, 1943–1949.

Hartman, G.D., Halczenko, W., Prugh, J.D., Smith, R.L., Sugrue, M.F., Mallorga, P.J., Michelson, S.R., Randall, W.C., Schwam, H., and Sondey, J.M. (1992) Thieno[2,3-b]furan-2-sulfonamides as topical carbonic anhydrase inhibitors. *Journal of Medicinal Chemistry* **35**, 3027–3033.

Heming, T.A., Geers, C., Gros, G., Bidani, A., and Crandall, E.D. (1986) Effects of dextran-bound inhibitors on carbonic anhydrase activity in isolated rat lungs. *Journal of Applied Physiology* **61**, 1849–1856.

Henry R.P. (1996) Multiple roles of carbonic anhydrase in cellular transport and metabolism. *Annual Reviews in Physiology* **58**, 523–538.

Ilies, M., Supuran, C.T., Scozzafava, A., Casini, A., Mincione, F., Menabuoni, L., Caproiu, M.T., Maganu, M., and Banciu, M.D. (2000) Carbonic anhydrase inhibitors: Sulfonamides incorporating furan-, thiophene- and pyrrole-carboxamido groups possess strong topical intraocular pressure lowering properties as aqueous suspensions. *Bioorganic and Medicinal Chemistry* **8**, 2145–2155.

Jain, A., Whitesides, G.M., Alexander, R.S., and Christianson, D.W. (1994) Identification of two hydrophobic patches in the active-site cavity of human carbonic anhydrase II by solution-phase and solid-state studies and their use in the development of tight-binding inhibitors. *Journal of Medicinal Chemistry* **37**, 2100–2105.

Jayaweera, G.D.S.A., MacNeil, S.A., Trager, S.F., and Blackburn, G.M. (1991) Synthesis of 2-substituted-1,3,4-thiadiazole-5-sulphonamides as novel water-soluble inhibitors of carbonic anhydrase. *Bioorganic and Medicinal Chemistry Letters* **1**, 407–410.

Katritzky, A.R., Caster, K.C., Maren, T.H., Conroy, C.W., and Bar-Ilan, A. (1987) Synthesis and physico-chemical properties of thiadiazolo[3,2-a]pyrimidinesulfonamides and thiadiazolo[3,2-a]triazinesulfonamides as candidates for topically effective carbonic anhydrase inhibitors. *Journal of Medicinal Chemistry* **30**, 2058–2062.

Kehayova, P.D., Bokinsky, G.E., Huber, J.D., and Jain, A. (1999) A caged hydrophobic inhibitor of carbonic anhydrase II. *Organic Letters* **1**, 187–188.

Korman, J. (1958) Carbonic anhydrase inhibitors. I. Benzothiazole derivatives. *Journal of Organic Chemistry* **23**, 1768–1771.

Krebs, H.A. (1948) Inhibition of carbonic anhydrase by sulfonamides. *Biochemical Journal* **43**, 525–528.

Lucci, M.S., Tinker, J.P., Weiner, I.M., and DuBose, T.D. (1983) Function of proximal tubule carbonic anhydrase defined by selective inhibition. *American Journal of Physiology* **245**, F443–F449.

Mann, T., and Keilin, D. (1940) Sulphanilamide as a specific carbonic anhydrase inhibitor. *Nature* **146**, 164–165.

Maren, T.H. (1967) Carbonic anhydrase: Chemistry, physiology and inhibition. *Physiological Reviews* **47**, 595–781.

Maren, T.H. (1995) The development of topical carbonic anhydrase inhibitors. *Journal of Glaucoma* **4**, 49–62.

Maren, T.H., Bar-Ilan, A., Caster, K.C., and Katritzky, A.R. (1987) Ocular pharmacology of methazolamide analogs: Distribution in the eye and effects on pressure after topical application. *Journal of Pharmacology and Experimental Therapeutics* **241**, 56–63.

Maren T.H., and Conroy, C.W. (1993) A new class of carbonic anhydrase inhibitor. *Journal of Biological Chemistry* **268**, 26233–26239.

Maren, T.H., Conroy, C.W., Wynns, G.C., and Godman, D.R. (1997) Renal and cerebrospinal fluid formation pharmacology of a high molecular weight carbonic anhydrase inhibitor. *Journal of Pharmacology and Experimental Therapeutics* **280**, 98–104.

Maren, T.H., Jankowska, L., Sanyal, G., and Edelhauser, H.F. (1983) The transcorneal permeability of sulfonamide carbonic anhydrase innhibitors and their effect on aqueous humor secretion. *Experimental Eye Research* **36**, 457–480.

Masereel, B., Rolin, S., Abbate, F., Scozzafava, A., and Supuran, C.T. (2002) Carbonic anhydrase inhibitors: Anticonvulsant sulfonamides incorporating valproyl and other lipophilic moieties. *Journal of Medicinal Chemistry* **45**, 312–320.

Menabuoni, L., Scozzafava, A., Mincione, F., Briganti, F., Mincione, G., and Supuran, C.T. (1999) Carbonic anhydrase inhibitors: Water-soluble, topically effective intraocular pressure lowering agents derived from isonicotinic acid and aromatic/heterocyclic sulfonamides — Is the tail more important than the ring? *Journal of Enzyme Inhibition* **14**, 457–474.

Miller, W.H., Dessert, A.M., and Roblin, R.O., Jr.. (1950) Heterocyclic sulfonamides as carbonic anhydrase inhibitors. *Journal of the American Chemical Society* **72**, 4893–4896.

Mincione, F., Menabuoni, L., Briganti, F., Mincione, G., Scozzafava, A., and Supuran, C.T. (1998) Carbonic anhydrase inhibitors: Inhibition of isozymes I, II and IV with *N*-hydroxysulfonamides — A novel class of intraocular pressure lowering agents. *Journal of Enzyme Inhibition* **13**, 267–284.

Mincione, G., Menabuoni, L., Briganti, F., Mincione, F., Scozzafava, A., and Supuran, C.T. (1999) Carbonic anhydrase inhibitors. Part 79. Synthesis and ocular pharmacology of topically acting sulfonamides incorporating GABA moieties in their molecule, with long-lasting intraocular pressure lowering properties. *European Journal of Pharmaceutical Sciences* **9**, 185–199.

Mincione, F., Starnotti, M., Menabuoni, L., Scozzafava, A., Casini, A., and Supuran, C.T. (2001) Carbonic anhydrase inhibitors: 4-Sulfamoyl-benzenecarboxamides and 4-chloro-3-sulfamoyl-benzenecarboxamides with strong topical antiglaucoma properties. *Bioorganic and Medicinal Chemistry Letters* **11**, 1787–1791.

Northey E.H. (1948) *The Sulfonamides and Allied Compounds*, Reinhold, New York, pp. 1–267.

Owa, T., and Nagasu, T. (2000) Novel sulphonamide derivatives for the treatment of cancer. *Expert Opinion on Therapeutic Patents* **10**, 1725–1740.

Pastorek, J., Pastorekova, S., Callebaut, I., Mornon, J.P., Zelnik, V., Opavsky, R., Zatovicova, M., Liao, S., Portetelle, D., Stanbridge, E.J., Zavada, J., Burny, A., and Kettmann, R. (1994) Cloning and characterization of MN, a human tumor-associated protein with a domain homologous to carbonic anhydrase as a putative helix-loop-helix DNA binding segment. *Oncogene* **9**, 2877–2888.

Pastorekova, S., Parkkila, S., Parkkila, A.K., Opavsky, R., Zelnik, V., Saarnio, J., and Pastorek, J. (1997) Carbonic anhydrase IX, MN/CA IX: analysis of stomach complementary DNA sequence and expression in human and rat alimentary tracts. *Gastroenterology* **112**, 398–408.

Pierce, W.M., Jr., Sharir, M., Waite, K.J., Chen, D., and Kaysinger, K.K. (1993) Topically active ocular carbonic anhydrase inhibitors: Novel biscarbonylamidothiadiazole sulfonamides as ocular hypotensive agents. *Proceedings of the Society of Experimental Biology and Medicine* **203**, 360–365.

Ponticello, G.S., Freedman, M.B., Habecker, C.N., Lyle, P.A., Schwam, H., Varga, S.L., Christy, M.E., Randall, W.C., and Baldwin, J.J. (1987) Thienothiopyran-2-sulfonamides: A novel class of water-soluble carbonic anhydrase inhibitors. *Journal of Medicinal Chemistry* **30**, 591–597.

Popescu, A., Simion, A., Scozzafava, A., Briganti, F., and Supuran, C.T. (1999) Carbonic anhydrase inhibitors: Schiff bases of some aromatic sulfonamides and their metal complexes: towards more selective inhibitors of carbonic anhydrase isozyme IV. *Journal of Enzyme Inhibition* **14**, 407– 423.

Prugh, J.D., Hartman, G.D., Mallorga, P.J., McKeever, B.M., Michelson, S.R., Murcko, M.A., Schwam, H., Smith, R.L., Sondey, J.M., Springer, J.B., and Sugrue, M.F. (1991) New isomeric classes of topically active ocular hypotensive carbonic anhydrase inhibitors: 5-Substituted thieno[2,3-b]thiophene-2-sulfonamides and 5-substituted thieno[3,2-b]thiophene-2-sulfonamides. *Journal of Medicinal Chemistry* **34**, 1805–1818.

Renzi, G., Scozzafava, A., and Supuran, C.T. (2000) Carbonic anhydrase inhibitors: Topical sulfonamide antiglaucoma agents incorporating secondary amine moieties. *Bioorganic and Medicinal Chemistry Letters* **10**, 673–676.

Roblin, R.O., Jr., and Clapp, J.W. (1950) The preparation of heterocyclic sulfonamides. *Journal of the American Chemical Society* **72**, 4890–4892.

Schoenwald, R.D., Eller, M.G., Dixson, J.A., and Barfknecht, C.F. (1984) Topical carbonic anhydrase inhibitors. *Journal of Medicinal Chemistry* **27**, 810–812.

Scholz, T.H., Sondey, J.M., Randall, W.C., Schwam, H., Thompson, W.J., Mallorga, P.J., Sugrue, M.F., and Graham, S.L. (1993) Sulfonylmethanesulfonamide inhibitors of carbonic anhydrase. *Journal of Medicinal Chemistry* **36**, 2134–2141.

Scolnick, L.R., Clements, A.M., Liao, J., Crenshaw, L., Hellberg, M., May, J., Dean, T.R., and Christianson, D.W. (1997) Novel binding mode of hydroxamate inhibitors to human carbonic anhydrase II. *Journal of the American Chemical Society* **119**, 850–851.

Scozzafava, A., Banciu, M.D., Popescu, A., and Supuran, C.T. (2000a) Carbonic anhydrase inhibitors: Synthesis of Schiff bases of hydroxybenzaldehydes with aromatic sulfonamides, and their reactions with arylsulfonyl isocyanates. *Journal of Enzyme Inhibition* **15**, 533–546.

Scozzafava, A., Banciu, M.D., Popescu, A., and Supuran, C.T. (2000b) Carbonic anhydrase inhibitors: Inhibition of isozymes I, II and IV by sulfamide and sulfamic acid derivatives. *Journal of Enzyme Inhibition* **15**, 443–453.

Scozzafava, A., Briganti, F., Ilies, M.A., and Supuran, C.T. (2000c) Carbonic anhydrase inhibitors: Synthesis of membrane-impermeant low molecular weight sulfonamides possessing *in vivo* selectivity for the membrane-bound versus the cytosolic isozymes. *Journal of Medicinal Chemistry* **43**, 292–300.

Scozzafava, A., Briganti, F., Mincione, G., Menabuoni, L., Mincione, F., and Supuran, C.T. (1999a) Carbonic anhydrase inhibitors: Synthesis of water-soluble, amino acyl/dipeptidyl sulfonamides possessing long lasting-intraocular pressure lowering properties via the topical route. *Journal of Medicinal Chemistry* **42**, 3690–3700.

Scozzafava, A., Menabuoni, L., Mincione, F., Briganti, F., Mincione, G., and Supuran, C.T. (1999b) Carbonic anhydrase inhibitors: Synthesis of water-soluble, topically effective, intraocular pressure-lowering aromatic/heterocyclic sulfonamides containing cationic or anionic moieties: Is the tail more important than the ring? *Journal of Medicinal Chemistry* **42**, 2641–2650.

Scozzafava, A., Menabuoni, L., Mincione, F., Briganti, F., Mincione, G., and Supuran, C.T. (2000d) Carbonic anhydrase inhibitors: Perfluoroalkyl/aryl-substituted derivatives of aromatic/heterocyclic sulfonamides as topical intraocular pressure lowering agents with prolonged duration of action. *Journal of Medicinal Chemistry* **43**, 4542–4551.

Scozzafava, A., Menabuoni, L., Mincione, F., Mincione, G., and Supuran, C.T. (2001) Carbonic anhydrase inhibitors: Synthesis of sulfonamides incorporating dtpa tails and of their zinc complexes with powerful topical antiglaucoma properties. *Bioorganic and Medicinal Chemistry Letters* **11**, 575–582.

Scozzafava, A., Menabuoni, L., Mincione, F., and Supuran C.T. (2002) Carbonic anhydrase inhibitors: A general approach for the preparation of water soluble sulfonamides incorporating polyamino-polycarboxylate tails and of their metal complexes possessing long lasting, topical intraocular pressure lowering properties. *Journal of Medicinal Chemistry* **45**, 1466–1476.

Scozzafava, A., Owa, T., Mastrolorenzo, A. and Supuran, C.T. (2003) Anticancer and antiviral sulfonamides. *Current Medicinal Chemistry* **10**, 1241–1260.

Scozzafava, A., and Supuran, C.T. (1998a) Carbonic anhydrase inhibitors. Part 48. Ureido and thioureido derivatives of aromatic sulfonamides possess increased affinities for isozyme I. A novel route to 2,5-disubstituted-1,3,4-thiadiazoles via thioureas, and their interaction with isozymes I, II and IV. *Journal of Enzyme Inhibition* **13**, 103–123.

Scozzafava, A., and Supuran, C.T. (1998b) Carbonic anhydrase inhibitors: Novel compounds containing S-NH moieties in their molecule — sulfenamido-sulfonamides, sulfenimido-sulfonamides and their interaction with isozymes I, II and IV. *Journal of Enzyme Inhibition* **13**, 419–442.

Scozzafava, A., and Supuran, C.T. (1999) Carbonic anhydrase inhibitors: Arylsulfonylureido and arylureido-substituted aromatic and heterocyclic sulfonamides — towards selective inhibitors of carbonic anhydrase isozyme I. *Journal of Enzyme Inhibition* **14**, 343–363.

Scozzafava, A., and Supuran, C.T. (2000a) Carbonic anhydrase inhibitors: Synthesis of N-Morpholyl-thiocarbonylsulfenylamino aromatic/heterocyclic sulfonamides and their interaction with isozymes I, II and IV. *Bioorganic and Medicinal Chemistry Letters* **10**, 1117–1120.

Scozzafava, A., and Supuran, C.T. (2000b) Carbonic anhydrase and matrix metalloproteinase inhibitors. Sulfonylated amino acid hydroxamates with MMP inhibitory properties act as efficient inhibitors of carbonic anhydrase isozymes I, II and IV, and N-hydroxysulfonamides inhibit both these zinc enzymes. *Journal of Medicinal Chemistry* **43**, 3677–3687.

Scozzafava, A., and Supuran, C.T. (2002) Carbonic anhydrase inhibitors: Preparation of potent sulfonamides inhibitors incorporating bile acid tails. *Bioorganic and Medicinal Chemistry Letters* **12**, 1551–1557.

Sharir, M., Pierce, W.M., Jr., Chen, D., and Zimmerman, T.J. (1994) Pharmacokinetics, acid-base balance and intraocular pressure effects of ethyloxaloylazolamide: A novel topically active carbonic anhydrase inhibitor. *Experimental Eye Research* **58**, 107–116.

Shepard, K.L., Graham, S.L., Hudcosky, R.J., Michelson, S.R., Scholz, T.H., Schwam, H., Smith, A.M., Sondey, J.M., Strohmaier, K.M., Smith, R.L., and Sugrue, M.F. (1991) Topically active carbonic anhydrase inhibitors. 4. [(Hydroxyalkyl)sulfonyl]benzene and [(Hydroxyalkyl)sulfonyl]-thiophenesulfonamides. *Journal of Medicinal Chemistry* **34**, 3098–3105.

Shinkai, I. (1992) A practical asymmetric synthesis of a novel topically active carbonic anhydrase inhibitor. *Journal of Heterocyclic Chemistry* **29**, 627–630.

Singh, J., and Wyeth, P. (1991) The enzyme-inhibitor approach to cell-selective labelling. III. Sulphonamide inhibitors of carbonic anhydrase as carriers for red cell labelling. *Journal of Enzyme Inhibition* **5**, 1–24.

Smith, G.M., Alexander, R.S., Christianson, D.W., McKeever, B.M., and Ponticello, G.S. (1994) Positions of His-64 and a bound water in human carbonic anhydrase II upon binding three structurally related inhibitors. *Protein Science* **3**, 118–125.

Sterling, D., Brown, N.J.D., Supuran, C.T., and Casey, J.R. (2002) The functional and physical relationship between the DRA bicarbonate transporter and carbonic anhydrase II. *American Journal of Physiology — Cell Physiology* **283**, C1522–C1529.

Supuran, C.T. (1994) Carbonic anhydrase inhibitors. In *Carbonic Anhydrase and Modulation of Physiologic and Pathologic Processes in the Organism*, Puscas, I., Ed., Helicon Press, Timisoara, Romania, pp. 29–111.

Supuran, C.T. (2003) Indisulam: An anticancer sulfonamide in clinical development. *Expert Opinion on Investigational Drugs*, **12**, 283–287.

Supuran, C.T., and Banciu, M.D. (1991) Carbonic anhydrase inhibitors. Part 9. Inhibitors with modified sulfonamido groups and their interaction with the zinc enzyme. *Revue Roumaine de Chimie* **36**, 1345–1353.

Supuran, C.T., Briganti, F., Menabuoni, L., Mincione, G., Mincione, F., and Scozzafava, A. (2000a) Carbonic anhydrase inhibitors. Part 78. Synthesis of water-soluble sulfonamides incorporating β-alanyl moieties, possessing long lasting-intraocular pressure lowering properties via the topical route. *European Journal of Medicinal Chemistry* **35**, 309–321.

Supuran, C.T., Briganti, F., and Scozzafava, A. (1997a) Sulfenamido-sulfonamides as inhibitors of carbonic anhydrase isozymes I, II and IV. *Journal of Enzyme Inhibition* **12**, 175–190.

Supuran, C.T., Briganti, F., Tilli, S., Chegwidden, W.R., and Scozzafava, A. (2001) Carbonic anhydrase inhibitors: Sulfonamides as antitumor agents? *Bioorganic and Medicinal Chemistry* **9**, 703–714.

Supuran, C.T., and Clare, B.W. (1995) Carbonic anhydrase inhibitors. Part 24. A quantitative structure-activity study of positively-charged sulfonamide inhibitors. *European Journal of Medicinal Chemistry* **30**, 687–696.

Supuran, C.T., Conroy, C.W., and Maren, T.H. (1996a) Carbonic anhydrase inhibitors: Synthesis and inhibitory properties of 1,3,4-thiadiazole-2,5-bissulfonamide. *European Journal of Medicinal Chemistry* **31**, 843–846.

Supuran, C.T., Ilies, M.A., and Scozzafava, A. (1998a) Carbonic anhydrase inhibitors. Part 29. Interaction of isozymes I, II and IV with benzolamide-like derivatives. *European Journal of Medicinal Chemistry* **33**, 739–752.

Supuran, C.T., Manole, G., Dinculescu, A., Schiketanz, A., Gheorghiu, M.D., Puscas, I., and Balaban, A.T. (1992) Carbonic anhydrase inhibitors. Part 5. Pyrylium salts in the synthesis of isozyme-specific inhibitors. *Journal of Pharmaceutical Sciences* **81**, 716–719.

Supuran, C.T., Manole, G., Schiketanz, A., Gheorghiu, M.D., and Puscas, I. (1991) Carbonic anhydrase inhibitors. Part 4. Bifunctional derivatives of 1,3,4-thiadiazole-2-sulfonamide. *Revue Roumaine de Chimie* **36**, 251–255.

Supuran, C.T., Nicolae, A., and Popescu, A. (1996b) Carbonic anhydrase inhibitors. Part 35. Synthesis of Schiff bases derived from sulfanilamide and aromatic aldehydes: the first inhibitors with equally high affinity towards cytosolic and membrane-bound isozymes. *European Journal of Medicinal Chemistry* **31**, 431–438.

Supuran, C.T., Popescu, A., Ilisiu, M., Costandache, A.. and Banciu, M.D. (1996c) Carbonic anhydrase inhibitors. Part 36. Inhibition of isozymes I and II with Schiff bases derived from chalkones and aromatic/heterocyclic sulfonamides. *European Journal of Medicinal Chemistry* **31**, 439–447.

Supuran, C.T., and Scozzafava, A. (2000a) Carbonic anhydrase inhibitors and their therapeutic potential. *Expert Opinion on Therapeutic Patents* **10**, 575–600.

Supuran, C.T., and Scozzafava, A. (2000b) Carbonic anhydrase inhibitors. Part 94. 1,3,4-Thiadiazole-2-sulfonamide derivatives as antitumor agents? *European Journal of Medicinal Chemistry* **35**, 867–874.

Supuran, C.T., and Scozzafava, A. (2000c) Carbonic anhydrase inhibitors: Aromatic sulfonamides and disulfonamides act as efficient tumor growth inhibitors. *Journal of Enzyme Inhibition* **15**, 597–610.

Supuran, C.T., and Scozzafava, A. (2001) Carbonic anhydrase inhibitors. *Current Medicinal Chemistry — Immunologic, Endocrine and Metabolic Agents*, **1**, 61–97.

Supuran, C.T., and Scozzafava, A. (2002a) Applications of carbonic anhydrase inhibitors and activators in therapy. *Expert Opinion on Therapeutic Patents* **12**, 217–242.

Supuran, C.T., and Scozzafava, A. (2002b) Matrix metalloproteinases (MMPs). In *Proteinase and Peptidase Inhibition: Recent Potential Targets for Drug Development*, Smith, H.J., and Simons, C., Eds., Taylor & Francis, New York, pp. 35–61.

Supuran, C.T., Scozzafava, A., and Briganti, F. (1999a) Carbonic anhydrase inhibitors: *N*-Cyanosulfonamides — a new class of high affinity isozyme II and IV inhibitors. *Journal of Enzyme Inhibition* **14**, 289–306.

Supuran, C.T., Scozzafava, A., and Casini, A. (2003) Carbonic anhydrase inhibitors. *Medicinal Research Reviews* **23**, 146–189.

Supuran, C.T., Scozzafava, A., Ilies, M.A., and Briganti, F. (2000b) Carbonic anhydrase inhibitors: Synthesis of sulfonamides incorporating 2,4,6-trisubstituted-pyridinium-ethylcarboxamido moieties possessing membrane-impermeability and *in vivo* selectivity for the membrane-bound (CA IV) versus the cytosolic (CA I and CA II) isozymes. *Journal of Enzyme Inhibition* **15**, 381–401.

Supuran, C.T., Scozzafava, A., Ilies, M.A., Iorga, B., Cristea, T., Briganti, F., Chiraleu, F., and Banciu, M.D. (1998b) Carbonic anhydrase inhibitors. Part 53. Synthesis of substituted-pyridinium derivatives of aromatic sulfonamides: the first non-polymeric membrane-impermeable inhibitors with selectivity for isozyme IV. *European Journal of Medicinal Chemistry* **33**, 577–594.

Supuran, C.T., Scozzafava, A., Jurca, B.C., and Ilies, M.A. (1998c) Carbonic anhydrase inhibitors. Part 49. Synthesis of substituted ureido- and thioureido derivatives of aromatic/heterocyclic sulfonamides with increased affinities for isozyme I. *European Journal of Medicinal Chemistry* **33**, 83–93.

Supuran, C.T., Scozzafava, A., Menabuoni, L., Mincione, F., Briganti, F., and Mincione, G. (1999b) Carbonic anhydrase inhibitors. Part 70. Synthesis and ocular pharmacology of a new class of water-soluble, topically effective intraocular pressure lowering agents derived from nicotinic acid and aromatic/heterocyclic sulfonamides. *European Journal of Medicinal Chemistry* **34**, 799–808.

Supuran, C.T., Scozzafava, A., Menabuoni, L., Mincione, F., Briganti, F., and Mincione, G. (1999c) Carbonic anhydrase inhibitors. Part 71. Synthesis and ocular pharmacology of a new class of water-soluble, topically effective intraocular pressure lowering sulfonamides incorporating picolinoyl moieties. *European Journal of Pharmaceutical Sciences* **8**, 317–328.

Supuran, C.T., Scozzafava, A., Popescu, A., Bobes-Tureac, R., Banciu, A., Creanga, A., Bobes-Tureac, G., and Banciu, M.D. (1997b) Carbonic anhydrase inhibitors. Part 43. Schiff bases derived from aromatic sulfonamides: towards more specific inhibitors for membrane-bound versus cytosolic isozymes. *European Journal of Medicinal Chemistry* **32**, 445–452.

Teicher, B.A., Liu, S.D., Liu, J.T., Holden, S.A., and Herman, T.S. (1993) A carbonic anhydrase inhibitor as a potential modulator of cancer therapies. *Anticancer Research* **13**, 1549–1556.

Tinker, J.P., Coulson, R., and Weiner, I.M. (1981) Dextran-bound inhibitors of carbonic anhydrase. *Journal of Pharmacology and Experimental Therapeutics* **218**, 600–607.

Vaughan, J.R., Eichler, J.A., and Anderson, G.W. (1956) Heterocyclic sulfonamides as carbonic anhydrase inhibitors, 2-acylamido and 2-sulfonamido-1,3,4-thiadiazole-5-sulfonamides. *Journal of Organic Chemistry* **21**, 700–771.

Woltersdorf, W., Schwam, H., Bicking, J.B., Brown, S.L., deSolms, S.J., Fishman, D.R., Graham, S.L., Gautheron, P.D., Hoffman, J.M., Larson, R.D., Lee, W.S., Michelson, S.R., Robb, C.M., Share, C.N., Shepard, K.L., Smith, A.M., Smith, R.L., Sondey, J.M., Strohmeyer, K.M., Sugrue, M.F., and Viader, M.P. (1989) Topically active carbonic anhydrase inhibitors. 1. O-Acyl derivatives of hydroxybenzothiazole-2-sulfonamide. *Journal of Medicinal Chemistry* **32**, 2486–2492.

Young, R.W., Wood, K.H., Vaughan, J.R., and Anderson, G.W. (1956) 1,3,4-Thiadiazole and thiadiazolinesulfonamides as carbonic anhydrase inhibitors. Synthesis and structural studies. *Journal of the American Chemical Society* **78**, 4649–4654.

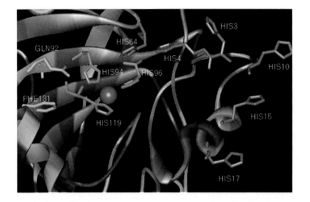

COLOR FIGURE 1.2 hCA II active site. The Zn(II) ion (central pink sphere) and its three histidine ligands (in green, His 94, His 96, His 119) are shown. The histidine cluster, comprising residues His 64, His 4, His 3, His 17, His 15 and His 10, is also shown, as this is considered to play a critical role in binding activators of the types **6** to **14** reported in the chapter as well as the carboxyterminal part of the anion exchanger AE1. The figure was generated from the x-ray coordinates reported by Briganti *et al.* (1997) (PDB entry 4TST). (Reproduced from Supuran, C.T., and Scozzafava, A. (2002) *Expert Opinion on Therapeutic Patents* **12**, 217–242. With permission from Elsevier.)

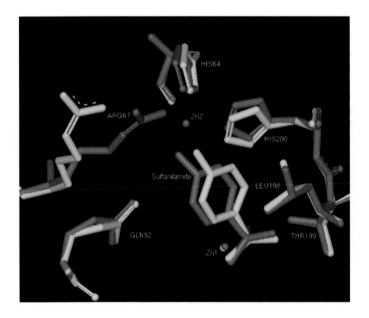

COLOR FIGURE 1.6 Least-squares superimposition of the most relevant active site residues of the natural mutant CA I Michigan 1 (in yellow) and the CA I Michigan 1 $(Zn)_2$ adduct (in red) involved in sulfonamide inhibitor binding, with bound sulfanilamide, as determined by x-ray crystallography. The catalytic zinc ion is Zn1. (Reproduced from Supuran, C.T. *et al.* (2003) *Medicinal Research Reviews* **23**, 146–189, John Wiley & Sons. With permission.)

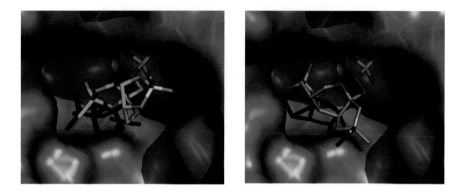

COLOR FIGURE 3.3 The binding mode of RWJ-37947 observed in the crystal structure (A) and manually rotated by 220° around the single bond connecting the ring system with the sulfamate anchor (B). (B) reveals that a binding mode as observed in the topiramate complex is sterically possible. The solvent accessible surface of the binding pocket of CAII hosting topiramate is shown (A, B).

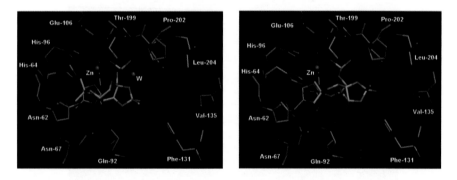

COLOR FIGURE 3.4 Binding mode of topiramate (A) and RWJ-37947 (B) in CA II. All amino acids in the binding site are highly conserved, except His 64, which adopts the *out* conformation in the topiramate complex (A) and the *in* conformation in complex with RWJ-37947 (B). All molecule representations are drawn with PyMOL. (DeLano, 2002.)

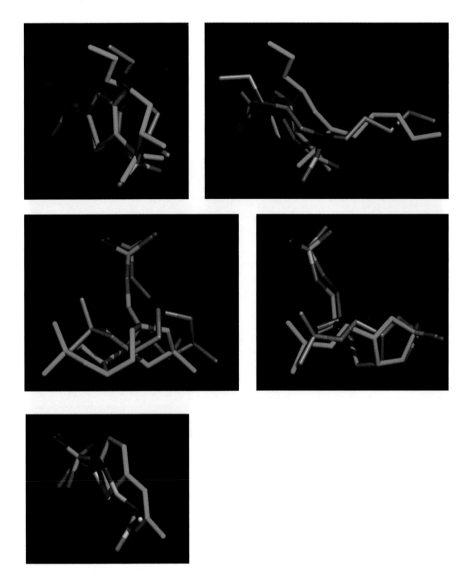

COLOR FIGURE 3.5 Dorzolamide (A), brinzolamide (B), topiramate (C), RWJ-37947 (D) and acetazolamide (E) docked into the binding site of CA II (PDB code 1cil). The first rank of the largest cluster is represented. X-ray crystal structures of the corresponding ligands are shown in green.

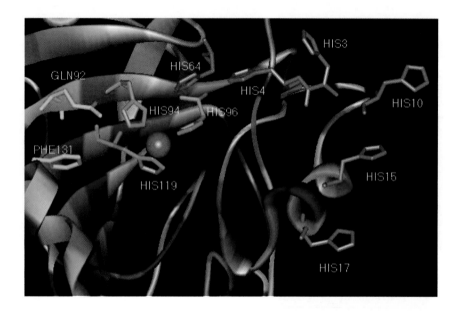

COLOR FIGURE 12.4 hCA II active site. Zinc ion (pink); its three histidine ligands —
His 94, His 96, His 119 (green); and the histidine cluster — His 64, His 4, His 3, His 17, His
15, His 10 (orange) are shown. The figure was generated from the diffraction coordinates
reported by Briganti *et al.* (1997a) (PDB entry 4TST). (Reproduced from Scozzafava and
Supuran, 2002a. With permission from Elsevier.)

5 QSAR Studies of Sulfonamide Carbonic Anhydrase Inhibitors

Brian W. Clare and Claudiu T. Supuran

CONTENTS

5.1 Introduction 149
5.2 Benzenoid Aromatics 151
    5.2.1 Electronic Correlates 151
    5.2.2 Steric Effects, Area, Volume and Polarizability 160
    5.2.3 Lipophilicity 160
    5.2.4 Quantum Measures Other Than Charge 162
    5.2.5 Topological Indices 163
    5.2.6 Direct Binding Studies 165
5.3 Heteroaromatics 166
    5.3.1 Charge and Dipole Moment 166
    5.3.2 Steric Variables and Polarizability 174
    5.3.3 Lipophilicity 174
    5.3.4 Quantum Measures Other Than Charge 174
    5.3.5 Topological Indices 174
    5.3.6 Direct Binding 174
5.4 Aliphatics 175
    5.4.1 Charge and Dipole Moment 175
5.5 Conclusions 178
Acknowledgements 179
References 179

5.1 INTRODUCTION

The enzyme carbonic anhydrase (CA, carbonate hydrolase, EC 4.2.1.1) is involved in a variety of physiological and physiopathological processes. [For a review see Maren (1967) and for updates see Maren (1984, 1987, 1991) and Maren et al. (1993, 1994).) Among its inhibitors, sulfonamides are important clinical agents, used to treat glaucoma (Supuran 1993; Maren 1984, 1987, 1991; Maren et al. 1993), gastroduodenal ulcers (Puscas 1984), certain neurological disorders (Lindskog and

0-415-30673-6/04/$0.00+$1.50

Wistrand 1987), motion and altitude sickness (Evans et al. 1976; Forster 1982) and many other conditions (Maren 1984, 1987, 1991; Maren et al. 1993; Lindskog and Wistrand 1987).

Many aromatic, heterocyclic, and, more recently, aliphatic sulfonamides (Maren and Conroy 1993) have been synthesized and tested for their CA inhibitory properties (Maren, 1984, 1987, 1991; Maren et al. 1993; Supuran 1993) and some of them are widely used clinical or investigational drugs, such as acetazolamide **5.1**, methazolamide **5.2**, ethoxzolamide **5.3** or the thienothiopyran derivative **5.4**, MK-972 which together with some closely related congeners are candidates for being clinically used as topical inhibitors to manage glaucoma. (Maren 1984, 1987, 1991; Maren et al. 1993; Blacklock 1993).

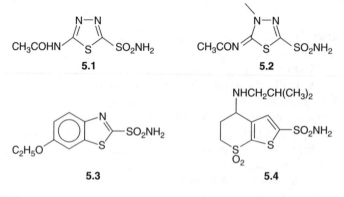

SCHEME 5.1

Early structure–activity correlations in this class of drugs were quite simple, leading Maren (1976) to state that perhaps "the criteria for activity in this class are the simplest in pharmacology." Indeed, all unsubstituted sulfonamides of the type RSO_2NH_2, where R is generally an aromatic/heterocyclic moiety (Supuran 1993), but can also be a polyhalogenoalkyl one (Evans et al. 1976; Forster 1982; Wright et al. 1983), show inhibitory properties toward this enzyme. When the number and variety of compounds examined increase, modifying influences come to light and the quantitative structure–activity relationships (QSARs) become more complex.

Sulfonamides bind in the ionized form (RSO_2NH-) to the zinc ion within the CA active site, as the fourth ligand, substituting the zinc-bound water molecule and perturbing in this way the entire catalytic cycle (Silverman and Lindskog 1998; Silverman 1983, 1990; Liljas et al. 1994).

Considering the clinical importance of these agents per se and because they constitute the starting point for obtaining other classes of widely used pharmacological agents, such as thiazide diuretics, high-ceiling diuretics (of the furosemide type) and certain antithyroid compounds (Silverman 1983, 1990), reviews on structure–activity correlations are available (Maren 1967, 1984, 1987, 1991; Maren et al. 1993, 1994; Supuran 1993; Lindskog and Wistrand 1987) and several important QSAR studies have also been published. This chapter reviews the QSAR work in the field of CA inhibitors, such as the different classes of compounds included, the types of correlations performed, the mathematical models obtained and their impact on designing novel inhibitors and the understanding of their mechanism of action at the molecular level. Some original work from our laboratories is also referenced.

5.2 BENZENOID AROMATICS

Derivatives of benzenesulfonamide are among the first known CA inhibitors and were the first for which QSAR results were obtained.

5.2.1 ELECTRONIC CORRELATES

The earliest QSAR studies on benzenesulfonamide derivatives indicated that the electric charge induced on the atoms of the sulfonamide moiety itself was the dominant determinant of CA inhibitory activity. Thus, empirical parameters such as the pK_a and ${}^1H$ NMR chemical shift (Kakeya et al. 1969a, 1969b, 1969c, 1970) of the sulfonamide protons, the Hammett σ constant of substituents on the benzene ring (Kakeya et al. 1969a, 1969b, 1969c, 1970; Hansch et al. 1985; Carotti et al. 1989) the force constant of the S–O bond, (Kakeya et al. 1969a, 1969b, 1969c, 1970) and calculated variables such as components of the dipole moment (Supuran and Clare 1998; Clare and Supuran 1999) and Mulliken and electrostatic potential-based charges (Supuran and Clare 1998; Clare and Supuran 1999; DeBenedetti and Menziani 1985; DeBenedetti et al. 1985) on the atoms of the sulfonamide group correlate with CA inhibitory activity. Another electronic parameter notionally related to lipophilicity and polarity is solvation energy, calculated as the difference in the heat of formation calculated as usual in an isolated molecule and that calculated in a medium with a high dielectric constant, using a continuum model for the solvent. Lipophilicity includes entropic contributions not present in the calculated solvation energy.

Kaketa et al. (1969a, 1969b, 1969c, 1970) gave the following equations for the set of 16 m- and p-substituted benzenesulfonamide inhibitors of CA (commercial, probably a mixture of CA I and CA II; Table 5.1):

$$\log (1/K_I) = 1.021\sigma + 0.474 \ (R = 0.938, \ s = 0.208, \ n = 16) \qquad (5.1)$$

$$\log (1/K_I) = 0.276\pi + 0.800\sigma + 0.413 \ (R = 0.965, \ s = 0.16, \ n = 16) \quad (5.2)$$

where R is the multiple correlation coefficient, s the standard error of estimate, n the number of compounds, σ the empirical Hammett constant and π the Hansch constant for the substituent.

Hansch et al. (1985) carried out similar calculations for the set of 29 monosubstituted benzenesulfonamide inhibitors of human CA II (Table 5.2) and obtained the following equation:

$$\log K = 1.55\sigma + 0.64 \log P - 2.07 \ I_1 - 3.28 \ I_2 + 6.94 \qquad (5.3)$$

$$(R = 0.991, \ s = 0.204, \ n = 29)$$

TABLE 5.1

Benzenesulfonamides in the Study by Kakeya et al.[a]

$R\text{-}C_6H_4\text{-}SO_2NH_2$

	R	$10^7 K_I$		R	$10^7 K_I$
a	p-MeNH	150	k	p-CN	11
b	p-NH$_2$	230	l	m-NO^2	13
c	p-MeO	45	m	p-NO$_2$	9
d	p-Me	38	n	3,4-Cl$_2$	4
e	m-Me	50	o	3-NO$_2$,4-Cl	1.7
f	H	61	p	3-CF$_3$,4-NO$_2$	1.4
g	p-Cl	19	q	o-Me	160
h	p-Br	12	r	o-Cl	30
i	m-Cl	23	s	o-NO$_2$	85
j	p-Ac	11			

[a] Kakeya, N. et al. (1969a) *Chemical and Pharmaceutical Bulletin* **17**, 1010–1018; Kakeya, N. et al. (1969b) *Chemical and Pharmaceutical Bulletin* **17**, 2000–2007; Kakeya, N. et al. (1969c) *Chemical and Pharmaceutical Bulletin* **17**, 2558–2564; Kakeya, N. et al. (1970) *Chemical and Pharmaceutical Bulletin* **18**, 191–194.

---

where P is the octanol–water distribution coefficient, and the indicator variables I_1 is 1 for m-substitution and 0 otherwise and I_2 is 1 for o-substitution and 0 otherwise. This implies that m-substituted compounds are 100 times and o-substituted compounds 2000 times weaker inhibitors than the corresponding p-substituted benzenesulfonamides.

Carotti et al. (1989) studied the series of m- and p-monosubstituted benzenesulfonamide inhibitors of bovine CA B (now designated CA I; Table 5.3) and obtained the following equations:

$$\log 1/K_i = 1.04(\pm 0.37)\sigma + 0.52(\pm 0.12)\pi - 0.65(\pm 0.25)I_1 + 5.98(\pm 0.16) \quad (5.4)$$

$$(n = 31, \, r = 0.898, \, s = 0.320, \, F_{4,27} = 37.3)$$

and

$$\log 1/K_i = 0.95(\pm 0.33)\sigma + 0.54(\pm 0.12)\pi - 0.35(\pm 0.11)B_{5,3} + 6.29(\pm 0.22) \quad (5.5)$$

$$(n = 31, \, R = 0.914, \, s = 0.294, \, F_{4,27} = 45.6)$$

TABLE 5.2
Sulfonamides used in QSAR calculations by Hansch et al.[a]

$X-C_6H_4-SO_2NH_2$

X		log K obs	X		log K obs
a	H	6.69	p	4-CONH-n-Bu	8.49
b	4-Me	7.09	q	4-CONH-n-Am	8.75
c	4-Et	7.53	r	4-CONH-n-He	8.88
d	4-n-Pr	7.77	s	4-CONH-n-Hp	8.93
e	4-n-Bu	8.30	t	3-COOMe	5.87
f	4-n-Am	8.86	u	3-COOEt	6.21
g	4-COOMe	7.98	v	3-COO-n-Pr	6.44
h	4-COOEt	8.50	w	3-COO-n-Bu	6.95
i	4-COO-n-Pr	8.77	x	3-COO-n-Am	6.86
j	4-COO-n-Bu	9.17	y	2-COOMe	4.41
k	4-COO-n-Am	9.39	z	2-COOEt	4.80
l	4-COO-n-He	9.39	α	2-COO-n-Pr	5.28
m	4-CONHMe	7.08	β	2-COO-n-Bu	5.76
n	4-CONHEt	7.53	γ	2-COO-n-Am	6.18
o	4-CONH-n-Pr	8.08			

[a] Hansch, C. et al. (1985) *Molecular Pharmacology* **27**, 493–498.

where I_1 is an indicator variable for m-substitution and $B_{5,3}$ the Sterimol parameter for the substituent, which is a steric measure. The lipophilicity π was measured by the shake flask method.

DeBenedetti et al. (1985) reported a CNDO/2 study of the 25 mono- and disubstituted benzenesulfonamide inhibitors of bovine CA B (CA I; Table 5.4), calculating the Mulliken charges on the atoms of the sulfonamide moiety and HOMO and LUMO energies. They found relationships among the lipophilicity, the force constant of the S–O bond, the Hammett σ constant and the pK_a and the chemical shift of the sulfonamide protons with q_{SA}, the total calculated charge on the SO_2NH_2 moiety. They gave the following equation:

$$\log II_{50} = 29.201(\pm 5.312)q_{SA} + 4.397(\pm 0.614) \tag{5.6}$$

$$(n = 23, \, r = 0.896, \, s = 0.292, \, F = 85.33)$$

Supuran and Clare (1998) in an AM1 quantum theoretical study of a series of Schiff's bases (Table 5.5) of sulfanilamide with substituted benzaldehyde and heteroaromatic aldehydes related CA inhibitory activity to a large number of theoretically calculated indices. They found that activities correlated marginally better for values calculated in solution by the COSMO technique and charges based on electrostatic potential distribution than those from a vacuum calculation and Mulliken population analysis.

TABLE 5.3
Activity Data Used by Carotti et al.[a] to Derive QSAR
for Benzenesulfonamides

$X\text{-}C_6H_4\text{-}SO_2NH_2$

	X	log $1/K_I$		X	log $1/K_I$
1	H	5.50	17	4-Br	6.70
2	3-Me	5.60	18	4-I	7.08
3	3-i-Pr	5.40	19	4-NH$_2$	4.82
4	3-Cl	6.34	20	4-OMe	5.92
5	3-Br	6.42	21	4-COMe	6.85
6	3-I	6.21	22	4-SO$_2$NH$_2$	7.04
7	3-NH$_2$	5.28	23	4-O-n-Bu	7.03
8	3-OMe	5.49	24	4-i-Pr	6.63
9	3-COMe	5.44	25	4-NO$_2$	6.43
10	3-CONH$_2$	4.90	26	4-CN	6.53
11	3-SO$_2$NH$_2$	5.79	27	4-Ph	7.19
12	3-O-n-Bu	5.86	28	4-OH	5.13
13	3-NO$_2$	6.28	29	4-OHx	7.20
14	3-CN	6.14	30	4-NHCOMe	5.43
15	4-Me	5.85	31	4-NHSO$_2$CH$_3$	5.32
16	4-Cl	6.76	32	3-NO$_2$-2,4-(OH)$_2$	7.56

[a] Carotti, A. et al. (1989) *QSAR* **8**, 1–10.

TABLE 5.4
Activities of the Group of Benzenesulfonamides Studied
by DeBenedetti et al.[a]

	R	log II_{50} obs		R	log II_{50} obs
1	4-NH$_2$	0.00	14	4-CN	1.32
2	4-NHMe	0.18	15	4-Cl	1.08
3	2-Me	0.16	16	3-Cl	1.00
4	3,4-Me$_2$	0.48	17	2-Cl	0.88
5	4-OMe	0.71	18	3,4-Cl$_2$	1.76
6	4-NHCOMe	1.48	19	4-NO$_2$	1.41
7	2-NH$_2$	0.66	20	3-NO$_2$	1.25
8	4-Me	0.78	21	3-SO$_2$NH$_2$	1.87
9	3-NH$_2$	0.40	22	4-SO$_2$NH$_2$	2.00
10	3-Me	0.66	23	3-CF$_3$-4-NO$_2$	2.21
11	H	0.58	24	3-NO$_2$-4-Cl	2.13
12	4-COMe	1.32	25	2-NO$_2$	0.43
13	3-Me-4-Cl	1.74			

[a] DeBenedetti et al. (1985) *QSAR* **4**, 23–28.

TABLE 5.5
Sulfanilamide Schiff's Bases Studied by Supuran and Clare[a]

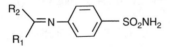

No.	R1	R2	C_lCA $I(\times 10^6$ $M)$	C_lCA II $(\times 10^8\ M)$
1	Phenyl	H	18	27
2	2-Hydroxyphenyl	H	35	41
3	2-Nitrophenyl	H	9	21
4	4-Chlorophenyl	H	25	, 28
5	4-Hydroxyphenyl	H	14	19
6	4-Methoxyphenyl	H	13	19
7	4-Dimethylaminophenyl	H	10	8
8	4-Nitrophenyl	H	13	5
9	4-Cyanophenyl	H	4	11
10	3-Methoxy-4-hydroxyphenyl	H	5	8
11	3,4-Dimethoxyphenyl	H	7	3
12	3-Methoxy-4-acetoxyphenyl	H	3	10
13	2,3-Dihydroxy-5-formylphenyl	H	4	2
14	2-Hydroxy-3-methoxy-5-formylphenyl	H	5	3
15	3,4,5-Trimethoxyphenyl	H	5	3
16	3-Methoxy-4-hydroxy-5-bromophenyl	H	12	4
17	2-Furyl	H	3	5
18	5-Methyl-2-furyl	H	3	4
19	Pyrol-2-yl	H	5	2
20	Imidazol-4(5)-yl	H	1	12
21	2-Pyridyl	H	2	9
22	3-Pyridyl	H	4	8
23	4-Pyridyl	H	4	5
24	Styryl	Me	14.4	0.39
25	4-Methoxystyryl	Me	9.4	0.12
26	4-Dimethylaminostyryl	Me	0.6	0.10
27	3,4,5-Trimethoxystyryl	Me	1.3	0.24
28	Styryl	Ph	20.9	0.56
29	4-Methoxystyryl	Ph	19.0	1.50
30	4-Dimethylaminostyryl	Ph	16.0	1.69
31	3,4,5-Trimethoxystyryl	Ph	10.7	2.35
32	3,4,5-Trimethoxystyryl	$4\text{-MeOC}_6\text{H}_4$	12.5	1.27
33	3-Nitrostyryl	$4\text{-MeOC}_6\text{H}_4$	6.3	0.65
34	3,4,5-Trimethoxystyryl	$4\text{-H}_2\text{NC}_6\text{H}_4$	10.6	0.85
35	3,4,5-Trimethoxystyryl	$4\text{-PhC}_6\text{H}_4$	25.0	2.48

[a] Supuran, C.T., and Clare, B.W. (1998) *European Journal of Medicinal Chemistry* **33**, 489–500.

Values of pK_a for the sulfonamide group calculated by the Pallas software were also included. They derived the following equation for CA II:

$$-\log K_{ii} = 0.0651(\pm 0.0212)D_x + 9916(\pm 3917)Q_H - 6614(\pm(2408)Q_H^2$$

$$-1.1921(\pm 0.2413)I_p + 603.5(\pm 57.8)\left(S_p^E - S_m^E/2\right) \tag{5.7}$$

$$+113.6(\pm 21.1)(Q_m + Q_p) - 4007.0(\pm 1592.5)$$

$$(n = 35, R^2 = 0.854, s = 0.274, F = 27.5)$$

where D_x is the component of the dipole moment in the direction of the benzene carbon–sulfur bond, Q_H the charge on the sulfonamide protons, I_p an indicator variable related to lipophilicity, $(S_p^E - S_m^E/2)$ the difference in electrophilic superdelocalizability between the p- and the average of the two m-atoms of the sulfanilamide benzene ring and $(Q_m + Q_p)$ the sum of the ESP-determined charges on those three atoms.

For CA I, their best equation was:

$$-\log K_i = 11.74(\pm 3.65)Q_S - 1.713(\pm 0.621)D_1 - 0.1956(\pm 0.0457)\log P$$

$$-1.0062(\pm 0.1571)pK_a - 172.8(\pm 62.0)S_p^E - 59.07(\pm 13.0) \tag{5.8}$$

$$(n = 35, R^2 = 0.762, s = 0.218)$$

where Q_S is the charge on the sulfonamide S, D_1 the local dipole index (a measure of separation of charge), $\log P$ the lipophilicity calculated by the program ClogP, pK_a the dissociation constant of the sulfonamide protons and S_p^E the electrophilic superdelocalizability of the p-carbon of the sulfanilamide benzene ring.

Clare and Supuran (1999) reported in an AM1 study of a series of 1,3- and 1,4-benzenedisulfonamides (Table 5.6) of the forms $RSO_2NHC_6H_4SO_2NH_2$ (m and p), $RSO_2NHCH_2C_6H_4SO_2NH_2$ and $RSO_2NHCH_2CH_2C_6H_4SO_2NH_2$ (both p) with one primary sulfonamide the QSARs for the inhibition of CA I, CA II and CA IV. For CA I, they obtained:

$$-\log IC_{50} = -5.79(\pm 2.12)Q_S - 4.26(\pm 1.01)Q_H + 0.0326(\pm 0.011)\mu_x \tag{5.9}$$

$$-0.299(\pm 0.108)E_H - 0.600(\pm 0.145)E_L + 2.12 \times 10^{-3}(-1.14 \times 10^{-3})\Pi_{yy}$$

$$+0.0209(\pm 0.0088)\Delta H_S - 3.13(\pm 0.96)Q_C + 10.30(\pm 5.58)$$

$$(n = 72, R^2 = 0.619, s = 0.22)$$

where Q_S is the charge on the primary sulfonamide S; Q_H the charge on the secondary sulfonamide H; E_H and E_L are the energies of the HOMO and LUMO, respectively;

TABLE 5.6
Benzenedisulfonamides Studied by Clare and Supuran[a]

Drug	Type	R	IC_{50} CA I	CA II	CA III
1	A	Me	125	64	98
2	A	$PhCH_2$	109	17	51
3	A	Me_2N	114	12	30
4	A	Ph	103	49	85
5	A	$4\text{-}Me\text{-}C_6H_4$	96	42	83
6	A	$2,4,6\text{-}Me_3\text{-}C_6H_2$	69	55	60
7	A	$4\text{-}F\text{-}C_6H_4$	40	9	15
8	A	$4\text{-}Cl\text{-}C_6H_4$	38	9	13
9	A	$4\text{-}Br\text{-}C_6H_4$	35	8	10
10	A	$4\text{-}MeO\text{-}C_6H_4$	44	30	14
11	A	$2\text{-}HOOC\text{-}C_6H_4$	26	7	10
12	A	$2\text{-}O_2N\text{-}C_6H_4$	29	11	18
13	A	$3\text{-}O_2N\text{-}C_6H_4$	33	10	21
14	A	$4\text{-}O_2N\text{-}C_6H_4$	31	10	22
15	A	$4\text{-}AcNH\text{-}C_6H_4$	245	82	164
16	A	$3\text{-}H_2N\text{-}C_6H_4$	168	43	114
17	A	$4\text{-}H_2N\text{-}C_6H_4$	164	46	129
18	A	C_6F_5	11	3	7
19	A	$4\text{-}Cl\text{-}3\text{-}O_2N\text{-}C_6H_3$	60	12	31
20	B	Me	176	79	125
21	B	$PhCH_2$	150	28	77
22	B	Me_2N	210	31	65
23	B	Ph	295	77	126
24	B	$4\text{-}Me\text{-}C_6H_4$	301	76	110
25	B	$2,4,6\text{-}Me_3\text{-}C_6H_2$	98	67	81
26	B	$4\text{-}F\text{-}C_6H_4$	65	10	22
27	B	$4\text{-}Cl\text{-}C_6H_4$	44	9	14
28	B	$4\text{-}Br\text{-}C_6H_4$	40	9	12
29	B	$4\text{-}MeO\text{-}C_6H_4$	59	40	48
30	B	$2\text{-}HOOC\text{-}C_6H_4$	32	7	20
31	B	$2\text{-}O_2N\text{-}C_6H_4$	33	10	26
32	B	$3\text{-}O_2N\text{-}C_6H_4$	35	9	22
33	B	$4\text{-}O_2N\text{-}C_6H_4$	36	10	24
34	B	$4\text{-}AcNH\text{-}C_6H_4$	298	101	188

TABLE 5.6 (continued)
Benzenedisulfonamides Studied by Clare and Supuran[a]

Drug	Type	R	IC$_{50}$		
			CA I	CA II	CA III
35	B	3-H$_2$N-C$_6$H$_4$	176	47	129
36	B	4-H$_2$N-C$_6$H$_4$	185	50	144
37	B	C$_6$F$_5$	14	3	10
38	B	4-Cl-3-O$_2$N-C$_6$H$_3$	74	15	46
39	C	Me	104	51	87
40	C	PhCH$_2$	96	10	45
41	C	Me$_2$N	90	8	21
42	C	Ph	81	40	64
43	C	4-Me-C$_6$H$_4$	85	31	60
44	C	2,4,6-Me$_3$-C$_6$H$_2$	60	52	57
45	C	4-F-C$_6$H$_4$	33	7	12
46	C	4-Cl-C$_6$H$_4$	29	7	10
47	C	4-Br-C$_6$H$_4$	24	5	8
48	C	4-MeO-C$_6$H$_4$	43	18	29
49	C	2-HOOC-C$_6$H$_4$	21	6	10
50	C	2-O$_2$N-C$_6$H$_4$	24	10	16
51	C	3-O$_2$N-C$_6$H$_4$	25	9	17
52	C	4-O$_2$N-C$_6$H$_4$	20	6	11
53	C	4-AcNH-C$_6$H$_4$	205	77	149
54	C	3-H$_2$N-C$_6$H$_4$	123	35	86
55	C	4-H$_2$N-C$_6$H$_4$	109	33	72
56	C	C$_6$F$_5$	10	2	4
57	C	4-Cl-3-O$_2$N-C$_6$H$_3$	58	11	27
58	D	Me	93	40	71
59	D	PhCH$_2$	81	8	38
60	D	Me$_2$N	75	6	13
61	D	Ph	69	28	39
62	D	4-Me-C$_6$H$_4$	70	27	54
63	D	2,4,6-Me$_3$-C$_6$H$_2$	46	40	45
64	D	4-F-C$_6$H$_4$	28	5	9
65	D	4-Cl-C$_6$H$_4$	24	5	8
66	D	4-Br-C$_6$H$_4$	19	3	7
67	D	4-MeO-C$_6$H$_4$	38	15	25

TABLE 5.6 (continued)
Benzenedisulfonamides Studied by Clare and Supuran[a]

				IC$_{50}$	
Drug	Type	R	CA I	CA II	CA III
68	D	2-HOOC-C$_6$H$_4$	15	5	9
69	D	2-O$_2$N-C$_6$H$_4$	22	6	9
70	D	3-O$_2$N-C$_6$H$_4$	22	8	13
71	D	4-O$_2$N-C$_6$H$_4$	20	4	10
72	D	4-AcNH-C$_6$H$_4$	187	75	119
73	D	3-H$_2$N-C$_6$H$_4$	101	30	84
74	D	4-H$_2$N-C$_6$H$_4$	95	30	72
75	D	C$_6$F$_5$	8	1	3
76	D	4-Cl-3-O$_2$N-C$_6$H$_3$	50	9	21

[a] Clare, B.W., and Supuran, C.T. (1999) *European Journal of Medicinal Chemistry* **34**, 463–474. With permission from Elsevier.

Π_{yy} is the intermediate diagonal component of the polarization tensor; ΔH_S the calculated solvation energy; and Q_C the charge on the carbon atom bound to the primary sulfonamide S.

For CA II they obtained:

$$-\log IC_{50} = -2.63 \times 10^{-3}(\pm 1.34 \times 10^{-3})\Pi_{yy} - 3.67 \times 10^{-3}(\pm 1.23 \times 10^{-3})\Pi_{zz} \quad (5.10)$$

$$-0.0239(\pm 0.0117)\mu - 0.644(\pm 0.173)E_L + 0.115(\pm 0.046)\log P$$

$$+4.16(\pm 1.14)Q_O + 0.592(\pm 0.095)I_3 + 7.50(\pm 2.26)$$

$$(n = 72, R^2 = 0.682, s = 0.24)$$

where Π_{zz} is the diagonal component of the polarizability tensor along the axis of highest inertia, μ the magnitude of the dipole moment, $\log P$ the lipophilicity calculated by ClogP, Q_O the sum of the charges on the secondary sulfonamide oxygens and I_3 is 1 if there are two carbon atoms between the secondary SO$_2$NH group and the primary benzenesulfonamide ring (rather than 1 or 0) and 0 otherwise.

For CA IV they obtained:

$$-\log \text{ IC}_{50} = 0.891(\pm 0.317)Q_N - 0.322(\pm 0.134)E_H - 0.377(\pm 0.166)E_L \quad (5.11)$$

$$+ 0.921(\pm 0.367)D_1 + 0.184(\pm 0.060)\log P + 2.885(1.108)Q_O$$

$$+ 0.35(\pm 0.101)I_3 - 0.45(\pm 2.16)$$

$$(n = 72, \; R^2 = 0.661, \; s = 0.25)$$

where Q_N is the charge on the secondary sulfonamide N and the other symbols are as defined for Equation 5.9 and Equation 5.10. Equation 5.9 to Equation 5.11 benefit from having a large number of compounds studied.

5.2.2 Steric Effects, Area, Volume and Polarizability

The area and volume of molecules tend to correlate so closely that usually only one can enter a QSAR equation without introducing unacceptable collinearity. Polarizability, either as the tensor components or as their average, also correlates with size but not so strongly so as to preclude the use of both. Equation 5.5 presents a correlation with the Sterimol parameters obtained by Carotti et al. (1989), and Equations 5.9 and 5.10 correlations with components of the polarizability obtained by Clare and Supuran (1999).

5.2.3 Lipophilicity

As with most other groups of drugs, lipophilicity enters the QSAR for CA inhibitors if a sufficiently large variety of drugs is examined. Thus, the results of Kakeya et al. (1969a, 1969b, 1969c, 1970) and Hansch et al. (1985) cited in Equations 5.2 and 5.3 involve lipophilicity, with higher lipophilicity favoring activity, and similarly in the work of Carotti et al. (1989) in Equations 5.4 and 5.5. Lipophilicity also plays a part in the more elaborate equations derived by Clare and Supuran (Clare and Supuran 1999; Supuran and Clare 1998) as shown in Equations 5.8, 5.10 and 5.11. Table 5.7 presents a study of the 4-, 5- and 6-substituted 1,3-benzenedisulfonamide inhibitors of chloroform CA (a mixture of CA I and CA II). Kishida and Manabee (1980) used only the substituent π values to explain activity. They gave the following equation:

$$-\log \text{ IC}_{50} = -0.2125\pi_4 - 2.1814\pi_5 + 2.8731(\pi_6 - 0.6193)^2 + 0.5674 \quad (5.12)$$

$$(R = 0.9486, \; n = 19, \; F = 50.9414)$$

This obviously cannot be valid, as it treats the equivalent 4- and 6-positions differently.

Altomare et al. (1991a) determined the lipophilicities of a number of benzenesulfonamide inhibitors of BCA B (Table 5.3) and gave the following equation:

TABLE 5.7
Activities and Structures
of Disulfonamides Studied
by Kishida and Manabe[a]

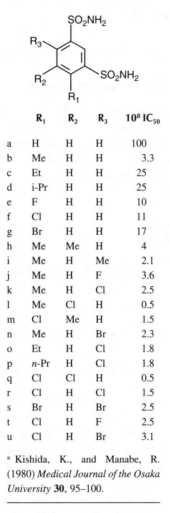

	R_1	R_2	R_3	$10^8 IC_{50}$
a	H	H	H	100
b	Me	H	H	3.3
c	Et	H	H	25
d	i-Pr	H	H	25
e	F	H	H	10
f	Cl	H	H	11
g	Br	H	H	17
h	Me	Me	H	4
i	Me	H	Me	2.1
j	Me	H	F	3.6
k	Me	H	Cl	2.5
l	Me	Cl	H	0.5
m	Cl	Me	H	1.5
n	Me	H	Br	2.3
o	Et	H	Cl	1.8
p	n-Pr	H	Cl	1.8
q	Cl	Cl	H	0.5
r	Cl	H	Cl	1.5
s	Br	H	Br	2.5
t	Cl	H	F	2.5
u	Cl	H	Br	3.1

[a] Kishida, K., and Manabe, R. (1980) *Medical Journal of the Osaka University* **30**, 95–100.

$$\log 1/K_i = 0.91(\pm 0.32)\sigma + 0.72(\pm 0.15)\pi^*_{ODP} - 0.35(\pm 0.11)B_{5,3} + 6.08(\pm 0.23)$$

$$(n = 31, \ r = 0.918, \ s = 0.287) \qquad (5.13)$$

where π^*_{ODP} is the new, chromatographically determined hydrophobic substituent constant. This equation is based on the same data as, and is comparable to, that in Equation 5.5.

Altomare et al. (1991b) used chromatographic octanol–water and chloro-form–water distribution coefficients to obtain the difference, $\Delta \log P_{oct-chf}$, which is a measure of hydrogen bond donor capacity of the substance. They derived for the series of benzenesulfonamide inhibitors of BCA B, of a subset given in Table 5.3, the following equation:

$$\log II_{50} = 0.65(\pm 0.30)\sigma + 0.66(\pm 0.34) \Delta \log P_{oct-chf} + 0.27(\pm 0.31) \quad (5.14)$$

$$(n = 19, R = 0.89, s = 0.305)$$

5.2.4 QUANTUM MEASURES OTHER THAN CHARGE

Frontier orbital energies were recognized early as predictors of CA inhibitory activity. The energies of the HOMO and LUMO are always to be considered as potential candidates for descriptors in QSAR, especially with aromatic compounds, and have been found to correlate with CA inhibitory activity in benzenesulfonamide derivatives. Thus, DeBenedetti and Menziani (1985) studied the series of 25 substituted benzenesulfonamide inhibitors of human CA II (Table 5.4), using the CNDO/2 semiempirical method and found for the neutral form the following relationship:

$$\log II_{50} = -2.038(\pm 0.348)E_{LUMO} + 1.629(\pm 0.138) \quad (5.15)$$

$$(n = 23, r = 0.907, s = 0.278, F = 96.95)$$

and for the anion:

$$\log II_{50} = -2.280(\pm 0.335)E_{HOMO} + 9.202(\pm 1.511) \quad (5.16)$$

$$(n = 23, r = 0.928, s = 0.244, F = 131.0)$$

A novel correlate of CA inhibitory activity in benzene derivatives, only recently detected, is the direction of the nodes in certain near-frontier, π-like orbitals (Clare and Supuran 2001). These orbitals can be regarded as arising from the degenerate HOMO and LUMO of benzene, and the resultant splitting of energy is also implicated in the nonbonded interaction between the ligand and some aromatic residue in the enzyme. It is clear that if two π-like orbitals, one on the ligand and one on the enzyme, are to interact, a requirement is that nodes in the orbitals largely coincide. We have a procedure to approximately quantify this coincidence. For human CA II:

$$-\log K_I = -18.75(\pm 17.44)Q_N - 3.252(\pm 1.381)D_1 + 0.2485(\pm 0.1854)\cos 2\Phi_H$$

$$- 0.0348(\pm 0.2659)\sin 2\Phi_H + 31.93(\pm 20.22) \quad (5.17)$$

$$(N = 27, R^2 = 0.706, Q^2 = 0.555)$$

where Q_N is the charge on the sulfonamide N, D_1 the local dipole index and Φ_H the angle the node in the HOMO makes with the sulfonamide group at the center of the benzene ring. The numbers in parentheses are the 95% confidence limits.

A quantum measure that attempts a comprehensive description of molecules is quantum similarity. This generates between each pair of compounds numbers analogous to correlation coefficients. Thus, for property $P(x,y,z)$ of drugs i and j, which are functions of the spatial coordinates x, y and z, a similarity measure can be defined, for example:

$$\rho_{i,j} = \int P_i P_j dr \Big/ \sqrt{\int P_i^2 dr \int P_j^2 dr}$$

where the integration is over all space and the two molecules are aligned to maximize ρ^2. The property P can be any spatial property of the molecule, such as electron density, an orbital magnitude or electrostatic potential, and the similarity measure is analogous to a Pearson correlation coefficient. The electron density is a property of special interest. It is tedious to compute any of these quantities as it requires optimization of both conformation and alignment.

Amat and Carbo-Dorca (1999) have applied the concept of self-similarity, defined as $Z_{AA} = \int |\rho_A(r)|^2 \, dr$, where ρ_A is the electron density, which avoids this problem, to CA inhibitors. Self-similarities can be substituted for classical descriptors such as Hansch π or Hammett σ. They treated the data of Hansch et al. (1985) and obtained the interesting equation, free from indicator variables, for inhibition of human CA C (now designated CA II):

$$\log K = 1.20000_{AA} - 2.31570_{AA}^{SO_2} + 2.51490_{AA}^{NH_2} - 0.62640_{AA}^{m-C} + 7.4441 \quad (5.18)$$

$$(n = 29, \ R^2 = 0.984, \ Q^2 = 0.976)$$

Note that Equation 24 in Amat and Carbo-Dorca (1999) appears to have been misprinted. Here, the scaled self-similarity θ_{AA} splits into contributions from fragments. The correlation indicates the splitting of SO_2NH_2 into two subfragments SO_2 and NH_2, suggesting that the SO_2NH_2 binds at two points: the SO_2 oxygen and the NH_2 hydrogen. Such calculations can throw light on the details of the mechanism of drug activity.

5.2.5 TOPOLOGICAL INDICES

Although topological and graph theoretical indices are very difficult to interpret physically they can be good predictors of activity, and compared to quantum theoretical indices they are very quick and easy to calculate. These two kinds of descriptors should not be regarded as mutually exclusive, but the topological indices can be applied on large data sets, perhaps derived from combinatorial synthesis and high throughput screening, to select data sets for more laborious methods, which give fits of comparable quality but those that are much more interpretable.

TABLE 5.8
Heterocyclic Sulfonamides Studied by Cash[a]

	X	Y	R	log II_{50}
1	N	N	$NHCOCH_2NH_2$	1.02
2	N	N	NH_2	1.64
3	N	N	NHMe	1.84
4	N	N	NHCHO	2.28
5	N	N	H	1.90
6	N	N	NHCOMe	2.52
7	N	N	NHC_6H_5	2.62
8	N	N	N(Me)COMe	2.66
9	N	N	$NHCOC_6H_5$	2.95
10	N	N	$NHSO_2C_6H_5$	3.38
11	N	N	$NHSO_2C_6H_4$-p-NHCOMe	3.16
12	N	N	$NHSO_2C_6H_4$-p-Cl	3.29
13	N	N	NHCOEt	2.62
14	N	N	$NHCOCH_2Cl$	2.36
15	N	N	$NHCOCF_3$	2.10
16	N	N	N(Me)COEt	2.60
17	N	N	N(Et)COMe	2.48
18	N	N	$N(C_6H_5)COMe$	2.52
19	N	N	NHCOOMe	2.36
20	N	N	C_6H_4-m-Cl	2.97
21	N	N	$NHSO_2C_6H_4$-p-Me	3.34
22	N	N	$NHSO_2C_6H_3$-3,4-Cl_2	3.18
23	N	N	$NHSO_2C_6H_4$-p-NH_2	2.92
24	CH	N	NH_2	0.48
25	CH	N	NHCOMe	1.18
26	N	CMe	H	1.70
27	N	CH	H	2.08
28	CH	CH	H	1.12
29	CH	CH	NO_2	3.04
30	CH	CH	NHCOMe	0.59

[a] Cash, G.G. (1995) *Structural Chemistry* **6**, 157–160.

For the pooled group of benzenesulfonamides and heteroaromatic sulfonamide inhibitors of bovine CA B (CA I) given in Table 5.8 and Table 5.9, Cash (1995) found correlations with molecular connectivity indices. Disregarding five of their heteroaromatics they obtained the following equation:

TABLE 5.9
Benzenesulfonamides Used in Connectivity Index QSAR by Cash[a]

$$X-C_6H_4-SO_2NH_2$$

	X	log II_{50}		X	log II_{50}
31	4-NH$_2$	0.00	42	4-CN	1.32
32	4-NHMe	0.18	43	4-Cl	1.08
33	3,4-Me$_2$	0.48	44	3-Cl	1.00
34	4-OMe	0.71	45	3,4-Cl$_2$	1.76
35	4-NHCOMe	1.48	46	4-NO$_2$	1.41
36	4-Me	0.78	47	3-NO$_2$	1.25
37	3-NH$_2$	0.40	48	3-SO$_2$NH$_2$	1.87
38	3-Me	0.66	49	4-SO$_2$NH$_2$	2.00
39	H	0.58	50	3-CF$_3$-4-NO$_2$	2.21
40	4-COMe	1.32	51	3-NO$_2$-4-Cl	2.13
41	3-Me-4-Cl	1.74			

[a] Cash, G.G. (1995) *Structural Chemistry* **6**, 157–160.

$$\log II_{50} = 0.38(\pm 0.05)\ ^5\chi_{PC} + 8.75(\pm 1.02)^5\chi_{CH} + 0.01 \tag{5.19}$$

$$(n = 45,\ r = 0.91,\ s = 0.39,\ F = 101)$$

Saxena and Khadikar (1999) found correlations of subsets of previously studied compounds (Supuran and Clare 1998; Equations 5.7 and 5.8; Table 5.5) with the Wiener index. Whereas five subsets gave good correlations for CA II and three subsets gave good correlations for CA I, the correlations for the complete sets were $r = 0.2338$ and 0.2512, respectively. This does not seem significant enough.

Mattioni and Jurs (2002) published a study of the large series of CAIs, including both benzenoid and heteroaromatic inhibitors of CA I, CA II and CA IV of Scheme 5.2, using a mixture of topological, electronic and geometric, and hybrid descriptors. They used mainly neural nets as their QSAR method, and although they obtained extremely good data fits (Figure 5.1), their use of neural nets and difficult-to-interpret descriptors precludes any physical interpretation of their results. These methods do, however, enable one to predict activities.

5.2.6 DIRECT BINDING STUDIES

In addition to approaches that consider the ligand only, there are those that deal with the interaction of the ligand with the enzyme on the molecular mechanics level. Menziani et al. (1989b) transformed *p*-chlorobenzenesulfonamide into benzene-sulfonamide in the presence of the enzyme in a molecular dynamics simulation, obtaining a $\Delta G°$ for the difference within 2.5 kJ mole^{-1} of the experimental value.

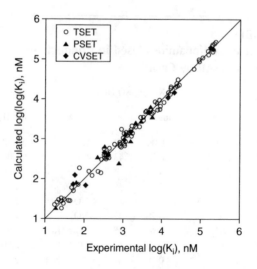

FIGURE 5.1 Fit of a large group of sulfonamide CA I inhibitors by using neural nets. TSET: training set, CVSET: cross-validation set and PSET: prediction set. (From Mattioni, B.E., and Jurs, P.C. (2002) *Journal of Chemical Information and Computer Science* **42**, 94–102. With permission.)

Menziani et al. (1989a) calculated energies for free molecules and CA-bound complexes for 20 benzenesulfonamides separated into a number of components: $BE = E_{CA-S} + E_S^D + E_{CA}^D$, where E_{CA-S} is the total interaction energy of enzyme with ligand and E_{CA}^D and E_S^D are distortion energies for the enzyme and ligand, respectively. The calculated binding energies were related to calculated AM1 MO descriptors, such as charges on the atoms of the sulfonamide moiety and the HOMO energy of the ligand.

5.3 HETEROAROMATICS

Many highly potent heterocyclic sulfonamide inhibitors of CA have been synthesized, and some are in clinical use as diuretics and intraocular pressure (IOP) lowering drugs and other types of drugs.

5.3.1 CHARGE AND DIPOLE MOMENT

The first significant study of the QSAR of heteroaromatics (Kishida 1978) used the Huckel theory to calculate for the series of nine amine *N*-substituted 5-amino-1,3,4-thiadiazole-2-sulfonamides given in Table 5.10 charges, atom–atom polarizabilities, HOMO and LUMO energies and electrophilic and nucleophilic superdelocalizabilities of the various atoms. These parameters and also the hydrophobicity π calculated by summing substituent contributions were correlated with CA inhibitory properties. It was found that in this series the electronic properties of the sulfonamide moiety varied little and there was no correlation with HOMO energy, but there was a strong correlation with LUMO energy ($R = 0.9439$). He obtained the following equation:

TABLE 5.10
Aminothiadiazole Sulfonamides
Studied by Kishida[a]

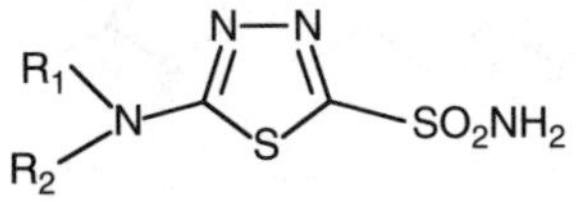

	R_1	R_2	$-pI_{50}$
a	H	H	−0.7109
b	H	Me	−0.5144
c	H	Ph	0.3841
d	H	Ac	0
e	Me	Ac	0.1727
f	Ph	Ac	0.1279
g	Et	Ac	0.0059
h	n-Bu	Ac	0.3765
i	Me	EtCO−	0.1308

[a] Kishida, K. (1978) *Chemical and Pharmaceutical Bulletin* **26**, 1049–1053.

$$-pI_{50} = -1.7313 + 6.3877Q_N + 0.1393\pi \tag{5.20}$$

$$(n = 9, R = 0.9438, F = 24.46)$$

where Q_N is the charge on the 5-nitrogen atom, suggesting that the atoms in the 5-position interact hydrophobically and electrostatically with the enzyme. Nine compounds, however, are rather few for a QSAR.

DeBenedetti's group (DeBenedetti et al. 1987; Menziani and DeBenedetti 1991a) used the same approach as that for the benzenoids. They (Menziani et al. 1989a) calculated with AM1 quantum chemical indices for the twelve 2-substituted 1,3,4-thiadiazole-5-sulfonamides given in Table 5.11. They calculated by molecular mechanics binding energies to bovine CA B and correlated CA inhibitory potency with binding energy as well as binding energy with quantum indices:

$$\log II_{50} = -0.199BE - 4.84 \tag{5.21}$$

$$(n = 12, r = 0.96, s = 0.22)$$

$$BE = 45.4E_H + 212.8 \tag{5.22}$$

$$(n = 11, r = 0.88, s = 1.64)$$

$$BE = -547.4Q_{SA} - 318.9 \tag{5.23}$$

$$(n = 11, r = 0.92, s = 1.32)$$

TABLE 5.11

Heteroaromatic Sulfonamides Studied by DeBenedetti et al.[a]

	Structure	R	log II_{50}
1	A	2-NHCOH	2.28
2	A	2-NHCOEt	2.62
3	A	2-NHCOCH$_2$Cl	2.36
4	A	2-NHCOCF$_3$	2.10
5	A	2-NHCOPh	2.95
6	A	2-NHCOCH$_2$NH$_2$	1.02
7	A	2-NHCOMe	2.52
8	A	2-N(Me)COMe	2.66
9	A	2-N(Me)COEt	2.60
10	A	2-N(Et)COMe	2.48
11	A	2-N(Ph)COMe	2.52
12	A	2-NHCOOMe	2.36
13	A	2-NH$_2$	1.64
14	A	2-NHMe	1.84
15	A	2-NHPh	2.62
16	A	2-C$_6$H$_4$(3-Cl)	2.97
17	A	2-NHSO$_2$Ph	3.38
18	A	2-NHSO$_2$C$_6$H$_4$(4-Me)	3.34
19	A	2-NHSO$_2$C$_6$H$_4$(4-Cl)	3.29
20	A	2-NHSO$_2$C$_6$H$_3$(3,4-Cl$_2$)	3.18
21	A	2-NHSO$_2$C$_6$H$_4$(4-NHCOMe)	3.16
22	A	2-NHSO$_2$C$_6$H$_4$(4-NH$_2$)	2.92
23	B	2-NH$_2$	0.48
24	B	2-NHCOMe	1.18
25	C	4-Me	1.70
26	C	5-H	2.08
27	D	2-H	1.12
28	D	2-NO$_2$	3.04
29	D	2-NHCOMe	0.59
30	D	2-CH$_2$CHMe$_2$	2.84

[a] De Benedetti, P.G. et al. *QSAR* **6**, 51–53.

where BE is the binding energy, E_H the energy of the HOMO and Q_{SA} the sum of the charges on the atoms of the SO$_2$NH– moiety.

Clare and Supuran (1997) in a quantum chemical study of the heterogeneous group of sulfonamides shown in Table 5.12, which included benzenoid, bicyclic and heteroaromatic compounds, used a set of descriptors that did not require that the

TABLE 5.12
Mixed Group of Sulfonamides Studied by Clare and Supuran[a]

Drug	Structure	$K_I \times 10^9 M$	$k_{on} \times 10^{-6}$ l/mole·sec^{-1}	$k_{off} \times$ sec^{-1}
A		10000	0.003	0.03
B		6000	0.002	0.01
C		3000	0.003	0.01
D		3000	0.003	0.01
E		900	0.04	0.04
F		800	0.01	0.008
G		17000	0.003	0.050
H		2350	0.013	0.030
I		750	0.033	0.024

TABLE 5.12 (continued)
Mixed Group of Sulfonamides Studied by Clare and Supuran[a]

Drug	Structure	$K_i \times 10^9\ M$	$k_{on} \times 10^{-6}$ l/mole·sec^{-1}	$k_{off} \times$ sec^{-1}
J		460	0.066	0.030
K		440	0.10	0.044
L		80	0.3	0.024
M		16	0.6	0.010
N		13	3.5	0.042
O		7	3.0	0.021
P		4	4.4	0.020
Q		0.9	31	0.028

TABLE 5.12 (continued)
Mixed Group of Sulfonamides Studied by Clare and Supuran[a]

Drug	Structure	$K_I \times 10^9\ M$	$k_{on} \times 10^{-6}$ l/mole·sec^{-1}	$k_{off} \times$ sec^{-1}
R	$C_5H_{11}O$ ... SO_2NH_2	0.8	15	0.012
S	Cl ... N—N ... S ... SO_2NH_2	0.8	30	0.024
T	C_2H_5O ... S ... N ... SO_2NH_2	0.7	14	0.010

[a] Clare, B.W., and Supuran, C.T. (1997) *European Journal of Medicinal Chemistry* **32**, 311–319. With permission from Elsevier.

series be congeneric, beyond including a sulfonamide moiety. They also used a statistical technique, ACE, which models nonlinearity. They obtained the following equation:

$$-\log K_I = 61.0(\pm 11.5)Q_N + 1.82(\pm 0.43)E_L + 0.592(\pm 0.091)A_x$$

$$+ 0.484(\pm 0.166)A_y' - 0.434(\pm 0.093)D - 0.919(\pm 0.242)\log P' \quad (5.24)$$

$$+ 8.41(\pm 3.01)$$

$$(n = 20,\ R^2 = 0.864,\ s = 0.678)$$

where Q_N is the charge on the sulfonamide N; E_L the LUMO energy; A_x and A_y the longest and second longest orthogonal dimensions of the molecule, respectively, determined from the moments of inertia; D the magnitude of the dipole moment; and $\log P$ the lipophilicity determined by using the program ClogP. The piecewise linear term A_y' is equal to A_y if $A_y > 6.0$ and equal to 6.0 otherwise; similarly, $\log P'$ is equal to $\log P$ if $\log P > 0.35$ and equal to 0.35 otherwise. This kind of nonlinear dependence on $\log P$ is well known, and can imply a kinetic effect of partitioning between compartments. The statistical significance of nonlinearity in A_y is not particularly good, but if it is valid implies that activity increases with increasing breadth, provided it is at least 6 Å wide.

Supuran and Clare (1999) carried out a quantum chemical QSAR study of the group of 40 (mainly aryl) 5-aminosulfonyl-substituted 1,3,4-thiadiazole- and thia-

TABLE 5.13
Aliphatic Sulfonamides Studied
by Maren and Conroy[a]

RSO_2NH_2

No.	R	pK_a	pK_1
1	CF_3	5.7	8.9
2	C_2F_5	5.9	8.1
3	C_4F_9	6.1	8.6
4	CCl_3	6.5	7.8
5	CHF_2	7.4	7.4
6	CF_3CH_2	9.0	6.1
7	CH_2Cl	9.2	5.4
8	Me	11.1	3.6
9	CH_2F	9.1	5.8

[a] Maren, T.H., and Conroy, C.W. (1993) A new class of carbonic anhydrase inhibitor. *Journal of Biological Chemistry* **268**, 26233–26239.

diazoline-2-sulfonamide inhibitors of CA I and CA II given in Table 5.13. For CA I, the results for the pooled data were not as good as those of the two groups separately and were affected by two highly influential cases. For the thiadiazoles, the best equation was:

$$-\log IC_{50} = -59.43(\pm 6.57)Q_S - 0.1359(\pm 0.0325)\mu_x + 0.0300(\pm 0.0116)\mu_z$$
$$+ 0.0204(\pm 0.0073)\Delta H_S - 98.87(\pm 10.30)Q_O - 27.83(\pm 10.39) \tag{5.25}$$

$$(n = 20, R^2 = 0.909, s = 0.18)$$

where Q_S and Q_O are the charges on the primary sulfonamide S and O atoms, respectively; μ_x and μ_z the components of the dipole moment along the ring C-sulfonamide S bond and that normal to the thiadiazole ring, respectively; and ΔH_S is the calculated solvation energy. For the thiadiazolines, the best equation for CA I inhibition was:

$$-\log IC_{50} = -0.00847(\pm 0.0014)\Pi_{yy} + 5.871(\pm 1.791)Q_S + 1.787(\pm 0.367)E_H$$
$$+ 1.575(\pm 0.329)E_L - 0.0501(\pm 0.0100)\Delta H_S \tag{5.26}$$
$$+ 82.3(\pm 17.76)Q_{O1} + 16.36(\pm 4.16)Q_{O2} + 182.6(\pm 32.8)$$

$$(n = 20, R^2 = 0.917, s = 0.21)$$

where Π_{yy} is the intermediate diagonal component of the polarizability tensor referred to the inertial axes; Q_S the charge on the secondary sulfonamide S; E_H and E_L the HOMO and LUMO energies, respectively; ΔH_S the solvation energy; and Q_{O1} and Q_{O2} the charges on the oxygen atoms of the primary and secondary sulfonamide groups, respectively.

For CA II inhibition by thiadiazoles the best equation was:

$$-\log \text{ IC}_{50} = -63.34(\pm 6.97)Q_S + 0.696(\pm 0.204)E_H - 0.1346(\pm 0.0345)\mu_x$$

$$+ 0.0504(\pm 0.0135)\mu_z - 0.0406(\pm 0.0077)\Delta H_S \qquad (5.27)$$

$$- 104.7(\pm 10.6)Q_O - 22.46(\pm 11.49)$$

$$(n = 20, \ R^2 = 0.902, \ s = 0.18)$$

where Q_S and Q_O are the charges on the primary sulfonamide S and O, respectively, and E_H, ΔH_S, μ_x and μ_z are as defined previously.

No satisfactory equation could be obtained for CA II inhibition by thiadiazolines. For CA IV inhibition by thiadiazoles the best equation obtained was:

$$-\log \text{ IC}_{50} = -0.00736(\pm 0.00104)\Pi_{xx} + 18.90(\pm 6.72)Q_N$$

$$- 0.1228(\pm 0.0452)\mu_x + 22.29(\pm 7.78) \qquad (5.28)$$

$$(n = 20, \ R^2 = 0.760, \ s = 0.29)$$

where Π_{xx} is the largest component of the polarizability tensor, Q_N the charge on the primary sulfonamide N and μ_x the component of the dipole moment as defined previously.

For CA IV inhibition by the thiadiazolines, the equation obtained was:

$$-\log \text{ IC}_{50} = -0.00729(\pm 0.00122)\Pi_{yy} + 1.628(\pm 0.384)E_H$$

$$- 0.0977(\pm 0.0226)\mu_x + 105.4(\pm 21.5)Q_O + 208.9(\pm 40.1) \qquad (5.29)$$

$$(n = 20, \ R^2 = 0.822, \ s = 0.25)$$

where Q_O is the charge on the primary sulfonamide oxygen and the other symbols are as defined previously. It should be noted that Equations 5.25 to 5.29 are good fits, but because a fairly large number of parameters was fitted to only 20 compounds they are not good predictors.

Equation 5.20 and Equations 5.23 to 5.29 contain at least one term involving the charge on one or more atoms on a primary sulfonamide group. Several also involve the dipole moment or its component along the C–S bond linking this group to the heterocyclic ring. This supports the long-held belief that electrostatic effects in the neighborhood of the sulfonamide group, and, in particular, the readiness of that group to deprotonate are of major importance in the QSAR of these substances.

5.3.2 STERIC VARIABLES AND POLARIZABILITY

Equation 5.24 for the compounds given in Table 5.12 involves the longest and second-longest linear dimensions of the molecule, and is the only equation in which steric effects were apparent. Both are in the direction of activity increasing with the size of the molecule, and A_x is of high statistical significance. Polarizability occurs in Equations 5.26 and 5.29 (the intermediate component) and Equation 5.28 (the largest component). In all three cases, a more polarizable molecule is less active.

5.3.3 LIPOPHILICITY

Only Equations 5.20 and 5.24 involve lipophilicity: in the first case high lipophilicity leads to high activity, whereas in the second it leads to low activity if log P exceeds 0.35.

5.3.4 QUANTUM MEASURES OTHER THAN CHARGE

Equations 5.23, 5.24, 5.26, 5.27 and 5.29 contain E_H, E_L or both. In each case, the activity of the drug increases with increasing energy. Solvation energy occurs in Equations 5.25 to 5.27. For Equation 5.25, the more solvated drugs are more active, whereas for Equation 5.26 and 5.27 they are less active.

5.3.5 TOPOLOGICAL INDICES

Gao and Bajorath (1999) computed binary and conventional QSARs for many sulfonamides, amides, alcohols and other compounds, the structures of which they did not give. They obtained good correlations, both conventional ($R^2 = 0.84$) and cross-validated ($Q^2 = 0.82$), with the atomic valence connectivity indices $^1\chi^v$ and $^0\chi^v$, the Kier shape indices $^1\kappa$ and $^2\kappa$, and the sum of atom–atom polarizabilities apol, an indicator variable for the presence of unsubstituted sulfonamide, and linear and quadratic terms in log P. The work by Mattioni and Jurs (2002) on a mixed series of CAIs has been mentioned previously (Section 5.2.5)

5.3.6 DIRECT BINDING

DeBenedetti's group (Menziani and DeBenedetti 1991b) used the AMBER molecular mechanics force field to calculate binding energies of both benzene and thiadiazole sulfonamides to HCA C, and related the calculated binding energies to total charge on SO_2NH^- N and CA inhibitory activity:

$$\log \text{II}_{50} = 106.82(\pm 12.66)q_O - 1.24(\pm 0.15)I + 57.90(\pm 6.58) \qquad (5.30)$$

$$(n = 48, r = 0.945, F = 183.5)$$

where I is an indicator variable for benzene. For the thiadiazoles:

$$\log II_{50} = -0.199(\pm 0.034)BE - 4.84(\pm 1.29) \qquad (5.31)$$

$$(n = 12, \; r = 0.96, \; F = 105.5)$$

5.4 ALIPHATICS

Until recently, the generally accepted idea was that aliphatic sulfonamides are very poor CA inhibitors (Maren 1984, 1987, 1991). Maren and Conroy (1993) showed that strong CA inhibitors can be designed and obtained from this class of sulfonamides too.

Figure 5.2 shows the linear correlation between pK_a and K_I for the nine aliphatic sulfonamides given in Table 5.14. Thus, the simplest derivative, $CH_3SO_2NH_2$, with a pK_a ca. 11 is an extremely weak CA inhibitor, whereas with polyhalogenoalkylsulfonamides increasing pK_as show an increasing potency of inhibitory power. $CF_3SO_2NH_2$, the most acidic sulfonamide ever reported, with a pK_a of 5.9, is one of the most potent CA inhibitors obtained, with a K_I of 2 nM (for CA II at 0°C; Maren and Conroy 1993). Some of these compounds are also topically active sulfonamides for lowering elevated IOP in glaucoma.

5.4.1 CHARGE AND DIPOLE MOMENT

In a series of aryl, alkyl and aralkyl-sulfonylmethanesulfonamides and -thiomethanesulfonamide inhibitors of human CA II (Scholz et al. 1993), the sulfonyl compounds were found to be much more active, probably because of the enhancement of acidity

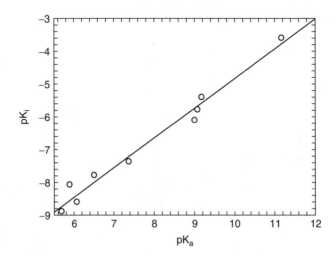

FIGURE 5.2 Plot of log inhibitory constant vs. pK_a for a group of aliphatic CA II inhibitors described by Maren, T.H., and Conroy, C.W. (1993) *Journal of Biological Chemistry* **268**, 26233–26239.

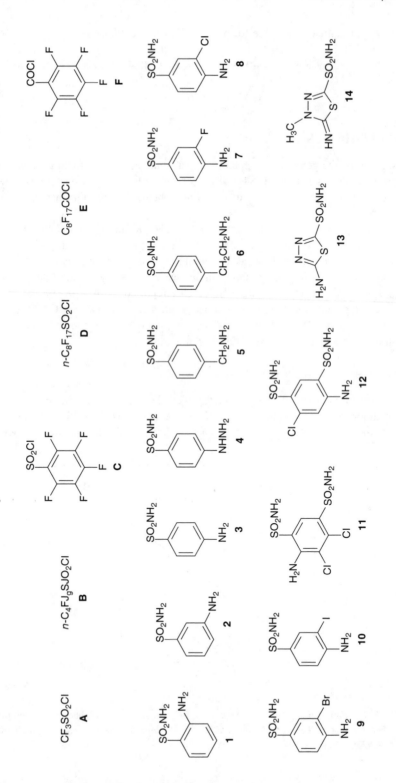

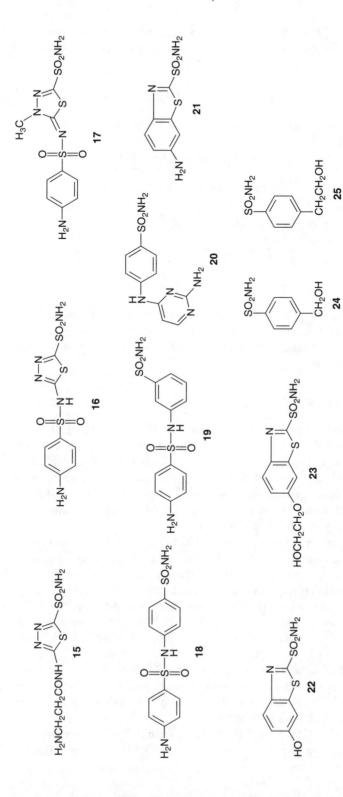

SCHEME 5.2

of the sulfonamide by the electronegative sulfonyl group. Increasing the size or hydrophobicity of the aryl, alkyl and aralkyl group also increased activity. Thus the equations:

$$-\log IC_{50} = -3.29 + 1.43(\pm 0.53)\Sigma \sigma_I \qquad (5.32)$$

$$(n = 13, \ r^2 = 0.40, \ s = 0.77, \ F = 7.34)$$

$$-\log IC_{50} = -3.78 + 2.85(\pm 0.66)\Sigma \sigma_I + 0.95(\pm 0.34)\Delta_{size} \qquad (5.33)$$

$$(n = 13, \ r^2 = 0.66, \ s = 0.60, \ F = 9.8)$$

$$-\log IC_{50} = -4.84 + 0.32(\pm 0.06)C\log P_R + 2.91(\pm 0.35)\Sigma \sigma_I$$
$$-1.28(\pm 0.19)\Delta_{size} \qquad (5.34)$$

$$(n = 13, \ r^2 = 0.91, \ s = 0.32, \ F = 31.89)$$

where $\Sigma \sigma_I$ is the sum of the Taft inductive parameters of substituents on the α carbon, Δ_{size} the increased size of the α substituents over hydrogen and $C\log P_R$ the hydrophobicity of the R substituent. Equations 5.33 and 5.34 differ in the sign of the correlation with size.

5.5 CONCLUSIONS

The two studies of DeBenedetti and coworkers (DeBenedetti et al. 1987; Menziani and DeBenedetti 1991a) described previously are the most suggestive and agree that in separate simple congeneric series, substituted benzenes and substituted aminothiadiazoles, good positive correlations are found with total charge on the sulfonamide in each case. This is also consistent with the studies of Kakeya et al. (1960a, 1969b, 1969c, 1970) that relate activity to pK_a of the SO_2NH_2 group, and influences thereon, such as Hammett σ value of substituents.

It is always easier to obtain a good relationship when the number of compounds is small and their variability is limited. Some of the more recent studies have pooled the different drug types and introduced more variation, and are correspondingly more difficult to interpret. They do, however, allow the examination of less significant influences on activity and also show some differences between the QSARs of the different isozymes. In some cases, even the sign of correlation with a descriptor varies from series to series. This can happen when for a particular descriptor there is an optimal value, and the range of values differs between the groups of drugs. Other anomalies might reflect genuine differences between isozymes. Thus, Equations 5.9 and 5.10 reflect a difference in dependence on polarizability between CA I and CA II by the benzenedisulfonamides, which is counter to that for the inhibition of CA I by thiadiazolines in Equation 5.26.

Equations 5.1–5.11, 5.13, 5.14, 5.17, 5.20 and 5.23–5.34 have terms involving either the charge on one or more atoms of the SO_2NH_2 group or a Hammett σ or

Taft inductive term. The charge on the sulfonamide H in these series is weakly positively correlated with that on S, and the two are strongly negatively correlated with those on N and O. The separation of charge in this group of atoms is a significant but not dominant contribution to the local dipole index D_1. More positive charge on the sulfonamide S and more negative charge on the N or O lead to more activity in most cases. Other quantities modify activity: in general, a larger size and polarizability and a greater solvation energy lead to lesser activity.

Equations 5.12, 5.18, 5.19 and 5.21 exclude charge and inductive terms by design, leaving only Equations 5.15, 5.16 and 5.22, which involve E_H or E_L. These energies are also involved in the more complex relationships of Equations 5.9, 5.11, 5.26, 5.27 and 5.29. Dependence of activity on E_H and E_L is negative for CA II (with the exception of Equation 5.27) and CA IV and positive for CA I. This suggests a way of achieving selectivity. Lipophilicity, an important consideration in many QSAR studies, enters into Equations 5.2–5.5, 5.10–5.14, 5.20, 5.24 and 5.34. The sign is positive in all cases but Equation 5.12, in which there is a complex dependence on three π values, and in Equation 5.24 (HCA II).

Consideration of the orientation of the nodes in the π-like orbitals on benzenoid compounds, as in Equation 5.17, leads to more readily interpretable QSAR, and this is particularly so if the symmetry of benzene is properly taken into account (Supuran and Clare, in press). This problem was avoided by the studies by Kakeya's and DeBenedetti's groups by their choice of compounds.

ACKNOWLEDGEMENTS

We are indebted to Elsevier for providing permission to reproduce Table 12 from Clare, B.W., and Supuran, C.T. (1007) *European Journal of Medicinal Chemistry* **32**, 311–319; Table 5 from Supuran, C.T., and Clare, B.W. (1998) *European Journal of Medicinal Chemistry* **33**, 489–500; and Table 6 from Supuran, C.T., and Clare, B.W. *European Journal of Medicinal Chemistry* **34**, 463–474. We also thank the publishing house of the Romanian Academy for providing permission to publish Tables 7 and 10 from the review Maren, T.H., Clare, B.W., and Supuran, C.T. (1994) Structure–activity studies of sulfonamide carbonic anhydrase inhibitors. *Romanian Chemical Quarterly Reviews* **2**, 259–282.

REFERENCES

Altomare, C., Tsai, R.-S., el Tayar, N., Testa, B., Carotti, A., Cellamare, S., and DeBenedetti, P.G. (1991a) Determination of lipophilicity and hydrogen bond donor acidity of bioactive sulfonyl-containing compounds by reversed phase HPLC and centrifugal partition chromatography and their application to structure–activity relationships. *Journal of Pharmacy and Pharmacology* **43**, 191–197.

Altomare, C., Tsai, R.-S., El Tayar, N., Testa, B., Carrupt, P.-A., Carotti, A., and DeBenedetti, P.G. (1991b) Assessment of hydrogen bond donor acidity of bioactive sulfonyl-containing compounds by CCCC. In *QSAR: Rational Approaches to the Design of Bioactive Compounds*, Silipo, C., and Vittoria, A. (Eds.), Elsevier, Amsterdam, pp. 139–142.

Amat, L., and Carbo-Dorca, R. (1999) Simple linear QSAR models based on quantum similarity measures. *Journal of Medicinal Chemistry* **42**, 5169–5180.

Blacklock, T.J., Sohar, P., Butcher, J.W., Lamanec, T., and Grabowski, E.J.J. (1993) An enantioselective synthesis of the topically active carbonic anhydrase inhibitor MK-0507: 5,6-dihydro-(S)-4-(ethylamino)-(S)-6-methyl-4H-thieno[2,3-b]thiopyran-2-sulfonamide 7,7-dioxide hydrochloride. *Journal of Organic Chemistry* **58**, 1672–1679.

Carotti, A., Raguseo, C., Campagna, F., Langridge R., and Klein, T.E. (1989) Inhibition of carbonic anhydrase by substituted benzenesulfonamides: A reinvestigation by QSAR and molecular graphics analysis. *QSAR* **8**, 1–10.

Cash, G.G. (1995) Prediction of inhibitory potencies of arenesulfonamides towards carbonic anhydrase using easily calculated molecular connectivity indices. *Structural Chemistry* **6**, 157–160.

Clare, B.W., and Supuran, C.T. (1997) Carbonic anhydrase inhibitors. Part 41. Quantitative structure–activity correlations involving kinetic rate constants of 20 sulfonamide inhibitors from a non-congeneric series. *European Journal of Medicinal Chemistry* **32**, 311–319.

Clare, B.W., and Supuran, C.T. (1999) Carbonic anhydrase inhibitors. Part 61. Quantum chemical QSAR of a group of benzenedisulfonamides. *European Journal of Medicinal Chemistry* **34**, 463–474.

DeBenedetti, P.G., Menziani, M.C., and Frassinetti, C. (1985) A quantum chemical QSAR study of carbonic anhydrase inhibitors: Quantum chemical QSAR. *QSAR* **4**, 23–28.

DeBenedetti, P.G., Menziani, M., Cocchi, M.C., and Frassineti, G. (1987) A quantum chemical QSAR analysis of carbonic anhydrase inhibition by heterocyclic sulfonamides. Sulfonamide carbonic anhydrase inhibitors: quantum chemical QSAR. *QSAR* **6**, 51–53.

Evans, W.O., Robinson, S.M., Houstman, D.H., Jackson R.E., and Weinkopf, R.B. (1976) Amelioration of the symptoms of acute mountain sickness by staging and acetazolamide. *Aviation, Space and Environmental Medicine* **47**, 512–516.

Forster, P. (1982) Methazolamide in acute mountain sickness. *Lancet* **1**, 1254.

Gao, H., and Bajorath, J. (1999) Comparison of binary and 2D QSAR analyses using inhibitors of human carbonic anhydrase II as a test case. *Journal of Molecular Diversity,* **4**, 115–130.

Hansch, C., McLarin, J., Klein, T., and Langridge, R. (1985) A quantitative structure–activity relationship and molecular graphics study of carbonic anhydrase inhibitors. *Molecular Pharmacology* **27**, 493–498.

Kakeya, N., Aoki, M., Kamada, A., and Yata, N. (1969a) Biological activities of drugs. VI. Structure–activity relation of sulfonamide carbonic anhydrase inhibitors. *Chemical and Pharmaceutical Bulletin* **17**, 1010–1018.

Kakeya, N., Yata, N., Kamada, A., and Aoki, M. (1969b) Biological activities of drugs VI. Structure–activity relation of sulfonamide carbonic anhydrase inhibitors. *Chemical and Pharmaceutical Bulletin* **17**, 2000–2007.

Kakeya, N., Yata, N., Kamada, A., and Aoki M. (1969c) Biological activities of drugs. VIII. Structure–activity relation of sulfonamide carbonic anhydrase inhibitors. *Chemical and Pharmaceutical Bulletin* **17**, 2558–2564.

Kakeya, N., Yata, N., Kamada, A., and Aoki, M. (1970) Biological activities of drugs. IX. Structure–activity relation of sulfonamide carbonic anhydrase inhibitors. 4. *Chemical and Pharmaceutical Bulletin* **18**, 191–194.

Kishida, K. (1978) 1,3,4-Thiadiazole-5-sulfonamides as carbonic anhydrase inhibitors: Relationship between their electronic and hydrophobic structures and their inhibitory activity. *Chemical and Pharmaceutical Bulletin* **26**, 1049–1053.

Kishida, K., and Manabe, R. (1980) The role of the hydrophobicity of the substituted groups of dichlorphenamide in the development of carbonic anhydrase inhibition. *Medical Journal of the Osaka University* **30**, 95–100.

Liljas, A., Hakansson, K., Jonsson, B.H., and Xue, Y. (1994) Inhibition and catalysis of carbonic anhydrase: Recent crystallographic analyses. *European Journal of Biochemistry* **219**, 1–10.

Lindskog, S., and Wistrand, P.J. (1989) Inhibitors of carbonic anhydrase. In *Design of Enzyme Inhibitors as Drugs*, Sandler, M., and Smith, H.J. (Eds.), Oxford University Press, Oxford, pp. 698–723.

Maren, T.H. (1967) Carbonic anhydrase: Chemistry, physiology, and inhibition. *Physiological Reviews* **47**, 595–782.

Maren, T.H. (1976) Relations between structure and biological activity of sulfonamides. *Annual Reviews in Pharmacology and Toxicology* **16**, 309–327.

Maren, T.H. (1984) The general physiology of reactions catalyzed by carbonic anhydrase and their inhibition by sulfonamides. *Annals of the New York Academy of Science* **429**, 568–579.

Maren, T.H. (1987) Carbonic anhydrase: General perspectives and advances in glaucoma research. *Drug Development Research* **10**, 255–276.

Maren, T.H. (1991) The links among biochemistry, physiology and pharmacology in carbonic anhydrase mediated systems. In *Carbonic Anhydrase*, Botrè, F., Gros, C.G., and Storey, B.T. (Eds.), VCH, Weinheim, pp. 186–207.

Maren, T.H., Clare, B.W., and Supuran, C.T. (1994) Structure–activity studies of sulfonamide carbonic anhydrase inhibitors. *Romanian Chemical Quarterly Reviews* **2**, 259–282.

Maren, T.H., and Conroy, C.W. (1993) A new class of carbonic anhydrase inhibitor. *Journal of Biological Chemistry* **268**, 26233–26239.

Maren, T.H., Wynns, G.C., and Wistrand, P.J. (1993) Chemical properties of carbonic anhydrase IV, the membrane-bound enzyme. *Molecular Pharmacology* **44**, 901–905.

Mattioni, B.E., and Jurs, P.C. (2002) Development of quantitative structure–activity relationship and classification models for a set of carbonic anhydrase inhibitors. *Journal of Chemical Information and Computer Science* **42**, 94–102.

Menziani, M.C., and De Benedetti, P.G. (1991a) Theoretical investigation on enzyme–inhibitor interactions: The binding of heterocyclic sulfonamides to carbonic anhydrase. In *Carbonic Anhydrase — from Biochemistry and Genetics to Physiology and Clinical Medicine*, Botrè, F., Gros, G.G., and Storey, B.T. (Eds.), VCH, Weinheim, pp. 126–129.

Menziani, M.C, and De Benedetti, P.G. (1991b) Direct and indirect theoretical QSAR modelling in sulfonamide carbonic anhydrase inhibitors. In *QSAR: Rational Approaches to the Design of Bioactive Compounds*, Silipo, C., and Vittoria, A. (Eds.), Elsevier, Amsterdam, pp. 331–334.

Menziani, M.C., DeBenedetti, P.G., Gago, F., and Richards, W.G. (1989a) The binding of benzenesulfonamides to carbonic anhydrase enzyme: A molecular mechanics study and quantitative structure–activity relationships. *Journal of Medicinal Chemistry* **32**, 951–956.

Menziani, M.C., Reynolds, C.A., and Richards, W.G. (1989b) Rational drug design: Binding free energy differences of carbonic anhydrase inhibitors. *Journal of the Chemical Society: Chemical Communications* 853–855.

Puscas, I. (1984) Treatment of gastroduodenal ulcers with carbonic anhydrase inhibitors. *Annals of the New York Academy of Science* **429**, 587–591.

Saxena, A., and Khadikar, P.V. (1999) QSAR studies on sulphanilamide Schiff's base inhibitors of carbonic anhydrase. *Acta Pharmaceutica* **49**, 171–179.

Scholz, T.H., Sondey, J.M., Randall, W.C., Schwam, H., Thompson, W.J., Mallorga, P.J., Sugrue, M.F., and Graham, S.L. (1993) Sulfonylmethanesulfonamide inhibitors of carbonic anhydrase. *Journal of Medicinal Chemistry* **36**, 2134–2141.

Silverman, D.N. (1983) Proton transfer in the catalytic mechanism of carbonic anhydrase. *CRC Critical Reviews in Biochemistry* **14**, 207–255.

Silverman, D.N. (1990) The catalytic mechanism of carbonic anhydrase. *Canadian Journal of Botany* **69**, 1070–1078.

Silverman, D.N., and Lindskog, S. (1988) The catalytic mechanism of carbonic anhydrase: implications of a rate-limiting protolysis of water. *Accounts of Chemical Research* **21**, 30–36.

Supuran, C.T. (1993) Carbonic anhydrase inhibitors: Syntheses, properties and physiological significance. *Romanian Chemical Quarterly Reviews* **1**, 77–116.

Supuran, C.T., and Clare, B.W. (1998) Carbonic anhydrase inhibitors. Part 47: Quantum chemical quantitative structure–activity relationships for a group of sulphanilamide Schiff base inhibitors of carbonic anhydrase. *European Journal of Medicinal Chemistry* **33**, 489–500.

Supuran, C.T., and Clare, B.W. (1999) Carbonic anhydrase inhibitors. Part 57. Quantum chemical QSAR of a group of 1,3,4-thiadiazole- and 1,3,4-thiadiazoline disulfonamides with carbonic anhydrase inhibitory properties. *European Journal of Medicinal Chemistry* **34**, 41–50.

Supuran, C.T., and Clare, B.W. (2001) Orbital symmetry in QSAR: Some Schiff's base inhibitors of carbonic anhydrase. *SAR and QSAR in Enviromental Research* **12**, 17–29.

Supuran, C.T. and Clare, B.W. (2004) Quantum theoretic QSAR of benzene derivatives: some enzyme inhibitors. *Journal of Enzyme Inhibition and Medicinal Chemistry,* in press.

Wright, A.D., Bradwell A.R., and Fletcher, R.F. (1983) Methazolamide and acetazolamide in acute mountain sickness. *Aviation, Space and Environmental Medicine* **54**, 619–621.

6 Metal Complexes of Heterocyclic Sulfonamides as Carbonic Anhydrase Inhibitors

Joaquín Borrás, Gloria Alzuet, Sacramento Ferrer and Claudiu T. Supuran

CONTENTS

6.1   Introduction 183
6.2   Acetazolamide Complexes..184
6.3   Methazolamide Complexes..192
6.4   Benzolamide Complexes..195
6.5   Complexes Containing Other Sulfonamide CAIs 200
6.6   Applications of Metal Complexes of Sulfonamides in Therapy................202
References..203

This chapter reviews the synthesis, characterization and inhibitory properties of complexes of heterocyclic sulfonamide carbonic anhydrase inhibitors (CAIs). The different chelating properties of acetazolamide, methazolamide and benzolamide are described on the basis of crystal structures of Cu(II), Zn(II), Ni(II) and Co(II) complexes. These complexes show interesting applications in the mechanistic studies of carbonic anhydrases (CAs) as well as in the development of clinically useful derivatives, because this enzyme is involved in critical physiological/physiopathological processes.

6.1 INTRODUCTION

The importance of sulfonamides as pharmacological agents was realized when Dogmack (1935) showed that sulfanilamide was the metabolite of the antibacterial drug Prontosil. Later, many sulfanilamide derivatives were synthesized, character-

183

ized and tested as antibacterial agents, with many such derivatives currently being used to treat bacterial infections (Mandell and Petri 1996). Sulfonamide derivatives, widely used in clinical medicine as pharmacological agents with a wide variety of biological actions, were designed from the simple sulfanilamide lead molecule (Scozzafava et al. 2003; Casini et al. 2002). In addition to the antibacterials mentioned previously, the unsubstituted aromatic/heterocyclic sulfonamides act as carbonic anhydrase inhibitors (CAIs) (Supuran and Scozzafava 2001; Supuran et al. 2003), whereas other types of derivatives show diuretic activity (high-ceiling diuretics or thiadiazine diuretics), hypoglycemic activity, anticancer properties (Supuran 2003) or inhibitory effects on the aspartic HIV protease, being used to treat AIDS and HIV infection (Scozzafava et al. 2003). Historically, the first sulfonamide metal complex reported was the silver(I) derivative of sulfanilamide, prepared from the sulfonamide sodium salt and silver nitrate by Braun and Towle (1941). The metal complexes of substituted sulfanilamides were then investigated in detail by Bult (1983). However, few crystal structures of such complexes were reported, whereas metal complexes of other types of sulfonamides (except the CAIs, discussed later) have not been investigated.

6.2 ACETAZOLAMIDE COMPLEXES

Among the sulfonamide CAIs, acetazolamide [(H_2acm); 5-acetamido-1,3,4,-thiadiazole-2-sulfonamide; see Scheme 6.1) has been extensively used clinically (under the trademark Diamox) as a diuretic drug, and is still used at present to treat glaucoma, epilepsy and other neuromuscular diseases and as a diagnostic tool (Maren 1967; Supuran and Scozzafava 2000; Supuran et al. 2003).

The crystal structure of H_2acm was reported by Mathew and Palenik (1974). Neutral sulfonamides are expected to behave as poor ligands toward metal ions,

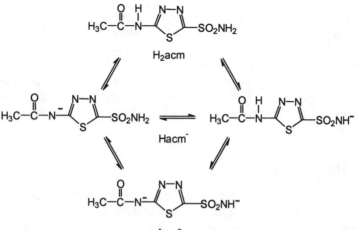

SCHEME 6.1

because of the withdrawal of electron density from the nitrogen atom onto the electronegative oxygen atoms. However, if the sulfonamide nitrogen atom bears a dissociable hydrogen atom, this same electron-withdrawing effect increases its acidity, and in the deprotonated form, sulfonamidate anions can act as effective σ-donor ligands for cations. Acetazolamide has several donor atoms that can bind metal ions; however, according to Mathew and Palenik (1974), H_2acm is not a good ligand, even though acetazolamide was earlier used for the gravimetric determination of Ag(I) (Malecki et al. 1984). As a consequence, complexes of sulfonamides have only recently been investigated in detail, mainly by our groups.

When does acetazolamide behave as a ligand through the thiadiazole ring and when through the sulfonamidate group? This depends on the presence of the two acidic groups in this ligand: the sulfonamido and the acetamido moieties. A search in the literature for research on the acidic character of acetazolamide indicates that this property was controversial until Ferrer et al. (1990a) studied potentiometrically the determination of acid constants of acetazolamide by using the SUPERQUAD program. As acetazolamide contains two acidic protons, four species can be present simultaneously in solution (Scheme 6.1). The pK_{a1} and pK_{a2} values determined in aqueous and in aqueous–ethanol media were 7.19 and 7.52, and 8.65 and 9.41, respectively. Which value corresponds to the monodeprotonated species through the sulfonamido group and which to the monodeprotonated species through the acetamido group? Probably the most acidic group is the sulfonamido one, as considered by several researchers, including Kimura et al. (1990), but there is little experimental evidence to confirm it. Independent of the assignment of pK_a values, the most important point is that at neutral pH the two monodeprotonated and dideprotonated species coexist in concentrations high enough to allow metal ions to bind to any of them (Ferrer et al. 1990a). Probably, the final result is not only a consequence of the affinity of the metal ion for this ligand but also an ensemble of factors such as crystal packing and hydrogen bonds of the coordination compound.

Acetazolamide is the sulfonamide that has been most extensively studied as a ligand (Alzuet et al. 1994c). The interest in its coordination chemistry comes from its properties as an inhibitor of CA and also from the multitude of coordination possibilities that it offers. Acetazolamide has several potential donor positions (at the heterocyclic ring, acetamide group, sulfonamide moiety), and, as mentioned previously, two acidic groups that afford four different species: (1) H_2acm, (2) *sulfonamidate*-Hacm⁻, (3) *acetamidate*-Hacm⁻ and (4) the dideprotonated species acm²⁻ (Scheme 6.1). Therefore, acetazolamide can act as a very versatile ligand (Ferrer et al., 1990a; Alzuet et al. 1994c).

Ferrer et al. (1987) reported the first complexes of acetazolamide, [Co(Hacm)₂(NH₃)₂] and [Zn(Hacm)₂(NH₃)₂]. Since then, many compounds with transition metal ions and a few with main-group metal ions have been reported (Alzuet et al. 1994a, 1994c; Ferrer et al. 1989a, 1989b; Supuran et al. 2003). Table 6.1 lists most of the acetazolamide complexes described in the literature. Most of them have been investigated by the research groups of Borrás and Supuran (previously reviewed by Alzuet et al. 1994c). In all the reported complexes, acetazolamide acts as a ligand in anionic form, either mono- or dideprotonated, which correlates well with the syntheses being performed: by adding a strong base (KOH, Bu₄NOH); or

TABLE 6.1
Reported Acetazolamide Complexes

Metal Ion	Compound[a]	Reference Wherein Reported
Co(II)	[Co(Hacm)$_2$(NH$_3$)$_2$]	Ferrer et al. 1987
	K$_6$Co(acm)$_4$ · 6H$_2$O	Alzuet et al. 1991
	(Hea)$_2$Co$_2$(acm)$_3$ · 6H$_2$O, (Hdea)$_2$Co$_2$(acm)$_3$ · 1/2H$_2$O, (Htea)$_2$Co$_2$(acm)$_3$ · 2H$_2$O	Alzuet et al. 1991
	[Co(Hacm)L](ClO$_4$) · 2H$_2$O	Alzuet et al. 1994a
Ni(II)	[Ni(Hacm)$_2$(NH$_3$)$_4$]	Ferrer et al. 1989b
	K$_2$Ni$_2$(acm)$_3$ · 5H$_2$O	Ferrer et al. 1989a
	(Hdea)$_2$Ni$_2$(acm)$_3$ · 3H$_2$O, (Htea)$_2$Ni$_2$(acm)$_3$ · 3H$_2$O	Alzuet et al. 1991
	[Ni(Hacm)(OH)(OH$_2$)$_2$]$_2$ · 6H$_2$O	Supuran et al. 1990
Cu(II)	[Cu(acm)(NH$_3$)$_2$(OH$_2$)$_2$]$_2$ · 2H$_2$O	Ferrer et al. 1990b
	Cu(acm)(NH$_3$)$_3$, Cu(acm)(NH$_3$) · H$_2$O, Cu$_3$(acm)(OH)$_4$ · 3H$_2$O	Ferrer et al. 1990b
	K$_6$Cu$_2$(acm)$_5$ · 2H$_2$O	Alzuet et al. 1991
	(Hea)$_6$Cu$_2$(acm)$_5$ · H$_2$O, (Hdea)$_2$Cu$_4$(acm)$_5$ · 3H$_2$O, (Htea)$_2$Cu$_4$(acm)$_5$ · 3H$_2$O	Alzuet et al. 1991
	[Cu(Hacm)$_2$(en)$_2$], [Cu(Hacm)$_2$(tn)$_2$]	Ferrer et al. 1992
	[Cu(Hacm)(dien)](ClO$_4$) · H$_2$O, [Cu(Hacm)(dipn)](ClO$_4$)	Alzuet et al. 1994a
	[Cu(Hacm)$_2$], [Cu(Hacm)Cl]$_2$	Supuran et al. 1990
d^{10}	[Zn(Hacm)$_2$(NH$_3$)$_2$]	Ferrer et al. 1987; Hartmann and Vahrenkamp 1991
	[Zn(Hacm)L'](ClO$_4$)	Kimura et al. 1990
	[Zn(Hacm)L](ClO$_4$)	Alzuet et al. 1994a
	[Zn(Hacm)$_2$]	Supuran et al. 1990
	Cd(acm) · 3/2H$_2$O	Ferrer et al. 1989a
	Hg(acm) · 1/2H$_2$O	Ferrer et al. 1989a
	Ag$_2$(acm), Ag$_2$(acm)(NH$_3$)	Ferrer et al. 1989a
Main group- M(II)	[Be(Hacm)$_2$]	Supuran et al. 1993a
	[Mg(Hacm)$_2$(OH$_2$)$_2$]	Supuran et al. 1990
	[Pb(Hacm)$_2$(OH$_2$)$_2$]	Supuran et al. 1990
M(III)	[M(Hacm)$_3$] × H$_2$O [M = Al, Ga, In, Tl, Fe, Ru, Rh, La, Ce, Pr, Nd, Sm, Eu, Gd, Tb, Dy, Ho, Er, Tm, Yb]	Supuran et al. 1992, 1993a; Supuran and Andruh 1994
	[Au(Hacm)$_2$]Cl	Manole et al. 1993; Supuran et al. 1993c
M(IV)	[VO(Hacm)$_2$]	Supuran 1993
	[M(Hacm)$_4$] · xH$_2$O [M = Ce, Th]	Supuran 1993
M(VI)	[UO$_2$(Hacm)$_2$(OH$_2$)] · H$_2$O	Supuran 1993

[a] ea: ethylamine, dea: diethylamine, tea: triethylamine, L: tris[2-(1-methylbenzimidazol-2-yl)ethyl]nitromethane, L': [12]aneN$_3$=1,5,9-triazacyclododecane, en: 1,2-ethanediamine, tn: 1,3-propanediamine, dien: diethylenetriamine, dipn: dipropylenetriamine.

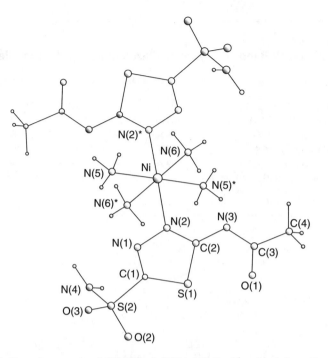

FIGURE 6.1 Crystal structure of [Ni(Hacm)$_2$(NH$_3$)$_4$] (**6.1**). (Reprinted from Ferrer, S. et al. (1989b) *Inorganic Chemistry* **28**, 160–163. With permission from ACS.)

in the presence of a weak one, such as ammonia, aliphatic monoamines (ea, dea and tea), chelating diamines (en and tn) or tridentate triamines (dien and dipn; see Table 6.1 for amine abbreviations); or by using the sodium salt of the ligand (NaHacm). The only exception is the silver complex Ag$_2$(acm), prepared directly from the acetazolamide without a base (Ferrer et al. 1989a).

Despite the extensive study of acetazolamide complexes, only five have been analyzed by crystallographic methods. The first published crystal structure corresponds to that of the purple compound [Ni(Hacm)$_2$(NH$_3$)$_4$] (**6.1**, Figure 6.1; Ferrer et al. 1989b). It consists of discrete units that contain a NiN$_6$ chromophore with the metal center in a slightly elongated rhombically distorted octahedral environment formed by four ammonia molecules in equatorial positions [average d(Ni – NH$_3$) = 2.105 Å] and two *trans* Hacm$^-$ ligands in the apical ones [d(Ni – N(2)) = 2.150 Å]. The structure shows acetazolamide deprotonated at the acetamido group and coordinating through N(2), the N-thiadiazole atom closest to the group with the negative charge. The changes in the ligand, mainly by the deprotonation rather than by the coordination, are clearly reflected on the IR spectrum of the complex (Table 6.2), with the carbonyl band shifted and split but with the sulfonamido bands not suffering significant modification as compared with the corresponding bands of the ligand.

The next compound for which the crystal structure was described was the dark blue [Cu(acm)(NH$_3$)$_2$(OH$_2$)$_2$]$_2$ · 2H$_2$O (**6.2**; Figure 6.2; Ferrer et al. 1990b). Its structure, made of dinuclear entities and crystallization water molecules, corresponds to

TABLE 6.2
Significant IR Bands (cm¹) for Acetazolamide and Its Complexes[a]

Compound	$\upsilon(C{=}O)$	$\upsilon(SO_2)_{asym}$	$\upsilon(SO_2)_{sym}$
H_2acm	1672s	1318s	1170s
[Ni(Hacm)$_2$(NH$_3$)$_4$] (**6.1**)	1681w, 1624m	1328s	1178s
[Cu(Hacm)$_2$(tn)$_2$] (**6.5**)	1580sh, 1540s-b	1355s	1170s
[Cu(Hacm)$_2$(en)$_2$] (**6.4**)	1620m	1330s	1160s
[Zn(Hacm)$_2$(NH$_3$)$_2$] (**6.3**)	1699s	1285s	1145s
[Cu(acm)(NH$_3$)$_2$(OH$_2$)]$_2$ ·2H$_2$O (**6.2**)	1617m	1296s	1145-1126s-d

[a] s: strong; m: medium; w: weak; sh: shoulder, b: broad, d: doublet.

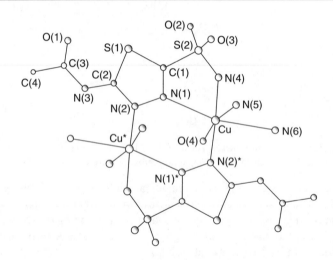

FIGURE 6.2 Crystal structure of [Cu(acm)(NH$_3$)$_2$(OH$_2$)]$_2$ · 2H$_2$O (**6.2**). (Reprinted from Ferrer, S. et al. (1990b) *Inorganic Chemistry* **29**, 206–210. With permission from ACS.)

the only polynuclear derivative of acetazolamide currently available. In the centro-symmetric dimer, the two Cu(II) ions are connected by a double $N(1) - N(2)$ thiadiazole bridge. The metallic centers form CuN$_3$O + N$_2$ chromophores with a tetragonally elongated octahedral geometry around each copper ion. Each acetazola-mide ligand, in its doubly deprotonated form (acm^{2-}), acts as a bidentate chelate toward one Cu(II) ion via the $N(1)$thiadiazole and the $N(4)$sulfonamidate atoms and as a monodentate ligand toward the other Cu(II) ion via the $N(2)$thiadiazole atom, with the following bonding distances: $d(Cu - N(2)) = 2.020$ Å, $d(Cu - N(4)) = 2.028$ Å and $d(Cu - N(1)) = 2.546$ Å. Despite its negative charge, the tridentate acm^{2-} does not bind through the acetamidate group. The IR spectrum of the complex is in agreement with the deprotonation of both the acetamido and sulfonamido groups, with significant shifts of the corresponding bands to lower frequencies (Table 6.2).

The third crystal structure, determined by Hartmann and Vahrenkamp (1991), was that of the complex [Zn(Hacm)$_2$(NH$_3$)$_2$] (**6.3**), already obtained by Ferrer et al.

(1987). The structure is formed by a mononuclear species that contains the Zn(II) ion tetrahedrically surrounded by four nitrogen atoms, two of them from two ammonia molecules, with d(Zn–N) = 2.021 Å, and the other two from two sulfonamidate groups, with d(Zn–N) = 1.977 Å. Despite the apparent similarity with compound **6.1**, [Ni(Hacm)$_2$(NH$_3$)$_4$], the behavior of acetazolamide as ligand in **6.3** is completely different. In the present case, acetazolamide is also monodeprotonated, but deprotonation takes place on a different group (at the sulfonamido moiety), and it also acts as a monodentate ligand, but through the nitrogen sulfonamidate atom. These differences are clearly indicated by the IR spectrum of **6.3** (Table 6.2), with shifts of the SO$_2$ bands to lower frequencies, which, together with the electronic spectrum of the [Co(Hacm)$_2$ (NH$_3$)$_2$] compound, allowed Borrás's group to predict the right structure before the x-ray study was available. (The only aspect not solved by the spectral evidences was whether the coordination was produced via the N or the O sulfonamidate atoms; Ferrer et al. 1987.) Finally, it is remarkable that the structural details of **6.3** compare very favorably to those of the acetazolamide bound to CA II, with similar findings for the coordination geometry and for the donor atom of the inhibitor (Vidgren et al. 1990).

The complexes described previously were synthesized working with an excess of ammonia. The next studies involved syntheses in the presence of chelating diamines and led to the isolation of two new compounds for which the crystal structures could be determined: the purple [Cu(Hacm)$_2$(en)$_2$] (**6.4**; en = 1,2-ethanediamine; Figure 6.3) and the dark blue [Cu(Hacm)$_2$(tn)$_2$] (**6.5**; tn = 1,3-propanediamine;

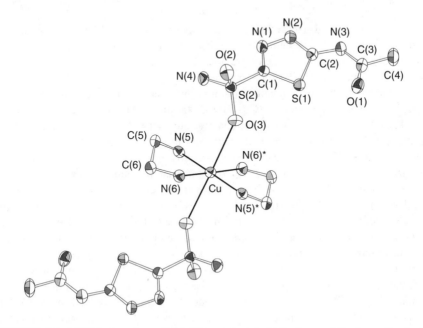

FIGURE 6.3 ORTEP drawing of [Cu(Hacm)$_2$(en)$_2$] (**6.4**) (en = 1,2-ethanediamine). (Reprinted from Ferrer, S. et al. (1992) *Inorganica Chimica Acta* **192**, 129–138. With permission from Elsevier.)

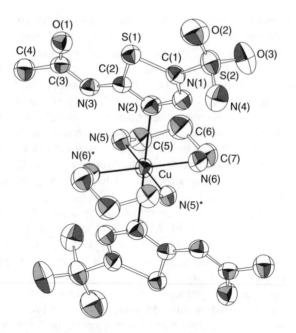

FIGURE 6.4 ORTEP drawing of [Cu(Hacm)$_2$(tn)$_2$] (**6.5**) (tn = 1,3-propanediamine). (Reprinted from Ferrer et al. (1992) *Inorganica Chimica Acta* **192**, 129–138. With permission from Elsevier.)

Figure 6.4; Ferrer et al. 1992). In both these mononuclear complexes, acetazolamide is deprotonated at the acetamido group, but its binding mode to the metal ion is completely different from those of previously investigated derivatives **6.1** to **6.3**. In both **6.4** and **6.5**, the Cu(II) ions, on the symmetry centers, exhibit an elongated octahedral geometry, with the four diamine nitrogen atoms in an approximately square coplanar arrangement [d(Cu–N) = 2.001 and 2.009 Å for **6.4**; d(Cu–N) = 2.044 and 2.048 Å for **6.5**]; two O-sulfonamido atoms in **6.4** and two $N(2)$-thiadiazole atoms in **6.5** (from two *trans* acetazolamidate ligands) complete the octahedra at longer distances [2.652 Å and 2.457 Å, respectively]. Again the donor positions of acetazolamide do not belong to the group that bears the deprotonation in either of the two complexes. In **6.5**, the ligand binds in the expected way; its structure is similar to that of **6.1** [although in **6.5** the distortion on apical positions is more pronounced, in agreement with the different nature of the metal ion, Cu(II) instead of Ni(II)]. For **6.4**, however, the coordination mode of acetazolamide is unexpected: it binds through one oxygen sulfonamido atom (and at a semicoordinating distance), even though the negative charge is on the other substituent of the ring (in **6.1** and **6.5**, the donor atom, $N(2)$, participates on the negative charge of $N(3)$ due to electronic delocalization). This unusual behavior of **6.4** must be related to the Cu(II) Jahn–Teller effect and to the existence of a network of strong hydrogen bonds that stabilize the excess of negative charge at $N(3)$ through pairing of acetazolamidate anions. Therefore, in the presence of these chelating amines, acetazolamide behaves as an ambidentate, unusually weak anionic ligand. The major IR changes are in

TABLE 6.3

Coordination Ways of Acetalozamide Determined by x-Ray Crystallography

Compound	Description of Structure	Ref.
[Ni(Hacm)$_2$(NH$_3$)$_4$] (**6.1**)	Mononuclear	Ferrer et al. 1989b, 1992
[Cu(Hacm)$_2$(tn)$_2$] (**6.5**)	Deprotonation at N(acetamido)	
	Monodentate via N(thiadiazole)	
[Cu(Hacm)$_2$(en)$_2$] (**6.4**)	Mononuclear	Ferrer et al. 1992
	Deprotonation at N(acetamido)	
	Monodentate via O(sulfonamido)	
[Zn(Hacm)$_2$(NH$_3$)$_2$] (**6.3**)	Mononuclear	Ferrer et al. 1987;
	Deprotonation at N(sulfonamido)	Hartmann and
	Monodentate via N(sulfonamido)	Vahrenkamp 1991
[Cu(acm)(NH$_3$)$_2$(OH$_2$)]$_2$ · 2H$_2$O (**6.2**)	Dinuclear	Ferrer et al. 1990b
	Deprotonation at N(acetamido)	
	and N (sulfonamido)	
	Bridge via N,N (thiadiazole) and	
	Chelate via N(1)(thiadiazole)	
	and N(4)(sulfonamido)	

agreement with the deprotonation of the acetamido group (**6.4, 6.5**); the weak O-sulfonamido coordination has small effects on the sulfonamide IR bands of **6.4** (Table 6.2).

Some conclusions were derived from these five structures of acetazolamide complexes. First, acetazolamide always binds in an anionic form, with three possibilities: as acetamidate, as sulfonamidate or dideprotonated. Second, coordination is not necessarily achieved through the deprotonated group: the acetamido/-ate group never interacts directly with the metal ion. Third, acetazolamide can exhibit at least four different ways of coordination, presented in Table 6.3, which confirm its versatility as a ligand. Finally, the IR spectrum can be used as a diagnostic tool to propose the binding mode in cases when a crystal structure is not available (Table 6.2).

Apart from the previously mentioned structures, the binding of acetazolamide to model complexes that mimic the CA active site has also been investigated. Kimura et al. (1990) reported the complex [Zn(Hacm)([12]aneN$_3$)](ClO$_4$) ([12]aneN$_3$ = 1,5,9-triazacyclododecane) as a model of the interaction of acetazolamide with the active metal center of CA. Although good crystals could not be obtained for x-ray studies, the spectroscopic investigation indicated that the monodeprotonated sulfonamide is coordinated to Zn(II) through the sulfonamido nitrogen atom. Casella's group prepared Zn(II), Co(II) and Cu(II) complexes with the ligand tris[2-(1-methylbenzimidazol-2-yl)ethyl]nitromethane (L), whose donor positions mimic the environment of the Zn(II) site of CA, and then studied the formation of adducts of such complexes with acetazolamide (Alzuet et al. 1994a). Two ternary complexes [Zn(Hacm)L](ClO$_4$) and [Co(Hacm)L](ClO$_4$) · H$_2$O were obtained. In the case of Cu(II), it was not possible to isolate the pure ternary complex, but mixed-ligand complexes with the triamines dien and dipn, with the formula [Cu(Hacm)(dien)](ClO$_4$) · H$_2$O and [Cu(Hacm)(dipn)](ClO$_4$), were obtained. The available spectral evidences indicated that in these cases acetazolamide binds in its monoanionic form through the deprotonated sulfonamido group. These ternary complexes can be considered analogues of the adducts formed by CA and the inhibitor acetazolamide. These complexes are the only acetazolamide compounds so far reported that contain a noncoordinating anion.

6.3 METHAZOLAMIDE COMPLEXES

From the coordination point of view, methazolamide [N-(4-methyl-2-sulfamoyl-Δ^2-1,3,4-thiadiazolin-5-ylidene)acetamide] (Hmacm) behavior is much simpler than that of the structurally related acetazolamide, because methylation on the thiadiazole N-2 atom (probably the best donor atom in acetazolamide) makes it unable to interact with metal ions. Furthermore, methazolamide incorporates only one acidic proton, at the sulfonamide group, with a pK_a of 7.4 (Alzuet et al. 1994c) and therefore can be used as a simpler model for understanding the interaction of sulfonamides with metal ions (Scheme 6.2). The synthesis, crystal structure and chemical characterization of several methazolamide metal complexes also containing ammonia or pyridine as ligands have been described (Table 6.4).

[Ni(macm)$_2$(NH$_3$)$_4$] (**6.6**) was the first reported metal complex of methazolamide (Figure 6.5; Alzuet et al. 1993a). The geometry around the metal ion can be described as a slightly elongated rhombic octahedron. Methazolamide occupies the axial positions,

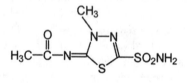

SCHEME 6.2

TABLE 6.4
Methazolamide Complexes with Structures Determined by x-Ray Crystallography

Compound	Description of Structure	Ref.
[Ni(macm)$_2$(NH$_3$)$_4$] (**6.6**)	Distorted octahedral Monodentate via N(sulfonamido)	Alzuet et al. 1993a
[Co(macm)(NH$_3$)(aib)$_2$] (NO$_3$)$_2$•2H$_2$O (**6.7**)	Octahedral Monodentate via N(sulfonamido)	Alzuet et al. 1993b
[Zn(macm)$_2$(NH$_3$)$_2$] (**6.8**)	Distorted tetrahedral Monodentate via N(sulfonamido)	Alzuet et al. 1995
[Ni(macm)$_2$(py)$_2$(OH$_2$)$_2$] (**6.9**)	Distorted octahedral Monodentate via N(sulfonamido)	Alzuet et al. 1992
[Co(macm)$_2$(py)$_2$(OH$_2$)$_2$] (**6.10**)	Distorted octahedral Monodentate via N(sulfonamido)	Alzuet et al. 1992
[Cu(macm)$_2$(py)$_2$(OH$_2$)$_2$] (**6.11**)	Distorted octahedral Monodentate via N(sulfonamido)	Alzuet et al. 1992

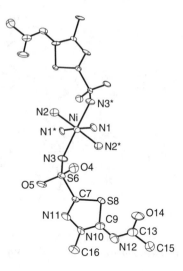

FIGURE 6.5 Crystal structure of $[Ni(macm)_2(NH_3)_4]$ (**6.6**). (Reprinted from Alzuet, G. et al. (1993a) *Inorganica Chimica Acta* **203**, 257–261. With permission from Elsevier.)

acting as a monodentate ligand coordinating through the nitrogen atom of the deprotonated sulfonamide group. A comparison with the related acetazolamide compound $[Ni(Hacm)_2(NH_3)_4]$ shows that the axial Ni-N(sulfonamide) distance (2.216Å) in the $[Ni(macm)_2(NH_3)_4]$ complex is slightly longer than that for Ni-N(thiadiazole) (2.150 Å) in $[Ni(Hacm)_2(NH_3)_4]$. This difference in axial distances has been connected with the different ligand fields created by the donor nitrogen atoms. The higher bond strength of the N(thiadiazole) atom seems to be the result of the strong delocalization in the thiadiazole ring of the Hacm⁻ anion.

The complex $[Co(macm)(NH_3)(aib)_2(NO_3)_2]\cdot 2H_2O$ (**6.7**; Figure 6.6) was prepared by reacting $Co(NO_3)_2$ with methazolamide and ammonia in acetone (Alzuet et al. 1993b). It contains two molecules of 2-methyl-2-amino-4-imino pentane (aib), which acts as a bidentate ligand. The aib ligand is obtained *in situ* by condensation of two acetone and two ammonia molecules in a template-type reaction described by Curtis (1963). In addition, the air spontaneous oxidation of Co(II) to Co(III) occurs. Co(III) exhibits a nearly regular octahedral geometry. Methazolamide exhibits its usual coordination behavior, binding to the metal ion through the nitrogen atom of the sulfonamidate group (Alzuet et al. 1993b).

More recently, the compound $[Zn(macm)_2(NH_3)_2]$ (**6.8**; Figure 6.7) has been shown to be a good model of the interaction of methazolamide with the metal center in the active site of CA (Alzuet et al. 1995). The coordination sphere of the Zn(II) ion is a distorted tetrahedron. The Zn–N(sulfonamido) bond distances and the coordination angles compare well with those reported for the hCAII–acetazolamide complex (Vidgren et al. 1990). Analysis based on extended Huckel calculations applied to this compound support the results obtained by Vidgren et al. (1990), which indicated the importance of interactions of acetamido and thiadiazole ring in the complex formation between enzyme and the sulfonamide bound within the active site. Furthermore the isostructural $[Co(macm)_2(NH_3)_2]$ derivative has been obtained

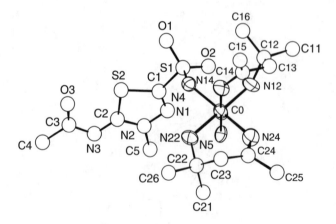

FIGURE 6.6 Crystal structure of $[Co(macm)(NH_3)(aib)_2(NO_3)_2] \cdot 2H_2O$ (**6.7**). (Reprinted from Alzuet, G. et al. (1993b) *Inorganica Chimica Acta* **205**, 79–84. With permission from Elsevier.)

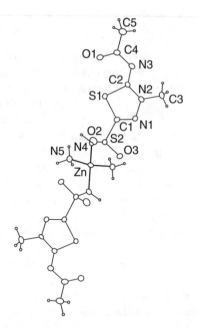

FIGURE 6.7 Crystal structure of $[Zn(macm)_2(NH_3)_2]$ (**6.8**). (Reprinted from Alzuet, G. et al. (1995) *Journal of Inorganic Biochemistry* **57**, 219–234. With permission from Elsevier.)

as a spectroscopic model for the binding of inhibitors to Co(II)-substituted CA (Alzuet et al. 1995).

The compounds $[M(macm)_2(py)_2(OH_2)_2]$ (M = Cu, Co, Ni; **6.9**, **6.10**, **6.11**; Figure 6.8) were reported to be octahedral (Alzuet et al. 1992). Methazolamide behaves again as a monodentate ligand, binding to the metal ion through the sulfonamidate nitrogen atom. Deprotonated methazolamide ligands occupy equatorial sites in a *trans* geometry in these compounds.

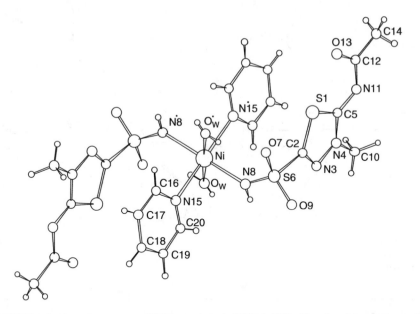

FIGURE 6.8 Crystal structure of [Ni(macm)$_2$(py)$_2$(OH$_2$)$_2$] (**6.9**). (Reprinted from Alzuet, G. et al. (1992) *Polyhedron* **22**, 2849–2856. With permission from Elsevier.)

In all these complexes, one relevant feature of methazolamide is the shortening of the sulfonamide S–N distance and the opposite lengthening of the contiguous S–C bond distance. The shortening of the S–N bond distance depends on the nature of the metal ion. The minor S–N bond length reduction was observed for the Co(III)–macm complex with respect to other M(II)–macm complexes, and it might be related to the trivalent nature of the metal center.

6.4 BENZOLAMIDE COMPLEXES

Benzolamide [(5-phenylsulfonamido-1,3,4-thiadiazole-2-sulfonamide), H$_2$bz] contains two dissociable protons similar to those in acetazolamide, incorporating multiple potential donor binding sites at the thiadiazole and the two sulfonamide moieties (Alzuet et al. 1998, 1999). Hence, as for acetazolamide, a multitude of coordination possibilities is expected (Scheme 6.3).

Our group reported ternary zinc and copper benzolamide complexes with amines [ammonia, diethylenetriamine (dien), dipropylenetriamine (dipt)] and the tripodal ligand *tris*(2-benzimidazolyl-methylamine) (L) (Alzuet et al. 1999). In these complexes, the coordination behavior of benzolamide was found to be different (Table 6.5).

In the first reported benzolamide complex [Cu(bz)(NH$_3$)$_4$] (**6.12**; Figure 6.9) for which the x-ray structure was determined, the coordination behavior of benzolamide is unexpected (Alzuet et al. 1998). Although dideprotonated, it acts as a monodentate ligand via the nitrogen atom of the primary sulfonamido group. This fact contrasts with the ligand behavior of acetazolamide (H$_2$acm) in the analogous [Cu(acm)(NH$_3$)$_2$(H$_2$O)]$_2$ · 2H$_2$O complex (Ferrer et al. 1990b), in which the doubly

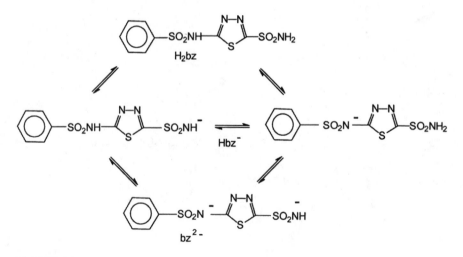

SCHEME 6.3

TABLE 6.5
Benzolamide Complexes with Structures Determined
by x-Ray Crystallography

Compound	Description of Structure	Ref.
[Cu(bz)(NH$_3$)$_4$] (**6.12**)	Square pyramidal	Alzuet et al.
	Dideprotonated at the two N sulfonamido atoms	1998
	Monodentate via N(unsubstituted sulfonamido)	
{[Zn$_2$(bz)$_2$(NH$_3$)$_4$]·2H$_2$O}∞ (**6.13**)	Tetrahedral	Alzuet et al.
	Dideprotonated at the two sulfonamido N atoms	1999
	Bridging via N(unsubstituted sulfonamido)/N(thiadiazole)	
[Zn(Hbz)(L)](ClO$_4$)·H$_2$O (**6.14**)	Trigonal bipyramidal	Alzuet et al.
	Monodeprotonated at substituted sulfonamido N atom	1999
	Monodentate via N(thiadiazole)	
[Cu(Hbz)$_2$(dien)] (**6.15**)	Square pyramidal	Alzuet et al.
	Monodeprotonated at substituted sulfonamido N atom	2000
	Monodentate via N(thiadiazole)	
[Zn(Hbz)$_2$(dien)] (**6.16**)	Square pyramidal	Alzuet et al.
	Monodeprotonated at substituted sulfonamido N atom	2000
	Monodentate via N(thiadiazole)	

---

ionized acetazolamide coordinates through the N(sulfonamido) and N(thiadiazole)
atoms. The reason for the different behaviors of the two sulfonamides can be inferred
from the stronger interaction between the thiadiazole ring and the acetamido group

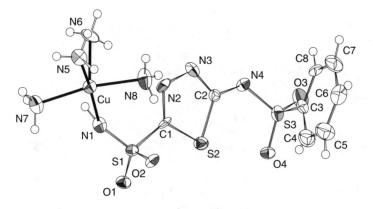

FIGURE 6.9 Crystal structure of $[Cu(bz)(NH_3)_4]$ (**6.12**). (Reprinted from Alzuet, G. et al. (1998) *Inorganica Chimica Acta* **273**, 334–338. With permission from Elsevier.)

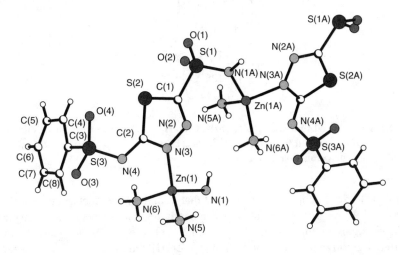

FIGURE 6.10 Crystal structure of $\{[Zn_2(bz)_2(NH_3)_4].2H_2O\}\infty$ (**6.13**). (Reprinted from Alzuet, G. et al. (1999) *Journal of Inorganic Biochemistry* **75**, 189–198. With permission from Elsevier.)

in acm^{2-} as compared to that of the thiadiazole ring and the sulfonamido group in bz^{2-}. From spectroscopic data, a similar coordination behavior has been proposed for the compounds $[Cu(bz)(dien)(OH_2)]$ and $[Cu(bz)(dipn)(OH_2)]$ (Alzuet et al. 2000).

The reaction of benzolamide, ammonia and Zn(II) led to the formation of $\{[Zn_2(bz)_2(NH_3)_4] \cdot H_2O\}\infty$ (**6.13**; Alzuet et al. 1999; Figure 6.10). In this case, the dianion of benzolamide acts as a bridge linking two metal centers through the nitrogen atom of the primary sulfonamido group and the thiadiazole nitrogen N(3). The structure consists of infinite chains linked together by hydrogens bonds. A similar polymeric structure was proposed for the (Hyt)Zn(NH₃)₂ compound described by Hartmann and Vahrenkamp (1994), where Hyt stands for the dideprotonated form of the *bis*-sulfonamide ligand hydrochlorothiazide. The Zn–*N*-sulfonamide bond

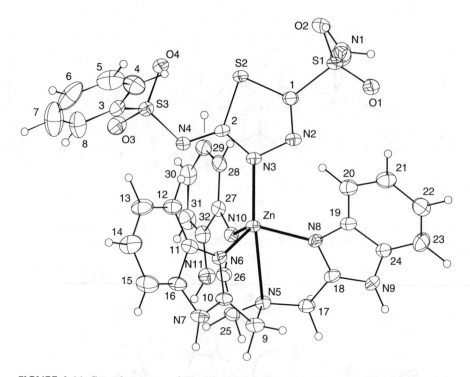

FIGURE 6.11 Crystal structure of [Zn(Hbz)L]ClO$_4$.H$_2$O (**6.14**) [L = *tris* (2-benzimidazolyl-methylamine)]. (Reprinted from Alzuet, G. et al. (1999) *Journal of Inorganic Biochemistry* **75**, 189–198. With permission from Elsevier.)

length of 1.978 Å compares well with that found in the acetazolamide–hCAII complex (1.90 Å). The Zn(II) ion adopts a nearly regular tetrahedral geometry with *N*–Zn–*N* angles ranging from 106.6 to 112.2°. These angles are close to those reported by Vidgren et al. (1990) for the Zn(II) tetrahedron in the acetazolamide–hCAII complex (*N*-sulfonamide–Zn–*N*–His94/119, 104° and 113°, respectively). Furthermore, this structure is reminiscent of those of [Zn(Hacm)$_2$(NH$_3$)$_2$] (Ferrer et al. 1987; Hartmann and Vahrenkamp 1991) and [Zn(macm)$_2$(NH$_3$)$_2$] (Alzuet et al. 1995), which are structural models of the sulfonamide interaction with the Zn(II) ion at the active site of CA.

For the synthesis of Zn-benzolamide complexes, a tripodal *tris*-benzimidazole ligand [*tris*(2-benzimidazolyl-methylamine) (L)] has been used, with the aim of mimicking the environment of the metal center in CA (Alzuet et al. 1999). The complex [Zn(Hbz)L](ClO$_4$) · H$_2$O (**6.14**; Figure 6.11) has been obtained, which consists of cationic [Zn(Hbz)L]$^+$ entities and no coordinating ClO$_4^-$ anions. The benzolamide anion interacts through the thiadiazole nitrogen atom contiguous to the deprotonated sulfonamido group. Angles around the Zn(II) ion show a distorted trigonal bipyramidal geometry.

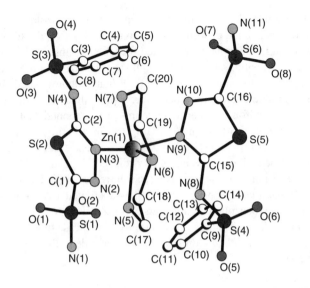

FIGURE 6.12 Crystal structure of [Zn(Hbz)₂(dien)] (**6.16**). (Reprinted from Alzuet, G. et al. (2000) *Polyhedron* **19**, 725–730. With permission from Elsevier.)

An interesting aspect is the different coordination behaviors exhibited by benzolamide and the closely related sulfonamide acetazolamide toward Zn(II) when these sulfonamides are monodeprotonated. Although both sulfonamides have the same potential donor atoms, acetazolamide interacts through the sulfonamide nitrogen atom in the [Zn(Hacm)₂(NH₃)₂] compound (Ferrer et al. 1987; Hartmann and Vahrenkamp 1991) whereas benzolamide coordinates through a thiadiazole nitrogen atom in the [Zn(Hbz)L](ClO₄)·H₂O complex.

The ternary complexes [M(Hbz)₂(dien)] (**6.15, 6.16**; M = Cu, Zn; Alzuet et al. 2000) were obtained by reacting the metal salt, the triamine and benzolamide in a molar ratio of 1:1:2. The ratio of reagents is important for obtaining these compounds, because addition of an equimolecular amount of triamine and benzolamide leads to the formation of the previously reported M(bz)(dien)(H₂O) (M = Cu, Zn) compounds, in which the benzolamide is present in the dideprotonated form.

In the [Cu(Hbz)₂(dien)] and [Zn(Hbz)₂(dien)] complexes (Figure 6.12), the metal ion is pentacoordinated by five nitrogen atoms, the structure being best described as a regular square pyramid. Both MN₅ chromophores are essentially square pyramidal, but with significant differences in the local molecular stereochemistry. For these two structures, the major changes in the angles surrounding the metal center are found in the N(5)-M-N(7) and N(6)-M-N(3), which expands from 149.8 and 153.4° [Zn(Hbz)₂(dien)] to 163.4 and 162.3° [Cu(Hbz)₂(dien)].

In these complexes, the coordination mode of benzolamide is determined by the deprotonation occurring at the secondary sulfonamide moiety. This makes the thiadiazole nitrogen closest to this sulfonamido group, N(3), the best donor atom. Structural determinations of all these complexes have demonstrated the coordinative versatility of benzolamide, a feature that it shares with acetazolamide. Although it is difficult to rationalize the coordination behavior of this ligand, our investigations have indicated that in the monodeprotonated form benzolamide behaves as a monodentate ligand coordinating the metal through the thiadiazole nitrogen atom closest to the substituted sulfonamido group, (N3). When dideprotonated, benzolamide shows a more variable coordination behavior, acting either as a bridge through the thiadiazole and free sulfonamido nitrogen atoms or as a monodentate ligand via the primary sulfonamido nitrogen. A comparison of the structures of the $\{[Zn_2(bz)_2(NH_3)_4] \cdot 2H_2O\} \infty$ and $[Cu(bz)(NH_3)_4]$ complexes indicates that the coordination mode of benzolamide, when doubly ionized, depends on the nature of the metal ion. A metal ion dependence has also been reported (Coleman 1975) for the binding of sulfonamides to Zn(II)- and Co(II)-substituted CAs. Furthermore, our results indicate that benzolamide interacts with Zn(II) through the sulfonamido nitrogen atom as a dinegative anion, so that only in this case this interaction can be considered as a model for the binding of the sulfonamide to the metal center within the CA active site.

6.5 COMPLEXES CONTAINING OTHER SULFONAMIDE CAIs

Besides metal complexes of acetazolamide, methazolamide and benzolamide, Supuran`s group prepared numerous metal complexes of other sulfonamides, such as ethoxzolamide (HEZA) or acetazolamide congeners incorporating other moieties in the 5-position of the thiadiazole ring, of types **6.20** to **6.25** (Andruh et al. 1991; Supuran et al. 1991, 1993a,b; Supuran 1992, 1995a, 1998a, 1998b, 1999; Almajan and Supuran 1997; Jitianu et al. 1997; Scozzafava et al. 2001, 2002; Briganti et al. 2000).

Ethoxzolamide **6.17** was shown to behave as a bidentate ligand in both the anion and the neutral form, binding the metal ions through the $N_{sulfonamide}$ and the benzothiazole nitrogen atoms (Andruh et al. 1991; Supuran et al. 1991). However, no crystal structures of ethoxzolamide complexes have been reported so far. The thienothiopyran sulfonamides dorzolamide **6.18** and MK-907 **6.19** (the first is a topically acting antiglaucoma agent in clinical use, and the second was a clinical candidate before dorzolamide; Supuran et al. 2003) have also been investigated for their complexation behavior (Supuran 1995b, 1996). These sulfonamides were shown to behave as bidentate ligands through the $N_{sulfonamide}$ and the $S_{heterocyclic}$ atoms (Supuran 1995b, 1996). Many complexes of di- and trivalent metal ions incorporating the various 1,3,4-thiadiazole-sulfonamides of types **6.20** to **6.25** were reported and assayed as CAIs against the major isozymes CA I and CA II (Supuran 1992, 1995a, 1998a, 1998b, 1999; Almajan and Supuran 1997; Jitianu et al. 1997; Scozzafava et al. 2001, 2002; Briganti et al. 2000). According to the IR spectra of these complexes, and taking into account the ligand behavior of acetazolamide and benzolamide discussed earlier, it has been proposed that these ligands bind metal ions through the $N_{sulfonamide}$ and the $N_{thiadiazole}$ atoms in a bidentate fashion (Supuran 1992,

1995a, 1998a, 1998b, 1999; Almajan and Supuran 1997; Jitianu et al. 1997; Scoz-zafava et al. 2001, 2002; Briganti et al. 2000).

Borja et al. (1998) described the crystal structure of $[Zn(ats)_2(NH_3)] \cdot H_2O$, where Hats is 5-amino-1,3,4-thiadiazole-2-sulfonamide **6.26** (the acetazolamide precursor). In this complex, the sulfonamide deprotonated ligand ats acts as a bridge between two Zn ions, binding them through the $N_{sulfonamide}$ and the $N_{thiadiazole}$ atoms. Later, Chufan et al. (2001) reported the crystal structures of $[Cu(ats^-)_2(dipn)]$ and $[Ni(dien)_2](ats^-)Cl \cdot H_2O$ (dipn = dipropylentriamine; dien = diethylenediamine). In the latter complex, ats acts as a counterion. Pedregosa et al. (1995) also reported the crystal structure of the $[Cu(B\text{-}ats^-)(NH_3)_2]_2$ complex, where B-ats$^-$ is the anion of the 5-tertbutyloxycarbonylamido-1,3,4-thiadiazole-2-sulfonamide **6.27**. This ligand behaves as a bidentate ligand through the $N_{sulfonamide}$ and the $N_{thiadiazole}$.

Scozzafava et al. (2001, 2002) have reported water-soluble metal complexes of sulfonamides incorporating polyamino-polycarboxylate tails of types **6.28** and **6.29** (as well as many congeners with tails other than those of EDTA and DTPA shown in these structures). It has been hypothesized that such sulfonamides behave as ligands coordinating the metal ion through the $O_{carboxylate}$ moieties of the polyamino-polycarboxylate tails, without the involvement of the sulfonamide moieties in the complexation, but no crystal structures of such derivatives are available at present.

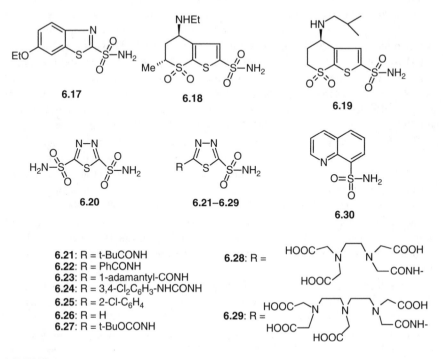

6.21: R = t-BuCONH
6.22: R = PhCONH
6.23: R = 1-adamantyl-CONH
6.24: R = 3,4-Cl$_2$C$_6$H$_3$-NHCONH
6.25: R = 2-Cl-C$_6$H$_4$
6.26: R = H
6.27: R = t-BuOCONH

6.28: R =
6.29: R =

SCHEME 6.4

Sumalan et al. (1996) described the crystal structure of the $[Zn(sa)(NH_3)] \cdot NH_3$ complex ([sa$^-$ = anion of the 8-quinolinsulfonamide **6.30**]). The ligand binds the metal ion bidentately through the N(sulfonamide) and the N(quinoline) atoms.

6.6   APPLICATIONS OF METAL COMPLEXES OF SULFONAMIDES IN THERAPY

The CA inhibitory properties of numerous metal sulfonamide complexes mentioned here have been investigated in detail, mostly against the red cell isozymes hCA I and hCA II, but in some cases also against the membrane-bound isozyme bCA IV (h = human, b = bovine isozyme; reviewed in Alzuet et al. 1994c and Supuran 1994). Such studies surprisingly indicated that the metal complexes are 10 to 100 times more potent inhibitors of these isozymes as compared to the corresponding parent sulfonamides, making them among the most potent CAIs ever reported, with inhibition constants in the low nanomolar–picomolar range (Alzuet et al. 1994c; Supuran 1994). It is believed that this powerful inhibition is due to a dual mechanism of action of the complexes, through sulfonamide anions and metal ions obtained in dilute solution by dissociation of the coordination compounds (Luca et al. 1991). Sulfonamidate anions formed this way then bind to the Zn(II) ion within the enzyme active site, whereas the metal ions block the proton shuttle residues of CA, for instance His 64 for hCA II (Alzuet et al. 1994c; Supuran 1994; Supuran et al. 2003).

As a consequence of these very powerful enzyme inhibitory properties, several interesting applications have been reported for some metal complexes of heterocyclic sulfonamides possessing strong CA inhibitory properties. Thus, some Zn(II) and Cu(II) complexes of heterocyclic sulfonamides of type **6.20** to **6.25** and **6.28**, **6.29** were very efficient intraocular pressure (IOP) lowering agents when administered topically in normotensive or glaucomatous rabbits, although most of the parent sulfonamides from which they were obtained do not show topical antiglaucoma activity (Supuran et al. 1998a, 1999; Briganti et al. 2000; Scozzafava et al. 2001, 2002). The observed topical activity has been explained by a modulation by the metal ion on the physicochemical properties of the complex, which in some cases becomes more polar and thus penetrates better though the cornea to inhibit ciliary processes CAs, thereby reducing elevated IOP in animal models of glaucoma (Supuran et al. 1998a, 1999; Briganti et al. 2000; Scozzafava et al. 2001, 2002).

Some Al(III) complexes, such as the benzolamide complex, act as efficient antisecretory agents in dogs. It has been proposed that the Zn(II), Mg(II) and Al(III) sulfonamide complexes might constitute a new class of antiulcer agents (Scozzafava et al. 2000).

It has been reported that copper(II) complexes of acetazolamide and methazolamide are potent anticonvulsant agents. Their activity is higher than that shown by the parent sulfonamides (Alzuet et al. 1994b).

Mastrolorenzo et al. (2000a, 2000b) reported that some silver(I) complexes of sulfonamide CAIs, structurally related to acetazolamide and benzolamide (possessing potent CA inhibitory properties), also show very strong antifungal properties, explaining the use of such pharmacological agents for the treatment of burns.

All these data show that the metal complexes of sulfonamide CAIs have a largely unexplored potential for the design of pharmacological agents with a host of biological activities. By choosing different metal ions and diverse sulfonamides, it is possible to fine tune the biological activity in a manner rarely attainable by other classical techniques of drug design.

REFERENCES

Almajan, L.G., and Supuran, C.T. (1997) Carbonic anhydrase inhibitors. Part 30. Complexes of 5-pivaloylamido-1,3,4-thiadiazole-2-sulfonamide with trivalent metal ions. *Revue Roumaine de Chimie* **42**, 593–597.

Alzuet, G., Casanova, J., Borrás, J., García Granda, S., Gutiérrez-Rodríguez, A., and Supuran, C.T. (1998) Copper complexes modelling the interaction between benzolamide and Cu-substituted carbonic anhydrase: Crystal structure of Cu(bz)(NH$_3$)$_4$ complex. *Inorganica Chimica Acta* **273**, 334–338.

Alzuet, G., Casanova, J., Ramírez, J.A, Borrás, J., and Carugo O. (1995) Metal complexes of the carbonic anhydrase inhibitor methazolamide: Crystal structure of the Zn(macm)$_2$(NH$_3$)$_2$. Anticonvulsant properties of Cu(macm)$_2$(NH$_3$)$_3$(H$_2$O). *Journal of Inorganic Biochemistry* **57**, 219–234.

Alzuet, G., Casella, L., Perotti A., and Borrás, J. (1994a) Acetazolamide binding to Zn(II), Co(II) and Cu(II) Model complexes of carbonic anhydrase. *Journal of the Chemical Society — Dalton Transactions*, 2347–2351.

Alzuet, G., Ferrer, S., and Borrás, J. (1991) Acetazolamide-M(II) [M(II) = Co(II), Ni(II) and Cu(II)] complexes with ethylamine, diethylamine, triethylamine and potassium hydroxide. *Journal of Inorganic Biochemistry* **42**, 79–86.

Alzuet, G., Ferrer, S., Borrás, J., Castiñeiras, A., Solans, X., and Font-Bardía M. (1992) Coordination compounds of methazolamide: Synthesis, spectroscopic studies and crystal Structures of [M(macm)$_2$(py)$_2$(OH$_2$)$_2$] [M=Co(II), Ni(II) and Cu(II)]. *Polyhedron* **22**, 2849–2856.

Alzuet, G., Ferrer, S., Borrás, J., Solans, X., and Font-Bardía, M. (1993a) Coordination behaviour of methazolamide [N(-4-methyl-2-sulfamoyl-Δ^2-1,3,4-thiadiazolin-5-ylidene)] acetamide, an inhibitor of carbonic anhydrase enzyme: Synthesis, crystal structure and properties of bi(methazolamidate)tetrammine nickel (II). *Inorganica Chimica Acta* **203**, 257–261.

Alzuet, G., Ferrer, S., Borrás, J., and Sorenson, J.R.J. (1994b) Anticonvulsant properties of copper acetazolamide complexes. *Journal of Inorganic Biochemistry* **55**, 147–151.

Alzuet, G., Ferrer, S., Borrás, J., and Supuran C.T. (1994c) Complexes of heterocyclic sulfonamides: A class of potent, dual carbonic anhydrase inhibitors. *Romanian Chemical Quarterly Reviews* **2**, 283–300.

Alzuet, G., Ferrer, S., Casanova, J., Borrás, J., and Castiñeiras, A. (1993b) A Co(III) complex of carbonic anhydrase inhibitor methazolamide and the amino-imino "aib" ligand formed by reaction of acetone and ammonia. *Inorganica Chimica Acta* **205**, 79–84.

Alzuet, G., Ferrer-Llusar, S., Borrás, J., and Martínez-Mánez, R. (2000) New Cu(II) and Zn(II) complexes of benzolamide with diethylentriamine: synthesis, spectroscopy and x-ray structures. *Polyhedron* **19**, 725–730.

Alzuet, G., Ferrer-Llusar, S., Borrás, J., Server-Carrio, J., and Martínez-Mánez, R. (1999) Co-ordinative versatility of the carbonic anhydrase inhibitor benzolamide in zinc and copper model compounds. *Journal of Inorganic Biochemistry* **75**, 189–198.

Andruh, M., Cristurean, E., Stefan, R., and Supuran, C.T. (1991) Carbonic anhydrase inhibitors. Part 6. Novel coordination compounds of Pd(II), Pt(II) and Ni(II) with 6-ethoxy-benzothiazole-2-sulfonamide. *Revue Roumaine de Chimie* **36**, 727–732.

Borja, P., Alzuet, G., Casanova, J., Server-Carrió, J., Borrás, J., Martinez-Ripoll, M., and Supuran, C.T. (1998) Zn complexes of carbonic anhydrase inhibitors: Crystal structure of [Zn(5-amino-1,3,4,-thiadiazole-2-sulfonamidate)$_2$(NH$_3$)] · H$_2$O. Carbonic anhydrase inhibitory activity. *Main Group Metal Chemistry* **21**, 279–292.

Braun, C.E., and Towle, J.L. (1941) N^1-silver(I) derivatives of sulfanilamide and some related compounds. *Journal of the American Chemical Society* **63**, 3523.

Briganti, F., Tilli, S., Mincione, G., Mincione, F., Menabuoni, L., and Supuran, C.T. (2000) Carbonic anhydrase inhibitors: Metal complexes of 5-(2-chlorophenyl)-1,3,4-thiadiazole-2-sulfonamide with topical intraocular pressure-lowering properties — The influence of metal ions upon the pharmacological activity. *Journal of Enzyme Inhibition* **15**, 185–200.

Bult, A. (1983) Metal Complexes of sulfanilamides. In *Metal Ions in Biological Systems*, Sigel, H. and Sigel, A. (Eds.), Marcel Dekker, New York, pp. 261–268 (and references cited therein).

Casini, A., Scozzafava, A., Mastrolorenzo, A., and Supuran, C.T. (2002) Sulfonamides and sulfonylated derivatives as anticancer agents. *Current Cancer Drug Targets* **2**, 55–75.

Chufán, E.E., García-Granda, S., Diaz, M.R., Borrás, J., and Pedregosa, J.C. (2001) Several coordination modes of 5-amino-1,3,4-thiadiazole-2-sulfonamide (Hats) with Cu(II), Zn(II) and Zn(II): Mimetic ternary complexes of carbonic anhydrase inhibitor. *Journal of Coordination Chemistry* **54**, 303–312.

Coleman, J.E. (1975) Carbonic Anhydrase. In *Inorganic Biochemistry*, Eichorn, G.L. (Ed.)., Elsevier, New York, pp. 488–544 (and references cited therein).

Curtis, N.F. (1963) Macrocyclic coordination compounds formed by condensation of metal-amine complexes with aliphatic carbonyl compounds. *Coordination Chemistry Reviews* **3**, 3–47.

Domagk, G. (1935) Ein Beitrag zur Chemotherapie der Bakteriellen Infektionen. *Deutsches Medizinische Wochenschricht* **61**, 250–253.

Ferrer, S., Alzuet, G., and Borrás, J. (1989a) Synthesis and characterization of acetazolamide complexes of Ni(II), Hg(II) and Ag(I). *Journal of Inorganic Biochemistry* **37**, 163–174.

Ferrer, S., Borrás, J., and García-España, E. (1990a) Complex formation equilibria between acetazolamide (5-acetamido-1,3,4-thiadiazole-2-sulfonamide), a potent inhibitor of carbonic anhydrase, and Zn(II), Co(II), Ni(II) and Cu(II) in aqueous and ethanol-aqueous solutions. *Journal of Inorganic Biochemistry* **39**, 297–306.

Ferrer, S., Borrás, J., Miratvilles, C., and Fuertes, A. (1989b) Coordination behavior of acetazolamide (5-acetamido-1,3,4-thiadiazole-2-sulfonamide): synthesis, crystal structure, and properties of bis(acetazolamidato)tetraamminenickel(II). *Inorganic Chemistry* **28**, 160–163.

Ferrer, S., Borrás, J., Miratvilles, C., and Fuertes, A. (1990b) Synthesis and characterization of copper(II)-acetazolamide (5-acetamido-1,3,4-thiadiazole-2-sulfonamide) complexes: Crystal structure of dimeric [Cu(Acm)(NH$_3$)$_2$(OH$_2$)]$_2$ ·2H$_2$O. *Inorganic Chemistry* **29**, 206–210.

Ferrer, S., Haasnoot, J.G., de Graaff, R.A.G., Reedijk, J., and Borrás, J. (1992) Synthesis, crystal structure and properties of two acetazolamide (5-acetamido-1,3,4-thiadiazole-2-sulfonamide) complexes: bis (5-acetamidato-1,3,4-thiadiazole-2-sulfonamide-*O*)bis(1,2-ethanediamine) copper(II) and bis (5-acetamidato-1,3,4-thiadiazole-2-sulfonamide-*N*)bis(1,3-propanediamine) copper(II); an unusually weak ambidentate ligand. *Inorganica Chimica Acta* **192**, 129–138.

Ferrer, S., Jiménez, A., and Borrás, J. (1987) Synthesis and characterization of the acetazolamide complexes of Co(II) and Zn(II). *Inorganica Chimica Acta* **129**, 103–106.

Hartmann, U., and Vahrenkamp H. (1991) A zinc complex of the carbonic anhydrase inhibitor acetazolamide (aaat): crystal structure of (aaa)$_2$ Zn (NH$_3$)$_2$. *Inorganic Chemistry* **30**, 4676–4677.

Hartmann, U., and Vahremkamp, H. (1994) Zinkcomplexe von sulfonamiden. *Zeitscrift Naturforshung* Teil **B49**, 1725–1730.

Jitianu, A., Ilies, M.A., Scozzafava, A., and Supuran, C.T. (1997) Complexes with biologically active ligands. Part 8. Synthesis and carbonic anhydrase inhibitory activity of 5-benzoylamido- and 5-(3-nitro-benzoylamido)-1,3,4-thiadiazole-2-sulfonamide and their metal complexes. *Main Group Metal Chemistry* **20**, 151–156.

Kimura, E., Shiota, T., Koike, T., Shiro, M., and Kodama, M. (1990) A zinc(II) complex of 1,5,9-triazacyclododecane ([12]aneN$_3$) as a model for carbonic anhydrase. *Journal of the American Chemical Society* **112**, 5805–5811.

Luca, C., Barboiu, M., and Supuran, C.T. (1991) Carbonic anhydrase inhibitors. Part 7. Stability constants of complex inhibitors and their mechanism of action. *Revue Roumaine de Chimie* **36**, 1169–1173.

Malecki, F., Staroscik, R., and Weiss-Gradzinska, W. (1984) Zur Komplexbildung in Kupfer(II)-Sulfonamid-Systemen. *Pharmazie* **39**, 158–160.

Mandell, G.L., and Petri, W.A. (1996) Antimicrobial agents. Sulfonamides, trimethoprim-sulfamethoxazole, and agents for urinary tract infections. In *Goodman's and Gilman's the Pharmacological Basis of Therapeutics*, 9th ed., Hardman, J.G., Limbird, L.E., Molinoff, P.B., Ruddon, R.W., and Gilman, A.G. (Eds.), McGraw-Hill, New York, pp. 1057–1072.

Manole, G., Maior, O., and Supuran, C.T. (1993) Carbonic anhydrase inhibitors. Part 17. Complexes of heterocyclic sulfonamides with Ru(III), Rh(III) and Au(III) are very strong dual inhibitors of isozymes I and II. *Revue Roumaine de Chimie* **38**, 475–480.

Maren, T.H. (1967) Carbonic anhydrase: chemistry, physiology and inhibition. *Physiological Reviews* **47**, 595–781.

Mastrolorenzo, A., Scozzafava, A., and Supuran, C.T. (2000a) Antifungal activity of Ag(I) and Zn(II) complexes of aminobenzolamide (5-sulfanilylamido-1,3,4-thiadiazole-2-sulfonamide) derivatives. *Journal of Enzyme Inhibition* **15**, 517–531.

Mastrolorenzo, A., Scozzafava, A., and Supuran, C.T. (2000b) Antifungal activity of silver and zinc complexes of sulfadrug derivatives incorporating arylsulfonylureido moieties. *European Journal of Pharmaceutical Sciences* **11**, 99–107.

Mathew, M., and Palenik, G.J. (1974) Crystal and molecular structure of acetazolamide (5-acetamido-1,3,4-thiadiazole-2-sulphonamide), a potent inhibitor of carbonic anhidrase. *Journal of the Chemical Society — Perkin Transactions* 2, 532–536.

Pedregosa, J.C., Casanova, J., Alzuet, G., Borrás, J., García-Granda, S., Diaz, M.R., and Gutierrez-Rodriguez, A. (1995) Metal complexes of 5-tertbutyloxycarbonylamido-1,3,4-thiadiazole-2-sulfonamide (B-H$_2$ats), a carbonic anhydrase inhibitor. Crystal structures of B-H$_2$ats and the [Cu(B-ats)(NH$_3$)$_2$]$_2$ dimer complex. *Inorganica Chimica Acta* **232**, 117–124.

Scozzafava, A., Ilies, M.A., and Supuran, C.T. (2000) Carbonic anhydrase inhibitors. Part 89. Metal complexes of benzolamide with strong enzyme inhibitory and putative antiulcer properties.*Revue Roumaine de Chimie* **45**, 771–778.

Scozzafava, A., Menabuoni, L., Mincione, F., Mincione, G., and Supuran, C.T. (2001) Carbonic anhydrase inhibitors: Synthesis of sulfonamides incorporating dtpa tails and of their zinc complexes with powerful topical antiglaucoma properties. *Bioorganic and Medicinal Chemistry Letters* **11**, 575–582.

Scozzafava, A., Menabuoni, L., Mincione, F., and Supuran, C.T. (2002) Carbonic anhydrase inhibitors: A general approach for the preparation of water soluble sulfonamides incorporating polyamino-polycarboxylate tails and of their metal complexes possessing long lasting, topical intraocular pressure lowering properties. *Journal of Medicinal Chemistry* **45**, 1466–1476.

Scozzafava, A., Owa, T., Mastrolorenzo, A., and Supuran, C.T. (2003) Anticancer and antiviral sulfonamides. *Current Medicinal Chemistry*, **10**, 925–953.

Sumalan, S.L., Casanova, J., Alzuet, G., Borrás, J., Castiñeiras, A., and Supuran, C.T. (1996) Metal complexes of carbonic anhydrase inhibitors: Synthesis and characterization of M(II)-8-quinolinsulfonamidato(sa) complexes (M=Co,Ni,Cu and Zn). Crystal structure of [Zn(sa)(NH₃)]. NH₃ complex. Carbonic anhydrase inhibitory properties. *Journal of Inorganic Biochemistry* **62**, 31–39.

Supuran, C.T. (1992) Carbonic anhydrase inhibitors. Part 13. Complex-type mechanism-based inhibitors. *Revue Roumaine de Chimie* **37**, 849–855.

Supuran, C.T. (1993) Carbonic anhydrase inhibitors. Part 16. Complex inhibitors containing metals in high oxidation states (V(IV); Ce(IV); Th(IV); U(VI)). *Revue Roumaine de Chimie* **38**, 229–236.

Supuran, C.T. (1994) Carbonic anhydrase inhibitors. In *Carbonic Anhydrase and Modulation of Physiologic and Pathologic Processes in the Organism*, Puscas, I. (Ed.), Helicon Press, Timisoara, Romania, pp. 29–111.

Supuran, C.T. (1995a) Metal complexes of 1,3,4-thiadiazole-2,5-disulfonamide are strong dual carbonic anhydrase inhibitors, although the ligand possesses very weak such properties. *Metal Based Drugs* **2**, 331–336.

Supuran, C.T. (1995b) Thienothiopyransulfonamides as complexing agents for the preparation of dual carbonic anhydrase inhibitors. *Metal Based Drugs* **2**, 327–330.

Supuran, C.T. (1996) Carbonic anhydrase inhibitors. Part 25. Thienothiopyransulfonamides — a novel class of complexing agents for the preparation of dual enzyme inhibitors. *Revue Roumaine de Chimie* **41**, 495–499.

Supuran, C.T. (2003) Indisulam: An anticancer sulfonamide in clinical development. *Expert Opinion on Investigational Drugs* **12**, 283–287.

Supuran, C.T., and Andruh, M. (1994) Carbonic anhydrase inhibitors. Part 18. Coordination compounds of heterocyclic sulfonamides with main group trivalent cations are potent isozyme II inhibitors. *Revue Roumaine de Chimie* **39**, 1229–1234.

Supuran, C. T., Andruh, M., and Puscas, I. (1990) Carbonic anhydrase inhibitors. Part 1. Metal complexes of sulfonamides — A novel class of carbonic anhydrase inhibitors. *Revue Roumaine de Chimie* **35**, 393–398.

Supuran, C.T., Loloiu, G., and Manole, G. (1993a) Carbonic anhydrase inhibitors. Part 15. Complex inhibitors containing main-group and transitional divalent cations. *Revue Roumaine de Chimie* **38**, 115–122.

Supuran, C.T., Manole, G., and Andruh, M. (1993b) Carbonic anhydrase inhibitors. Part 11. Coordination compounds of heterocyclic sulfonamides with lanthanides are potent inhibitors of isozymes I and II. *Journal of Inorganic Biochemistry* **49**, 97–103.

Supuran, C.T., Manole, G., and Manzatu, I. (1992) Carbonic anhydrase inhibitors. Part 12. Lanthanide complexes with acetazolamide as dual inhibitors. *Revue Roumaine de Chimie* **37**, 739–744.

Supuran, C.T., Mincione, F., Scozzafava, A., Briganti, F., Mincione, G., and Ilies, M.A. (1998a) Carbonic anhydrase inhibitors. Part 52. Metal complexes of heterocyclic sulfonamides: A new class of strong topical intraocular pressure-lowering agents with potential use as antiglaucoma drugs. *European Journal of Medicinal Chemistry* **33**, 247–254.

Supuran, C.T., Olar, R., Marinescu, D., and Brezeanu, M. (1993c) Enzyme mimics. Part 1. Mimicking the interaction of inhibitors with the carbonic anhydrase active site. *Romanian Chemical Quarterly Reviews* **1**, 193–210.

Supuran, C.T., and Scozzafava, A. (2000) Carbonic anhydrase inhibitors and their therapeutic potential. *Expert Opinion on Therapeutic Patents* **10**, 575–600.

Supuran, C.T., and Scozzafava, A. (2001) Carbonic anhydrase inhibitors. *Current Medicinal Chemistry — Immunologic, Endocrine and Metabolic Agents* **1**, 61-97 (and references cited therein).

Supuran, C.T., Scozzafava, A., and Casini, A. (2003) Carbonic anhydrase inhibitors. *Medicinal Research Reviews* **23**, 146–189.

Supuran, C.T., Scozzafava, A., Mincione, F., Menabuoni, L., Briganti, F., Mincione, G., and Jitianu, M. (1999) Carbonic anhydrase inhibitors. Part 60. The topical intraocular pressure-lowering properties of metal complexes of a heterocyclic sulfonamide: influence of the metal ion upon biological activity. *European Journal of Medicinal Chemistry* **34**, 585–594.

Supuran, C.T., Scozzafava, A., Saramet, I., and Banciu, M.D. (1998b) Carbonic anhydrase inhibitors. Inhibition of isozymes I, II and IV with heterocyclic mercaptans, sulfenamides, sulfonamides and their metal complexes. *Journal of Enzyme Inhibition* **13**, 177–194.

Supuran, C.T., Stefan, R., Manole, G., Puscas, I., and Andruh, M. (1991) Carbonic anhydrase inhibitors. Part 8. Complexes of ethoxzolamide with lanthanides are powerful inhibitors of isozymes I and II. *Revue Roumaine de Chimie* **36**, 1175–1179.

Vidgren, J., Liljas, A., and Walker, N.P. (1990) Refined structure of the acetazolamide complex of human carbonic anhydrase II at 1.9Å. *International Journal of Biological Macromolecules* **12**, 342–348.

7 Nonsulfonamide Carbonic Anhydrase Inhibitors

Marc Antoniu Ilies and Mircea Desideriu Banciu

CONTENTS

7.1　Introduction 209
7.2　Inhibition of CAs by Anions 210
7.3　Other Types of Nonsulfonamide Inhibitors 228
　　　7.3.1　Inhibitors of the Proton Shuttle 228
　　　7.3.2　Organic Sulfamates and Hydroxamates as CAIs 229
References 233

The study of carbonic anhydrase inhibitors (CAIs) followed and completed the biochemical and physiological investigations on the enzyme after its discovery. Besides sulfonamides, which act as powerful CAIs, inorganic anions inhibit the metalloprotein with sufficient potency to generate significant physiological consequences. Their diverse inhibition properties against different carbonic anhydrase isozymes are discussed, from both the kinetical and structural points of view. Other classes of nonsulfonamide CAIs, some of them recently discovered, are also reviewed.

7.1 INTRODUCTION

The inhibition of carbonic anhydrase (CA) was investigated almost immediately after its discovery (Meldrum and Roughton 1933; Forster 2000), and was decisive in understanding its functioning (Maren and Sanyal 1983). Since then, many carbonic anhydrase inhibitors (CAIs) have been discovered or invented; the field has been periodically reviewed (Bertini et al. 1982; Coleman 1975; Dodgson et al. 1991; Mansoor et al. 2000; Maren 1967, 2000; Maren and Sanyal 1983; Pocker and Sarkanen 1978; Silverman and Lindskog 1988; Supuran 1994; Supuran and Manole 1999; Supuran and Scozzafava 2000, 2002b) for the different applications of such inhibitors in either understanding the CA catalytic mechanism and its physiological role or for their therapeutic use.

The catalytic mechanism of CA has been the subject of debate for a long time (Lindskog 1982; Lindskog and Coleman 1973; Silverman and Lindskog 1988), mainly regarding the nature of the active species within the enzyme cavity. At present, it is generally accepted (Lindskog and Silverman 2000; Silverman and Lindskog 1988) that the Zn-coordinated hydroxyl ion (in the basic form of the enzyme) represents the nucleophilic species that attacks the substrate (CO_2), yielding a bicarbonate ion. The HCO_3^- is then displaced by a water molecule, generating the acidic form of the enzyme, which is catalytically inactive. The acidic form is converted back into the active, basic form by transferring a proton to the external buffer via a proton shuttle [identified as His-64 in CA II (Tu et al. 1981)] and probably via an interlocking histidine cluster in this and other isozymes (Briganti et al. 1997).

CAIs interfere with the catalytic mechanism in different ways, depending on the nature of the inhibitor (its electronic properties), the properties of the environment in which the enzyme and the inhibitor interact (pH, ionic strength, etc.), and the isoenzyme type (Bertini and Luchinat 1983; Bertini et al. 1982; Coleman 1975; Mansoor et al. 2000; Maren and Sanyal 1983; Supuran 1994; Supuran and Manole 1999; Supuran and Scozzafava 2000, 2002b). Fourteen different CA isozymes or CA-related proteins have been described to date in higher vertebrates, including humans, with different localizations, functions and inhibition properties (Chegwidden and Carter 2000; Supuran and Scozzafava 2000). Among them, the most studied are isozymes CA I, CA II, CA III and CA IV; therefore, most of the data available concern these isoforms.

The large majority of CAIs fall into two categories: monovalent anions and sulfonamides. Despite major structural differences, they share common inhibition kinetics, behaving as noncompetitive inhibitors with the natural substrate, carbon dioxide (Bertini et al. 1982; Lindskog and Coleman 1973; Maren and Sanyal 1983). They do not dock within the hydrophobic pocket and usually are bound directly to the Zn(II) ion into the active site of the enzyme.

7.2 INHIBITION OF CAs BY ANIONS

Anions were discovered to act as inhibitors when the enzyme was first discovered (Meldrum and Roughton 1933) and systematic investigations followed thereafter (Kernohan 1964, 1965; Lindskog 1966; Pocker and Stone 1965, 1967, 1968; Roughton and Booth 1946; Maren and Sanyal 1983). The common conclusion from these studies was that the anionic inhibition is noncompetitive for both the hydrase and the esterase activity of the enzyme (Pocker and Stone 1968), but competitive with regard to the anionic product bicarbonate (Lindskog et al. 1971). However, the isozyme susceptibility for anionic inhibitors is different. CA I is more susceptible to inhibition by anions than is CA II (Maren et al. 1976), which shares a common behavior with CA IV (Baird et al. 1997; Maren and Conroy 1993) and V (Heck et al. 1994). CAs III, VI and IX have an intermediate position toward anion inhibition (Engberg and Lindskog 1984; Murakami and Sly 1987; Rowlett et al. 1991; Wingo et al. 2001).

The inhibition mechanism is a complex topic, both in terms of kinetic aspects and in coordination behavior (Bertini and Luchinat 1983; Bertini et al. 1982; Coleman 1975; Maren and Sanyal 1983; Supuran and Manole 1999). The first important issue is that the association of the inhibitor (I) to the active site of the enzyme is pH dependent, because the enzyme exists in two forms — acidic and basic — which are in equilibrium. The equilibrium is characterized by pK_{aEnz}, which is specific for each isozyme: 7.0 for CA I (Pocker and Sarkanen 1978), 7.0 for CA II (Baird et al. 1997), 8.0 for CA III (Rowlett et al. 1991), 7.1 for CA IV (Baird et al. 1997), 7.4 for CA V (Heck et al. 1994), 6.3 for CA IX (Wingo et al. 2001). These two tetra-coordinate active species are in equilibrium with a five-coordinate adduct (Bertini and Luchinat 1983):

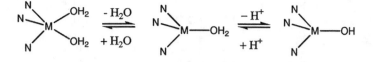

This is another important aspect for the binding of inhibitors for both kinetic and structural considerations (Bertini and Luchinat 1983).

Considering the interaction of the enzyme with the inhibitor

$$E + I \rightleftharpoons EI \qquad (7.1)$$

characterized by the inhibition constant $K_I = [EI]/[E][I]$, one can calculate the apparent affinity constant of the inhibitor (K_{app}) by Equation 7.2 (Bertini et al. 1982):

$$K_{app} = \frac{K_I}{1 + \dfrac{K_{aEnz}}{[H^+]}} \qquad (7.2)$$

The apparent affinity constant allows the correction of K_I for the pH variations during the enzyme–inhibitor interaction, taking into account the contribution of the enzyme and the intrinsic acidity of the inhibitor. For the inhibitors with K_{aIn} within 5.5 to 10.5 pH units, an additional correction must be introduced (Bertini et al. 1982; Pocker and Sarkanen 1978):

$$K_{app} = \frac{K_I}{1 + \dfrac{K_{aEnz}}{[H^+]}} \frac{1}{1 + \dfrac{[H^+]}{K_{aIn}}} \qquad (7.3)$$

Table 7.1 presents the K_{app} values for the most important anionic inhibitors against erythrocyte CA I and CA II isozymes.

The data of Table 7.1 reveal that for hCA II, HO^- has the highest affinity for the Zn(II) ion, reaching the potency of sulfonamide inhibitors. It is followed by CN^-,

TABLE 7.1
Apparent Affinity Constants of Some Anionic Inhibitors for hCA I and II, bCA II and Co(II)-substituted CA II

Anion	$pK_{app} = -lg\ K_{app}$			
	hCA I	hCA II	bCA II	Co-CA II
HO^-	—	—	6.5	—
HS^-	—	—	4.95	5.8
F^-	0.40	—	–0.08	1.6
Cl^-	1.30	1.17	0.72	1.7
Br^-	1.64	1.57	1.18	2.1
I^-	2.70	2.51	2.06	3.04
CN^-	—	—	5.58	>5
CNO^-	—	—	3.96	5.15
SCN^-	3.15	3.09	4.04	3.8
N_3^-	—	—	3.23	3.6
HCO_3^-	—	—	1.58	1.0
HSO_3^-	—	—	1.52	>5
NO_3^-	1.82	1.74	1.32	3.52
ClO_4^-	2.89	2.82	1.80	—
$HCOO^-$	—	1.74	—	2.68
CH_3COO^-	1.58	1.47	1.07	2.1
FCH_2COO^-	—	0.89	—	1.92
F_2CHCOO^-	—	1.30	—	2.15
F_3CCOO^-	—	1.96	—	1.96
$C_2O_4^{2-}$	—	1.74	—	2.5

Source: From Bertini, I. et al. (1981) *Journal of the American Chemical Society* **103**, 7784–7788; Bertini, I. et al. (1978a) *Journal of the American Chemical Society* **100**, 4873–4877; Bertini, I. et al. (1982) *Structure and Bonding* **48**, 45–92; Supuran, C.T., and Manole, G. (1999) *The Carbonic Anhydrase Inhibitors: Syntheses, Reactions and Therapeutical Applications* (in Romanian), Romanian Academy Publishing House, Bucharest.

HS^-, SCN^-, CNO^- and N_3^-, all class B ligands, known to have a high affinity for complexing transitional metal ions. Interestingly, class A ligands, such as ClO_4^- and NO_3^- still have a good inhibitory activity against these CA isozymes, particularly for isozyme I (Supuran and Manole 1999).

A wider picture of the CA isozyme susceptibility to anions could be compiled (in terms of K_I) only for some anionic species due to limited availability of data (results in Table 7.2).

As a general trend, it can be observed that the membrane-anchored CA IV is more resistant to anionic inhibition than cytosolic isozyme CA II. In contrast,

TABLE 7.2
Inhibition Constants of Some Anionic Inhibitors against Different CA Isozymes for the CO_2 Hydration Reaction Catalyzed by These Isozymes

Anion	K_I (mM)						
	hCA I	hCA II	mCA III	hCA IV	mCA V	hCA VI	hCA IX
F⁻	>300[a]	>300[a]	—	—	—	—	—
Cl⁻	6[b]	200[b]	6[b]	36[c]	—	14[g]	—
Br⁻	4[a]	63[d]	—	52[c]	—	—	—
I⁻	0.3[b]	26[b] (35)[i]	1.1[b]	11[c]	56[g]	—	—
CNO⁻	0.0007[e]	0.03[e]	—	0.03[e]	0.03[f]	—	0.046[h]
SCN⁻	0.2[a]	1.6[i]	—	—	—	—	—
N₃⁻	—	1.5[i]	—	—	—	—	—
HCO₃⁻	12[a]	85[c]	—	44[c]	—	—	—
HPO₄²⁻	3[a]	36[a]	<100[j]	27[c]	—	—	—
HCOO⁻	—	20[d]	—	6[c]	—	—	—
CH₃COO⁻	7[a]	79[d]	—	22[c]	—	—	—
SO₄²⁻	—	>200[c]	1.1[e]	44[c]	—	—	—
ClO₄⁻	3.6[a]	1.3[a]	—	—	—	—	—
NO₃⁻	7[a]	35[a]	—	—	—	—	—

[a] From Maren, T.H. et al. (1976) *Science* **191**, 469–472.

[b] From Maren, T.H., and Sanyal, G. (1983) *Annual Review of Pharmacology and Toxicology* **23**, 439–459.

[c] From Baird, T.T., Jr. et al. (1997) *Biochemistry* **36**, 2669–2678.

[d] From Liljas, A. et al. (1994) *European Journal of Biochemistry* **219**, 1–10.

[e] From Rowlett, R.S. et al. (1991) *Journal of Biological Chemistry* **266**, 933–941.

[f] From Heck, R.W. et al. (1994) *Journal of Biological Chemistry* **269**, 24742–27446.

[g] From Murakami, H., and Sly, W.S. (1987) *Journal of Biological Chemistry* **262**, 1382–1388.

[h] From Wingo, T. et al. (2001) *Biochemical and Biophysical Research Communications* **288**, 666–669.

[i] From Tibell, L. et al. (1984) *Biochimica et Biophysica Acta* **789**, 302–310.

[j] From Paranawithana, S.R. et al. (1990) *Journal of Biological Chemistry* **265**, 22270–22274.

isozymes CA III and CA I are more susceptible to be inhibited by halides than is CA II. Mitochondrial isozyme CA V behaves similar to CA II and slightly lower than tumoral isozyme CA IX against CNO⁻, whereas CA I is 100 times more susceptible to be inhibited with this ion (data for CAs V, VI and IX very limited, as seen from Table 7.2). Other notable facts are the resistance of CA III (mainly located in liver and muscles, organs with very dynamic metabolism) to be inhibited by phosphate (but not by the isosteric dianionic sulfate), and the relatively good resistance of salivary isozyme CA VI to chloride and iodide, which are common ingredients of the diet.

Taking into account the pH dependence for the binding of inhibitors to native and Co(II)-substituted CA, Bertini and Luchinat (1983) proposed a classification of CAIs into three classes:

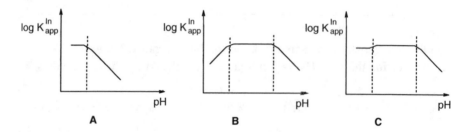

FIGURE 7.1 The pH dependence of the apparent affinity constants of classes A, B and C carbonic anhydrase inhibitors; the dashed lines represent the enzyme pK_a (left) and the inhibitor pK_a (right). (Adapted from Bertini, I., and Luchinat, C. 1983. *Accounts of Chemical Research* **16,** 272–279.)

Class A Inhibitors have a pH dependence of K_{app} as in Figure 7.1A and include mononegative anions that are conjugated bases of strong acids (X^-, NO_3^-, ClO_4^-, CNO^-/NCO^-, SCN^-, etc.; Coleman 1975). They formally bind the low pH form of the enzyme, their decrease in affinity at high pH being the result of the formal competition with the hydroxide ion. A similar behavior is shown by aniline (Bertini et al. 1977a), *N*-methylimidazole (Alberti et al. 1981; bind the low-pH form of the enzyme as neutral species) and by the bicarbonate ion (Bertini and Luchinat 1983).

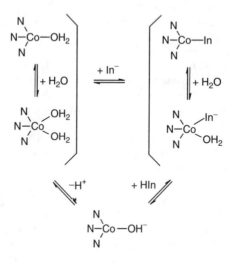

SCHEME 7.1

Class B Inhibitors display a bell-shaped curve for log K_{app} vs. pH (Figure 7.1B.). The anionic form of the inhibitor binds the low-pH form of the enzyme, or, alternatively, the inhibitor can bind the high-pH form of the enzyme in a neutral form (Scheme 7.1). This class includes CN^- (Thorslund and Lindskog 1967), HS^- (Pocker and Stone 1968; Thorslund and Lindskog 1967), the sulfonamides (Kernohan 1966; Lindskog and Thorslund 1968;

Taylor et al. 1970) and also the hydrated trichloroacetaldehyde (Bertini et al. 1979a) and α-aminoacids in their zwitterionic form (Bertini et al. 1977b).

Class C Inhibitors have an intermediate pH dependence of K_{app} (between Class A and Class B; Figure 7.1C). They can bind to the active site over a wide range of pH values, owing to their particular structural and ionizing properties (Scheme 7.2.). This class includes imidazole, 1,2,4-triazole and 1,2,3-triazole (Alberti et al. 1981). The behavior of these ligands clearly reveals the direct influence of the binding mode on the inhibitory properties of the compound.

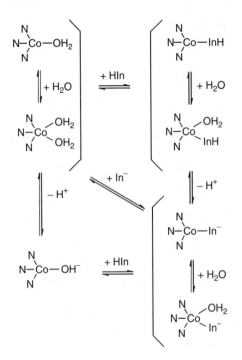

SCHEME 7.2

In general, both anions and sulfonamides bind to CA (as well as to Co-CA) in the stoichiometric ratio of 1:1 (Bertini and Luchinat 1983; Bertini et al. 1982). On binding, the electronic and magnetic properties of both the metal ion and the inhibitor are altered. Consequently, coordination behavior can be revealed by electronic, NMR-, EPR- and UV-VIS spectra (Bertini and Luchinat 1983). Ultimate details are provided by x-ray diffraction experiments on enzyme–inhibitor adducts, but because of the inherent difficulties associated with this technique, data are available for only a limited number of such inhibitors.

A basic fact revealed by ^{1}H-NMR studies on the cobalt protein is that the histidines remain coordinated on binding of inhibitors (Bertini et al. 1981). As a direct consequence, inhibitors can give rise to tetrahedral species, to five-coordinated species or to equilibrium between the two (Bertini and Luchinat 1983):

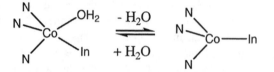

The position of the equilibrium can be evaluated by electronic spectroscopy, by monitoring the intensity of the overall absorption spectrum, or the presence or absence of a weak band, in the range of 13 to 15×10^3 cm^{-1} for Co-CA, characteristic of the five-coordinated complexes (Bertini et al. 1980a, 1978a). Similar results were obtained from EPR spectra (Bencini et al. 1981) performed on Cu(II)-CA (Taylor and Coleman 1971), V(IV)-CA (Bertini et al. 1979b; Fitzgerald and Chasteen 1974) and Mn(II)-CA (Bertini et al. 1978b; Lanir et al. 1973; Meirovitch and Lanir 1978). Also, a very powerful technique for studying the inhibition mechanism of CA is nuclear magnetic resonance spectroscopy. The large majority of the studies were done on Co(II)-CA (Alberti et al. 1981; Banci et al. 1990, 1989; Bertini et al. 1981; Luchinat et al. 1990; Moratal et al. 1992a), Ni(II)-CA (Moratal et al. 1992b), Cd(II)-CA (Armitage et al. 1978; Jarvet et al. 1989), as well as on the native enzyme (Jarvet et al. 1989).

The correlated results of all these studies allow a good overview (Bertini and Luchinat 1983) of the coordination behavior of different anionic and neutral inhibitors to the Zn(II) ion of the active site. In light of these experiments, anions such as CN$^-$, NCO$^-$, SH$^-$, as well as ligands such as aniline, anthranilate, trichloroacetaldehyde, thiadiazole, imidazole (at high pH), 1,2,4-triazole, tetrazole and all the sulfonamides generate pseudotetrahedral species. The anions SCN$^-$, HSO$_3^-$, NO$_3^-$, I$^-$, Au(CN)$_2^-$, Ag(CN)$_2^-$, formate, acetate, bromoacetate, oxalate, malonate, succinate, glutarate and ligands such as glycine, L(+)-alanine, D(–)-alanine, 2,4-pentanedione and 1,2,3-triazole yield five-coordinated species.

Between these two extremes are HCO$_3^-$, F$^-$, Cl$^-$, Br$^-$, N$_3^-$, phosphate and benzoate anions and imidazole at low pH, all generating four- to five-coordinated species, as shown by the electronic spectra and NMR experiments on the Co-CA adducts (Banci et al. 1989; Bertini et al. 1981, 1978a, 1979b, 1977a, 1982; Bertini and Luchinat 1983, 1984), which can be extrapolated to the native enzyme (Bertini and Luchinat 1983; Supuran and Manole 1999).

These results are supported, in general terms, by x-ray crystallographic data. Pseudotetrahedral coordination was confirmed for HS$^-$ (Mangani and Haakansson 1992), 1,2,4-triazole (Mangani and Liljas 1993) and sulfonamides (Eriksson et al. 1988b; Haakansson and Liljas 1994; Vidgren et al. 1990, 1993). Thus, for HS$^-$, Mangani and Haakansson (1992) showed that the strong inhibitor hydrosulfide coordinates the zinc atom of hCA II at a distance of 2.2 Å, replacing the metal-bound water/hydroxide (Wat-263) of the native enzyme (Figure 7.2). The binding mode is further stabilized by a hydrogen bond donated to the Oγ of Thr-199 (similar to the case of sulfonamide binding). The zinc ion maintains its tetrahedral coordination geometry, and differences from the native structure are minimal. The structured water network within the active site is also little affected by the inhibitor binding (Mangani and Haakansson 1992).

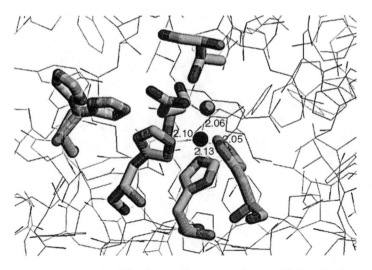

FIGURE 7.2 Coordination detail for the complex between human CA II and hydrogen sulfide (From Mangani, S. and Haakansson, K. 1992. *European Journal of Biochemistry* **210,** 867–871.); the structure was generated from the pdb file (1CAO) by using the program RasWin for Windows, version 2.7.1.1.

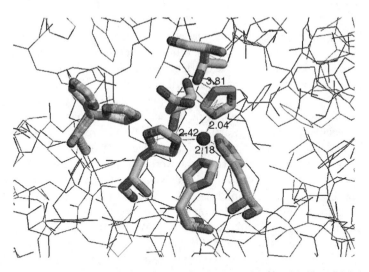

FIGURE 7.3 Coordination detail for the complex between human CA II and 1,2,4-triazole (From Mangani, S. and Liljas, A. 1993. *Journal of Molecular Biology* **232,** 9–14.); the structure was generated from the pdb file (1CRA) by using the program RasWin for Windows, version 2.7.1.1.

A related pattern is valid for the adduct of 1,2,4-triazole with hCA II (Mangani and Liljas 1993). The five-member ring of triazole coordinates to zinc through the N4 atom at a distance of 2.05 Å, displacing the water/hydroxyl ion, and generates a distorted tetrahedral geometry for the metal ion (Figure 7.3). It also forms two bent hydrogen bonds with the Oγ of Thr-200 (by N1) and the amide nitrogen atom

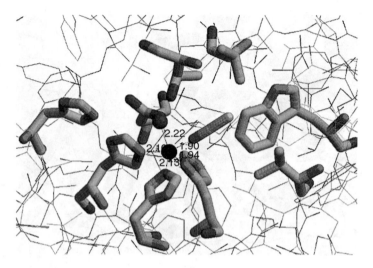

FIGURE 7.4 Coordination detail for the complex between human CA II and thiocyanate anion (From Eriksson, A.E. et al. 1988b. *Proteins: Structure, Function, and Genetics* **4**, 283–293.); the structure was generated from the pdb file (2CA2) by using the program RasWin for Windows, version 2.7.1.1.

of Thr-199 (with N2). Despite the smaller bond energy, the presence of the N2-Thr199 hydrogen bond accounts for the different coordination pattern observed for the enzyme with 1,2,4-triazole (clearly tetrahedral) as compared with the case of imidazole (tetra- to pentacoordinated; Alberti et al. 1981; Kannan et al. 1977; Luchinat et al. 1990; Mangani and Liljas 1993). In contrast to the hydrosulfide ion, the larger size of the 1,2,4-triazole molecule causes substantial changes in the network of water molecules present in the active-site cavity. Besides replacing the zinc-bound water, the inhibitor displaces the deep water molecule (Wat-338) and also another two water molecules from the active-site network. The positive entropic contribution arising from the release of these molecules also accounts in the total free energy of binding for the 1,2,4-triazole molecule (Mangani and Liljas 1993). Mention should be made that the location of this inhibitor overlaps the binding site for bicarbonate (Haakansson and Wehnert 1992; Xue et al. 1993) as well as the proposed binding site for carbon dioxide (Lindahl et al. 1993). This fact can explain the observed competitive binding of 1,2,4-triazole with respect to CO_2/HCO_3^- under equilibrium conditions (Tibell et al. 1985).

Alternatively, the binding of SCN^- ion to hCA II was shown to be a pure pentacoordinated one (Eriksson et al. 1988b). The three histidines, a water molecule and the SCN^- ion coordinate the zinc ion. The nitrogen atom of the thiocyanate is 1.9 Å away from the metal and shifted 1.3 Å with respect to the hydroxyl ion in the native structure, being at van der Waals distance from the Oγ of Thr-199 (Figure 7.4; Eriksson et al. 1988b).

The authors assigned this conformation to the inability of the Oγ1 atom of Thr-199 to serve as a hydrogen bond donor, thus repelling the nonprotonated nitrogen.

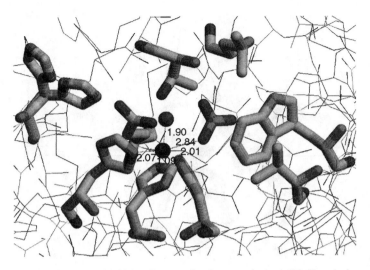

FIGURE 7.5 Coordination detail for the complex between human CA II and nitrate (From Mangani, S. and Haakansson, K. 1992. *European Journal of Biochemistry* **210**, 867–871.); the structure was generated from the pdb file (1CAN) by using the program RasWin for Windows, version 2.7.1.1.

The zinc-bound water was found to be located 2.2 Å away from the metal and 2.4 Å from the nitrogen atom of the thiocyanate ion. This coordinated water is hydrogen bonded to the Oγ1 atom of Thr-199 and to another water molecule (Wat-318). It was also shown that the sulfur atom of the SCN⁻ displaces the deep water molecule of the native enzyme and makes van der Waals interactions with Val-143, Leu-198 and Trp-209 side chains. Besides these modifications, the hydrogen network in the active site is almost identical to the native structure (Eriksson et al. 1988a, 1988b).

Mangani and Haakansson (1992) reported the structure of the hCA II–NO$_3^-$ complex and showed that this anion is also bound in a pentacoordinated geometry at the zinc ion. It shares a similar position with SCN⁻ within the active site of the enzyme, being located in the deep-water pocket of the protein (delimited by residues Val-121, Leu-141, Val-143, Leu-198, Thr-199, Val-207 and Trp-209). However, the interaction of the nitrate anion [in a monodentate fashion, which was calculated (Kumar and Marynick 1993) to be more stable than the bidentate form] is weaker than in the case of thiocyanate ion, the distance between the zinc and the ligand being 2.8 Å (Figure 7.5). The coordinated water (Wat-263) is placed 1.9 Å away from the zinc ion, slightly moved from its original position.

The same major characteristics can be found in the complexes of formate and acetate with hCA II (Haakansson et al. 1992, 1994). The formate anion is bound to zinc in a pentacoordinated manner, the other ligand besides the three histidines being a water molecule (Figure 7.6). The HCOO⁻ ion is linked in a monodentate fashion to the metal, with its O$_2$ atom 2.5 Å away from the zinc. The other oxygen of the formate ion forms a hydrogen bond with the Thr-199 amide nitrogen. The coordinated water is 2.2 Å from the zinc ion, only 0.7 Å away from its native position in the active protein.

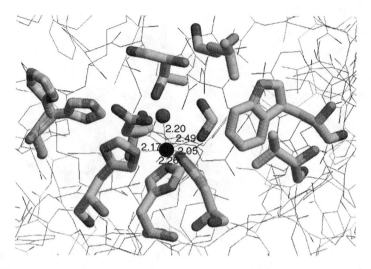

FIGURE 7.6 Coordination detail for the complex between human CA II and formate (From Haakansson, K. et al. 1992. *Journal of Molecular Biology* **227,** 1192–1204.); the structure was generated from the pdb file (2CBC) by using the program RasWin for Windows, version 2.7.1.1.

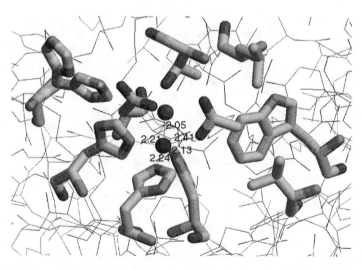

FIGURE 7.7 Coordination detail for the complex between human CA II and acetate (From Haakansson, K. et al. 1994. *Acta Crystallographica, Section D: Biological Crystallography* **D50,** 101–104.); the structure was generated from the pdb file (1CAY) by using the program RasWin for Windows, version 2.7.1.1.

In the case of acetate (Haakansson et al. 1994), one carboxylate oxygen is coordinated to the zinc at a distance of 2.4 Å, 1.7 Å away from the position of the zinc-bound water in the native protein. The zinc water itself is displaced 0.8 Å from its native position (Figure 7.7). The other carboxylate oxygen atom is hydrogen bonded to the Thr-199 backbone NH (3.0 Å away). The similarity with the formate

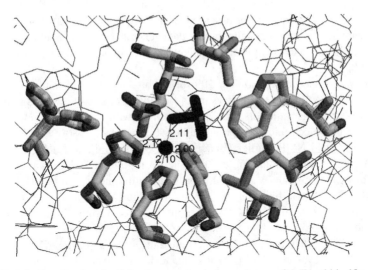

FIGURE 7.8 Coordination detail for the complex between human CA II and bisulfite (From Haakansson, K. et al. 1992. *Journal of Molecular Biology* **227,** 1192–1204.); the structure was generated from the pdb file (2CBD) by using the program RasWin for Windows, version 2.7.1.1.

and nitrate ions bonding is obvious. The methyl group of the inhibitor makes van der Waals interactions with the hydrophobic pocket of the protein, being in contact with Val-143 (3.3 Å), Leu-198 (3.9 Å) and Trp-209 (3.6 Å; Haakansson et al. 1994). Haakansson et al. (1992) also analyzed, in conjunction with the formate adduct, the complex of CA II with the bisulfite ion. Although bisulfite differs geometrically from bicarbonate, being pyramidal instead of planar as is the latter, it was assumed that its binding is analogous to the binding of the bicarbonate substrate (Haakansson et al. 1992). The results of the study showed that, in contrast to formate, the bisulfite anion is coordinated to the zinc active site in a tetrahedral geometry (Figure 7.8).

The explanation resides in the ability of bisulfite to act as a hydrogen-bond donor to the Oγ of Thr-199 and also to establish another hydrogen bond to the amide nitrogen atom of the same residue (2.9 Å away), similar to the cases of HS⁻ and the sulfonamides (Eriksson et al. 1988b; Haakansson et al. 1992; Haakansson and Liljas 1994; Mangani and Haakansson 1992; Vidgren et al. 1990, 1993). These additional H bonds account for sufficient energy to fully stabilize a tetrahedral geometry at the zinc ion. From this point of view, the Thr-199, along with H-bonded Glu-106, was considered to act as a door-keeper to the zinc ion, selecting protonated atoms for the water position on the zinc ion and excluding nonprotonated atoms from this coordination site (Eriksson et al. 1988b; Liljas et al. 1994; Lindskog and Liljas 1993; Lindskog and Silverman 2000; Merz 1990). Other structural characteristics of the bisulfite–hCA II complex are very similar to those of the formate ion. The HSO_3^- also displaces the deep water molecule and Wat-338 (similar to the cases of 1,2,4-triazole- and NO_3^- binding), thus interfering with the residues Val-121, Val-143, Leu-198, Thr-199 and Trp-209. The ligand, obviously bound in the monodentate mode, is 3.1 Å away from the Zn ion. Again, as in the case of formate, no changes in the peptide chain conformation are observed (Haakansson et al. 1992). The results are consistent with the Laue crystallographic data generated by Lindahl et al. (1992) for the same ligand; they are pH independent, as determined previously (Yachandra et al. 1983).

The very important role of Thr-199 for the binding mode of inhibitors into the active site of CA II (e.g., bisulfite vs. nitrate or formate) is also revealed in the case of adducts of strong inhibitors cyanide and cyanate with CA II. A common conclusion of several electronic absorption studies (Bertini et al. 1980b; Lindskog 1963, 1966) and ligand NMR studies (Banci et al. 1990; Bertini and Luchinat 1983) on Co-CA, carried out in solution, was that the strong inhibitors NCO^- and CN^- will substitute the zinc-bound water and coordinate to the metal in a tetrahedral mode (Banci et al. 1990; Bertini et al. 1978a), whereas the SCN^- ion, a weaker inhibitor, will coordinate as a fifth ligand, thus forming a distorted trigonal bipyramid. Eventually, this was proved to be true in the case of SCN^- (Eriksson et al. 1988b).

However, Lindahl et al. (1993) determined the structure of the hCA II inhibited by cyanide and cyanate. Their puzzling conclusion was that the two inhibitors bind in the close vicinity of the zinc ion, displacing the deep water molecule and forming a hydrogen bond with NH of Thr-199 backbone; the coordination of the metal ion was left unaltered, with a molecule of water coordinated in a tetrahedral fashion. The shortest distance between the zinc and the cyanate ion was found to be 3.2 Å, whereas the shortest Zn–CN-distance was a little longer –3.3 Å (Lindahl et al. 1993). It was hypothesized that this coordination behavior of the two ions is because of their inability to form a donor H bond with the Oγ1 of Thr-199, which can stabilize a tetrahedral coordination as in the case of hydrosulfide-, 1,2,4-triazole- and sulfonamide binding (Lindahl et al. 1993). In response to these findings, Bertini's group (Bertini et al. 1992) reinvestigated the binding of NCO^-, CN^- and SCN^- by spectroscopic techniques on Co-CA and [67]Zn-CA (which has a large quadrupolar moment). They arrived at the same conclusion that they considered previously, that the CN^- and CNO^- bind, in solution, the metal ion in bCA II. The discrepancies between their findings and the results of the crystallographic study were attributed to the existence of two different free-energy minima accessible to the system in solution and in the solid state.

These contradictory reports prompted the hypothesis (Supuran 1992; Supuran et al. 1994, 1995) that cyanate, which is isoelectronic and isosteric with the physiological substrate of the enzyme, CO_2, might behave as a poor substrate for CA (Supuran et al. 1997). Theoretical calculations by Merz's group (Peng et al. 1993) for the binding of cyanate, cyanide and thiocyanate to CA suggested that cyanide and cyanate might act as substrates; they tried to solve the controversy between the two approaches — the crystallographic methods and the spectroscopic ones — by suggesting that crystallographic methods observe the hydrated complexes of HCN and HNCO whereas the spectroscopic methods do not (Peng et al. 1993). Supuran et al. (1997) investigated whether different CA isozymes were able to catalyze the hydrolysis of cyanide or cyanate and found that a clear-cut answer was difficult to provide. They affirmed that either the process is unfavored kinetically or the two anions are actually hydrolyzed by CA but the corresponding products (carbamic acid for cyanate, formamide for cyanide) are very tightly bound to the enzyme and

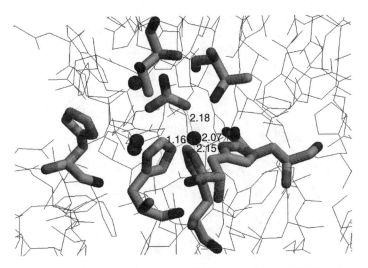

FIGURE 7.9 Coordination detail for the complex between human CA II and urea (From Briganti, F. et al. 1999. *Journal of Biological Inorganic Chemistry* **4**, 528–536.); the structure was generated from the pdb file (1BV3) by using the program RasWin for Windows, version 2.7.1.1.

do not dissociate from the active site (Supuran et al. 1997). The latter hypothesis is sustained by the recent finding (Briganti et al. 1999; Guerri et al. 2000) that cyanamide, which is also isoelectronic with CO_2, acts as a potent suicide substrate for CA. The urea formed after the hydrolytic reaction of the cyanate ion remains blocked within the enzyme active site, where it is directly coordinated to the zinc ion through a protonated nitrogen atom (Figure 7.9; Briganti et al. 1999). The hCA II–urea adduct was found to be quite a regular tetrahedron, with the distances and angles around the zinc atom similar to those of the native metalloprotein. The position of the ureate in the active site is further stabilized by hydrogen bonding with Thr-199, Thr-200 and several water molecules (Figure 7.9). These strong interactions at the active site explain why the zinc-bound ureate cannot be displaced even by high concentrations of very potent sulfonamide inhibitors (Briganti et al. 1999). The authors also provided a reaction mechanism to explain the cyanamide hydration, based on electronic spectroscopy and kinetic and x-ray crystallographic studies on the system.

The precise binding mode of cyanamide to the enzyme was clarified in a subsequent investigation by the same group (Guerri et al. 2000), using cryocrystallographic techniques. The crystal structure, obtained at 100 K and after soaking the protein crystal for a limited period in cyanamide solutions, shows that two different adducts are formed under the experimental conditions, with different occupancy in the crystal. The high-occupancy form consists of a binary hCA II–cyanamide complex, wherein the substrate has replaced the zinc-bound hydroxide/water, maintaining the tetrahedral geometry around the metal ion (Figure 7.10). The second, low-occupancy form consists of a hCA II–cyanamide–water ternary complex, wherein the catalytic zinc ion is still coordinated by a cyanamide molecule, but is also approached by a molecule of water. In the high-occupancy form, the zinc ion, besides the three

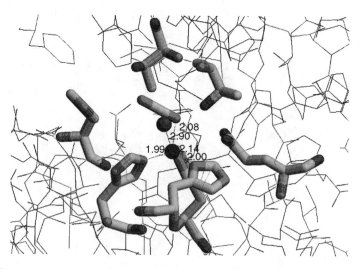

FIGURE 7.10 Coordination detail for the complex between human CA II and cyanamide (Wat-51 right) (From Guerri, A. et al. 2000. *Biochemistry* **39**, 12391–12397.); the structure was generated from the pdb file (1F2W) by using the program RasWin for Windows, version 2.7.1.1.

histidines, binds one cyanamide molecule at 2.09 Å; the water molecule (Wat-51) approaches zinc at 2.91 Å and can be considered as a zinc-outer sphere fifth ligand, generating a 4 + 1 coordination type for the metal ion. The deep water molecule (Wat-338) usually occupying the hydrophobic pocket of the enzyme is absent in this structure. The Wat-51 is closer to cyanamide (~2.5 Å) than to zinc (2.91 Å). The interaction of cyanamide with the CA II active site is completed by two strong hydrogen bonds. The first one involves the NH attached to zinc and the Oγ1 of Thr-199 proved to stabilize a tetrahedral geometry in the case of HS$^-$, 1,2,4-triazole, HSO$_3^-$ and sulfonamides. Another hydrogen bond is formed with the side chain of Thr-200. Except for seven water molecules displaced on binding of the ligand, the remaining water molecules and the overall structure of CA II remain unaltered on interaction with cyanamide. The low-occupancy form of the hCA II–cyanamide adduct is a ternary hCA II–cyanamide–water (Wat-51) complex, which can be interpreted as a frozen intermediate state of the hCA II-catalyzed conversion of cyanamide to urea. Wat-51 is considered to be added to the Zn coordination sphere, generating a five-coordination geometry. The coincidence of this water molecule position with that of the zinc-bound oxygen and nitrogen atoms of the five-coordinated adducts with nitrate and thiocyanate favors this hypothesis (Guerri et al. 2000).

Thus, the possibility of the existence of equilibria between four- and five-coordinate species in CA (Bertini and Luchinat 1983; Bertini et al. 1982) is once again confirmed. Moreover, formation of a pentacoordinated adduct in the presence of cyanamide was anticipated on the basis of spectroscopic investigations on Co-CA (Briganti et al. 1999).

If the ternary complex represents a snapshot along the reaction coordinate of the hCA II-catalyzed cyanamide hydration, this reaction will follow a different

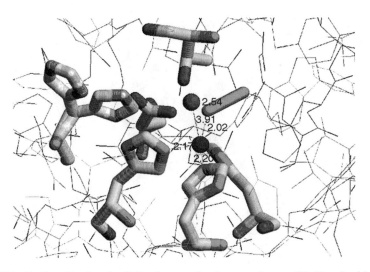

FIGURE 7.11 Coordination detail for the complex between human CA II and azide (Wat-318 also shown) (From Joensson, B.M. et al. *FEBS Letters* **322,** 186–190.); the structure was generated from the pdb file (1RAY) by using the program RasWin for Windows, version 2.7.1.1.

mechanism with respect to CO_2 hydration; the very slow rate of the cyanamide reaction is easily explained by considering that the attacking water molecule (Wat-51) is only partially activated by the interaction with the metal (Guerri et al. 2000).

As predicted by spectroscopic techniques (Bertini and Luchinat 1983; Bertini et al. 1982) another species to generate four- to five-coordinated species would be N_3^-, the halogen and bicarbonate ions. To confirm these facts, Joensson et al. (1993) determined the structure of CA II in complex with bromide and azide. They found that both azide and bromide ions bind into the CA II active site and replace the zinc-bound water and the deep water (Figure 7.11 and Figure 7.12), generating a distorted tetrahedral coordination for the metal ion. The inhibitors cannot bind close to Thr 199 Oγ1 because of their lack of protons. In the case of azide ion, the N3 atom is 2.0 Å away from the zinc ion and 3.2 Å away from Oγ1 of Thr-199. The N1 atom is positioned 3.4 Å from the Thr-199 backbone N and 3.7 Å from the Thr-199 Oγ1. The ligand forms a hydrogen bond with Wat-318 (positioned 2.6 Å away from N3 and 3.4 Å away from N2; Figure 7.11).

A similar pattern is observed in the structure of the CA II–bromide adduct; Br⁻ replaces the zinc-bound water and the deep water and coordinates 2.5 Å away from the zinc ion, 3.6 Å away from the Thr-199 Oγ1 and 3.8 Å from the Thr-199 backbone N (Figure 7.12). The difference between the position of the bromide ion and the azide N2 is only 0.2 Å. Also, there are no significant differences between the structure of the bromide adduct at pH 6.0 and at pH 7.8; however, at higher pH, the hydroxide ions compete for coordination to the metal center (Joensson et al. 1993). Iodide is reported to act in a similar way (Brown et al. 1977; Joensson et al. 1993). This behavior can be explained by their softness (greater polarizability) as compared with oxygen-complexing ligands, which might alleviate the repulsion between its electron

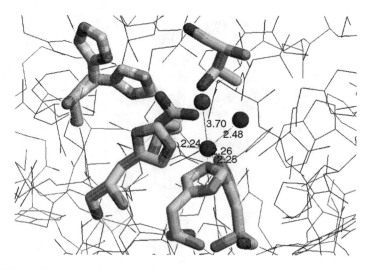

FIGURE 7.12 Coordination detail for the complex between human CA II and bromide (Wat-318 also shown) (From Joensson, B.M. et al. *FEBS Letters* **322,** 186–190.); the structure was generated from the pdb file (1RAZ) by using the program RasWin for Windows, version 2.7.1.1.

orbitals and the lone pair of Thr-199 Oγ1, as confirmed by more recent experiments (Lim et al. 1998; Merz and Banci 1996).

The structure of the native CA II–bicarbonate complex is not available to date, despite the efforts of several crystallographic groups. The insights of bicarbonate coordination were first revealed by using Co-CA II (Haakansson and Wehnert 1992). The ligand replaces the deep water molecule, binding at the cobalt ion in a pentaco-ordinate fashion, with the O3 atom 2.39 Å away from the metal. The O1 atom accepts a hydrogen bond from the Thr-199 backbone NH, whereas the O2 atom is 2.4 Å from the cobalt ion (Figure 7.13). The positions of O1 and O2 are very similar to the corresponding oxygen atoms of the formate–CA II complex (Haakansson et al. 1992). The metal-bound water (Wat-263) is moved 1.7 Å away from its native position, being located at 2.33 Å from the cobalt ion (Figure 7.13; Haakansson and Wehnert 1992).

In the adduct hCA I–HCO_3^- (Kumar and Kannan 1994), the bicarbonate ion occupies approximately the same position in the active site of the enzyme but coordinates as an essentially monodentate ligand, as proposed previously (Williams and Henkens 1985). An intermediate position between these two extremes is repre-sented by HCO_3^- coordination in its complex with the Thr-200 His hCA II mutant (Xue et al. 1993), in which a pseudobidentate coordination is observed. Here, the coordination geometry of the zinc ion is somewhere between tetrahedral and trigonal bipyramidal. The O3 atom is 2.2 Å away from the zinc ion and 2.6 Å from the Oγ1 of Thr-199; the O2 atom is 2.5 Å away from the metal, whereas the O1 atom is 3.0 Å from the Thr-199 backbone NH (Figure 7.14).

Thus, the crystallographic data confirm that bicarbonate interaction with CA is strongly influenced by the characteristics of the active site of each isozyme, as predicted by theoretical calculations (Krauss and Garmer 1991; Liang and Lipscomb

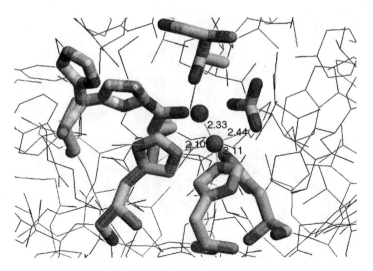

FIGURE 7.13 Coordination detail of the complex between Co-CA II and bicarbonate ion (From Haakansson, K. and Wehnert, A. 1992. *Journal of Molecular Biology* **228,** 1212–1218.); the structure was generated from the pdb file (1CAH) by using the program RasWin for Windows, version 2.7.1.1.

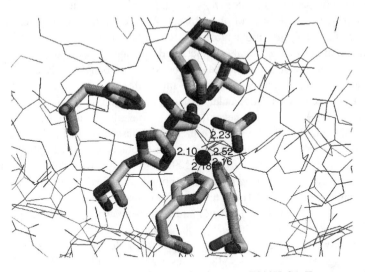

FIGURE 7.14 Coordination detail of the complex between T200H CA II mutant and bicarbonate ion (From Xue, Y. et al. 1993. *Proteins: Structure, Function, and Genetics* **15,** 80–87.); the structure was generated from the pdb file (1BIC) by using the program RasWin for Windows, version 2.7.1.1.

1987; Pullman 1981; Vedani et al. 1989; Zheng and Merz 1992) and by kinetic determinations (Pocker and Sarkanen 1978).

In addition to the x-ray crystallographic studies of different anions binding into the CA active site (all placed in the zinc coordination sphere), the crystal structure of the complex between CA II and phenol was reported (Nair et al. 1994) — its

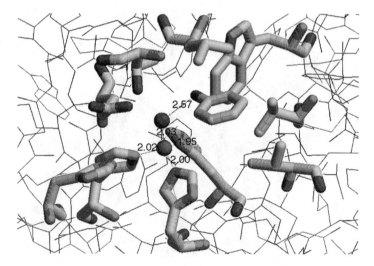

FIGURE 7.15 Coordination detail of the complex between CA II and phenol (From Nair, S.K. et al. 1994. *Journal of the American Chemical Society* **116**, 3659–3660.); the structure was generated from the pdb file using the program RasWin for Windows, version 2.7.1.1.

only known competitive inhibitor [K_I = 10 mM at pH 8.7 (Simonsson et al. 1982)]. Unexpectedly, two molecules of phenol were observed in the electron density maps of the complex. The first molecule binds to a hydrophobic patch, ca. 15 Å away from zinc, making van der Waals contacts with residues Leu-57, Ile-91and Pro-237 and hydrogen bonds with Asp-72 and with backbone carbonyl of Gly-235. Although this location does not correspond to the molecular trajectory or secondary CO_2 binding sites predicted in molecular dynamics calculations (Liang and Lipscomb 1990; Merz 1990, 1991), the second phenol molecule binds in the hydrophobic pocket of the enzyme, the known CO_2 precatalytic association site. It displaces the deep water molecule, making van der Waals contacts with hydrophobic pocket residues Val-121, Val-143, Leu-198, Trp-209. Its hydroxyl group does not coordinate to zinc but is hydrogen bonded to the zinc-bound water/hydroxyl ion (2.6 Å away) and also with the backbone NH of Thr-199 (3.2 Å away; Figure 7.15; Nair et al. 1994.

Considering that the CA II–phenol adduct crystals were prepared at pH 10 and also the differences between the pK_a values of phenol and zinc-bound water (10 and 7, respectively), it was assumed that the phenol molecule is ionized (Nair et al. 1994), as suggested also by ^{13}C-NMR studies carried out with this inhibitor (Khalifah et al. 1991).

7.3   OTHER TYPES OF NONSULFONAMIDE INHIBITORS

7.3.1   Inhibitors of the Proton Shuttle

A different class of CAIs is represented by the metal ions Cu^{2+} and Hg^{2+}. The copper ion was found to be a weak inhibitor of CA II (IC_{50} = 0.5 µM at pH 7.3; Eriksson et al. 1988b; Magid 1967). Subsequent [18]O exchange experiments done by Tu et al.

(1981) have shown that Cu^{2+} does not affect the rate of either substrate binding or product release. Moreover, lowering the pH to 6 completely abolishes the inhibitory effect. The same effects were observed for Hg^{2+} ions. The authors suggested that the metals bind to the proton shuttle of the enzyme — His-64 — and inhibit intramolecular proton transfer (Tu et al. 1981). X-ray crystallographic investigations by Ericksson et al. (1988b) confirmed this hypothesis. In the electron density chart of the Hg^{2+}–CA II complex, both nitrogens of His-64 were found to bind mercury at about half occupancy, suggesting that the two imidazolic nitrogen atoms bind the metal with about equal affinity, thereby eliminating the possibility for His-64 to participate in proton transfer (Eriksson et al. 1988b; Eriksson and Liljas 1991).

Linked with these facts is the common finding of different crystallographers that the use of mercuric ions or mercury-containing compounds, such as $HgCl_2$, 4-chloro-benzenesulfonic acid or 4-hydroxymercuribenzoate, to avoid dimerization of cysteine residues of CA II also enhances the hCA II crystal quality (Briganti et al. 1998, 1999; Eriksson et al. 1988b; Mangani and Liljas 1993; Tilander et al. 1965; Xue et al. 1993), probably by blocking the very mobile His-64 in a fixed conformation.

7.3.2 Organic Sulfamates and Hydroxamates as CAIs

Besides the shuttle inhibitors, other nonsulfonamide CAIs were discovered or invented in the quest for topical antiglaucoma drugs (Maren 2000; Supuran and Scozzafava 2000, 2002b) or by analogy with other zinc enzymes (Scolnick et al. 1997).

Thus, Lo et al. (1992), following a previous report that topiramate (**7.1**) and other related alkyl sulfamates can act as weak CAIs (Maryanoff et al. 1987), synthesized a series of imidazolylphenyl- and imidazolylphenoxyalkyl sulfamates, which were evaluated as topical antiglaucoma CAIs. The best results were obtained with the sulfamic acid ester **7.2** (AHR-16329, $K_I = 7$ nM against CA II), which displayed a peak 2.5 mmHg decrease in intraocular pressure (IOP) 1 h after being applied as a 2 to 5% solution to the eye (Brechue and Maren 1993; Lo et al. 1992).

Continuing their program on antiepileptic drugs, Maryanoff et al. (1998) conducted an extensive structure–activity study on topiramate-related compounds. Besides investigating their anticonvulsant properties, the authors also tested the drugs as CAIs. The cyclic sulfamate **7.3** (RWJ-37947) exhibited eight times greater anticonvulsant activity than did topiramate and also proved to be a potent inhibitor of CA ($IC_{50} = 36$ nM). Because of its rather unusual structure for a very efficient CAI, RWJ-37947 was further investigated by x-ray crystallography in complex with CA II (Recacha et al. 2002). The results showed that the sulfamate group of **7.3**, which has a pK_a value of 8.51 (Dodgson et al. 2000), coordinates to zinc, displacing the water/hydroxide ion and making a hydrogen bond with the Oγ1 of Thr-199, similar to sulfonamides. One sulfamate oxygen accepts a hydrogen bond from the Thr-199 backbone NH, the other 3.1 Å away from the zinc ion. Taking into account the distances and angles of the latter oxygen, it was considered weakly coordinated to the metal. The six-member tetrahydropyran ring of the inhibitor retains the skew conformation ($^3S_0$) found in solution (Dodgson et al. 2000; Figure 7.16). The cyclic sulfate group forms a weak hydrogen bond with Pro-202 and Leu-198. No evidence

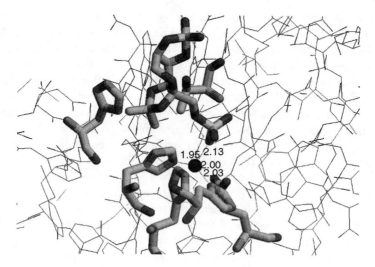

FIGURE 7.16 Coordination detail of the complex between CA II and RWJ-37947 (From **7.3**; Recacha, R. et al. 2002. *Biochemical Journal* **361,** 437–441.); the structure was generated from the pdb file (1EOU) by using the program RasWin for Windows, version 2.7.1.1.

of irreversible binding was found and the structure of CA II exhibited only minor differences in conformation when compared to the native protein. The coordination pattern is similar, in general terms, to those found for sulfamide [**7.4**, K_I = 1.13 mM (Briganti et al. 1996)] and sulfamic acid [**7.5**, K_I = 390 µM (Briganti et al. 1996)] in their adducts with CA (Abbate et al. 2002).

However, a very recent crystallographic study of the complex CA II–topiramate (Casini et al. 2003) revealed that this drug has a totally different conformation when bound to the active site of the enzyme, as compared with that of RWJ-37947. Although the sulfamate group is anchored in essentially the same way (Figure 7.17), the conformation of the backbone is entirely changed and strongly stabilized by an extended network of hydrogen bonds between the inhibitor and some amino acid residues inside the cavity: Asn-62–Nε1 to O4 of topiramate, Gln-92–Nε2 to O6 of topiramate, Wat-1134 to O3 of topiramate and to Oγ1 of Thr-200. Moreover, a full SAR study of the interaction of CA isozymes I, II and IV with related sugar sulfamates/sulfamides was reported and the main finding was that topiramate **1** is actually a very potent CAI, three orders of magnitude more efficient than previously reported ($K_{I\ hCA\ I}$ = 250 nM, $K_{I\ hCA\ II}$ = 5 nM, $K_{I\ hCA\ IV}$ = 54 nM). It is comparable in effect with acetazolamide, a fact sustained not only by the tightly bound crystal adduct with the CA II, but also by the side effects observed (Sabers and Gram 2000) in many patients treated with this antiepileptic drug, which are typical for strong systemic CAIs (Supuran and Scozzafava 2000, 2002b).

Hence, the use of sulfamide and sulfamic acid moieties as new anchoring groups for designing CAIs has a solid theoretical base. Other recent investigations (Scozzafava

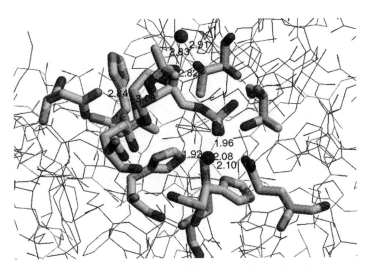

FIGURE 7.17 The complex between CA II and topiramate (From **7.1**; Casini, A. et al. 2003. *Bioorganic and Medicinal Chemistry Letters* **13**, 837–840.); the structure was generated from the pdb file by using the program RasWin for Windows, version 2.7.1.1.

et al. 2000, 2001) on these types of compounds confirmed these findings. Thus, when aromatic/heterocyclic scaffolds were used in conjunction with sulfamide and sulfamate anchoring groups, nanomolar levels of CA inhibition were obtained, e.g., **7.6** [$K_{I\ hCA\ II}$ = 16 nM, $K_{I\ hCA\ IV}$ = 50 nM (Scozzafava et al. 2001)], **7.7** [$K_{I\ hCA\ II}$ = 21 nM, $K_{I\ hCA\ IV}$ = 54 nM (Scozzafava et al. 2001)], **7.8** [$K_{I\ hCA\ I}$ = 4.6 nM, $K_{I\ hCA\ II}$ = 1.1 nM, $K_{I\ hCA\ IX}$ = 18 nM (Winum et al. 2003)], associated also by a good selectivity against different CA isozymes, such as in the case of compounds **7.9** [$K_{I\ hCA\ II}$ = 10 nM, $K_{I\ hCA\ IV}$ = 0.8 μM (Scozzafava et al. 2000)] and **7.10** [$K_{I\ hCA\ II}$ = 20 nM, $K_{I\ hCA\ IV}$ = 0.4 μM (Scozzafava et al. 2000)].

Christianson's group (Scolnick et al. 1997) showed that hydroxamic acids, known to be efficient inhibitors of zinc enzymes such as thermolysin (Jin and Kim 1998; Powers and Harper 1986) and matrix metalloproteinases (Supuran and Scozzafava 2002a), can act as CAIs. The three-dimensional structures of CA II complexes with acetohydroxamic acid (**7.11**, IC_{50} = 47 μM) and its fluorinated congener (**7.12**, IC_{50} = 3.8 μM) revealed an interesting binding mode: instead of forming a chelate with the CO and OH groups of the metal, they bind through the ionized nitrogen atom directly to the zinc ion, displacing the coordinated water/hydroxyl ion. The inhibitors also displace the deep water molecule and form a donor hydrogen bond with the Oγ1 of Thr-199. The complexes are further stabilized by a hydrogen bond between the Thr-199 backbone NH and the C=O group, and by van der Waals contacts with the hydrophobic pocket (Val-121, Val-143, Leu-198, Trp-209; Figure 7.18). Moreover, in the case of trifluoroacetohydroxamic acid **7.12**, a weakly polar C-F → Zn^{2+} interaction can be observed; the fluorine and ionized nitrogen atom of the ligand form a five-membered chelate with zinc (Scolnick et al. 1997).

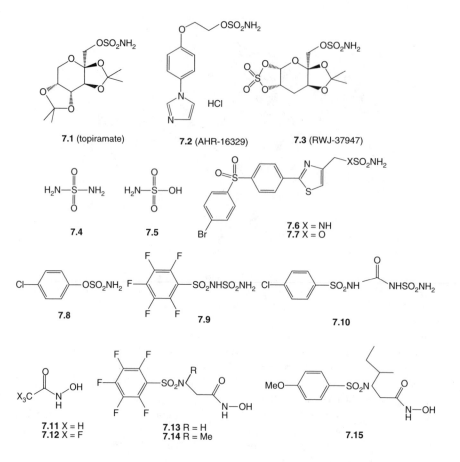

7.1 (topiramate)   7.2 (AHR-16329)   7.3 (RWJ-37947)

7.4   7.5   7.6 X = NH
                      7.7 X = O

7.8   7.9   7.10

7.11 X = H   7.13 R = H   7.15
7.12 X = F   7.14 R = Me

This finding prompted Scozzafava and Supuran (2000a) to investigate whether the sulfonylated amino acid hydroxamates synthesized within their MMP inhibitors program (Scozzafava and Supuran 2000b, 2000c; Supuran et al. 2000; Supuran and Scozzafava 2001) could act as efficient CAIs. The study confirmed this possibility and revealed some very potent inhibitors bearing this anchoring groups, such as **7.13** ($K_{I\ hCA\ I} = 7\ nM$, $K_{I\ hCA\ II} = 8\ nM$, $K_{I\ hCA\ IV} = 10\ nM$), **7.14** ($K_{I\ hCA\ I} = 7\ nM$, $K_{I\ hCA\ II} = 8\ nM$, $K_{I\ hCA\ IV} = 10\ nM$) and **7.15** [$K_{I\ hCA\ I} = 50\ nM$, $K_{I\ hCA\ II} = 5\ nM$, $K_{I\ hCA\ IV} = 39\ nM$; Scozzafava and Supuran 2000a]. This new anchoring group was considered recently in a successful virtual screening experiment for identifying novel inhibitors of CA (Gruneberg et al. 2002).

All these new insights on the coordination patterns of different inhibitors within the active site of CAs contribute to a better understanding of the role of CAs in different organisms and provide a solid ground for the design of drugs with improved selectivity and reduced side effects.

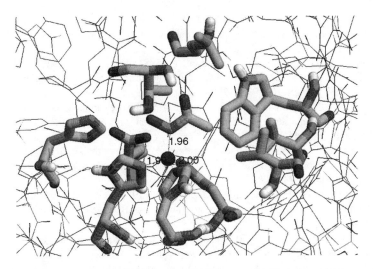

FIGURE 7.18 Coordination detail of the complex between CA II and acetohydroxamic acid (From **7.10**; Scolnick, L.R. et al. 1997. *Journal of the American Chemical Society* **119**, 850–851.); the structure was generated from the pdb file (1AM6) by using the program RasWin for Windows, version 2.7.1.1.

REFERENCES

Abbate, F., Supuran, C.T., Scozzafava, A., Orioli, P., Stubbs, M.T., and Klebe, G. (2002) Nonaromatic sulfonamide group as an ideal anchor for potent human carbonic anhydrase inhibitors: Role of hydrogen-bonding networks in ligand binding and drug design. *Journal of Medicinal Chemistry* **45**, 3583–3587.

Alberti, G., Bertini, I., Luchinat, C., and Scozzafava, A. (1981) A new class of inhibitors capable of binding both the acidic and alkaline forms of carbonic anhydrase. *Biochimica et Biophysica Acta* **668**, 16–26.

Armitage, I.M., Schoot Uiterkamp, A.J.M., Chlebowski, J.F., and Coleman, J.E. (1978) Cadmium-113 NMR as a probe of the active sites of metalloenzymes. *Journal of Magnetic Resonance* **29**, 375–392.

Baird, T.T., Jr., Waheed, A., Okuyama, T., Sly, W.S., and Fierke, C.A. (1997) Catalysis and inhibition of human carbonic anhydrase IV. *Biochemistry* **36**, 2669–2678.

Banci, L., Bertini, I., Luchinat, C., Donaire, A., Martinez, M.J., and Moratal Mascarell, J.M. (1990) The factors governing the coordination number in the anion derivatives of carbonic anhydrase. *Comments on Inorganic Chemistry* **9**, 245–261.

Banci, L., Bertini, I., Luchinat, C., Monnanni, R., and Moratal Mascarell, J. (1989) Proton NMR spectra of cobalt(II)-substituted carbonic anhydrase isoenzymes. *Gazzetta Chimica Italiana* **119**, 23–29.

Bencini, A., Bertini, I., Canti, G., Gatteschi, D., and Luchinat, C. (1981) The EPR spectra of the inhibitor derivatives of cobalt carbonic anhydrase. *Journal of Inorganic Biochemistry* **14**, 81–93.

Bertini, I., Borghi, E., Canti, G., and Luchinat, C. (1979a) Investigation of the system cobalt(II) bovine carbonic anhydrase B-trichloroacetaldehyde. *Journal of Inorganic Biochemistry* **11,** 49–56.

Bertini, I., Canti, G., Luchinat, C., and Mani, F. (1981) Hydrogen-1 NMR spectra of the coordination sphere of cobalt-substituted carbonic anhydrase. *Journal of the American Chemical Society* **103,** 7784–7788.

Bertini, I., Canti, G., Luchinat, C., and Romanelli, P. (1980a) Cyanometallates and cobalt(II) bovine carbonic anhydrase B: Five coordination with dicyanoaurate(I). *Inorganica Chimica Acta* **46,** 211–214.

Bertini, I., Canti, G., Luchinat, C., and Scozzafava, A. (1978a) Characterization of cobalt(II) bovine carbonic anhydrase and of its derivatives. *Journal of the American Chemical Society* **100,** 4873–4877.

Bertini, I., Canti, G., Luchinat, C., and Scozzafava, A. (1979b) Characterization of oxovanadium(IV) substituted bovine carbonic anhydrase B. *Inorganica Chimica Acta* **36,** 9–12.

Bertini, I., and Luchinat, C. (1983) Cobalt(II) as a probe of the structure and function of carbonic-anhydrase. *Accounts of Chemical Research* **16,** 272–279.

Bertini, I., and Luchinat, C. (1984) The structure of cobalt(II)-substituted carbonic anhydrase and its implications for the catalytic mechanism of the enzyme. *Annals of the New York Academy of Sciences* **429,** 89–98.

Bertini, I., Luchinat, C., Pierattelli, R., and Vila, A.J. (1992) A multinuclear ligand NMR investigation of cyanide, cyanate, and thiocyanate binding to zinc and cobalt carbonic anhydrase. *Inorganic Chemistry* **31,** 3975–3979.

Bertini, I., Luchinat, C., and Scozzafava, A. (1977a) Interaction of cobalt(II) bovine carbonic-anhydrase with aniline, benzoate, and anthranilate. *Journal of the American Chemical Society* **99,** 581–583.

Bertini, I., Luchinat, C., and Scozzafava, A. (1977b) Interactions between alpha-amino acids and cobalt(II) bovine-carbonic anhydrase. *Bioinorganic Chemistry* **7,** 225–231.

Bertini, I., Luchinat, C., and Scozzafava, A. (1978b) Evidence of exchangeable protons in the acidic form of manganese(II) bovine carbonic anhydrase B. *FEBS Letters* **87,** 92–94.

Bertini, I., Luchinat, C., and Scozzafava, A. (1980b) The acid-base equilibriums of carbonic anhydrase. *Inorganica Chimica Acta* **46,** 85–89.

Bertini, I., Luchinat, C., and Scozzafava, A. (1982) Carbonic anhydrase: An insight into the zinc binding site and into the active cavity through metal substitution. *Structure and Bonding* **48,** 45–92.

Brechue, W.F., and Maren, T.H. (1993) Carbonic anhydrase inhibitory activity and ocular pharmacology of organic sulfamates. *Journal of Pharmacology and Experimental Therapeutics* **264,** 670–675.

Briganti, F., Iaconi, V., Mangani, S., Orioli, P., Scozzafava, A., Vernaglione, G., and Supuran, C. T. (1998) A ternary complex of carbonic anhydrase: x-ray crystallographic structure of the adduct of human carbonic anhydrase II with the activator phenylalanine and the inhibitor azide. *Inorganica Chimica Acta* **275/276,** 295–300.

Briganti, F., Mangani, S., Orioli, P., Scozzafava, A., Vernaglione, G., and Supuran, C.T. (1997) Carbonic anhydrase activators: x-ray crystallographic and spectroscopic investigations for the interaction of isozymes I and II with histamine. *Biochemistry* **36,** 10384–10392.

Briganti, F., Mangani, S., Scozzafava, A., Vernaglione, G., and Supuran, C.T. (1999) Carbonic anhydrase catalyzes cyanamide hydration to urea: is it mimicking the physiological reaction? *Journal of Biological Inorganic Chemistry* **4,** 528–536.

Briganti, F., Pierattelli, R., Scozzafava, A., and Supuran, C.T. (1996) Carbonic anhydrase inhibitors. Part 37. Novel classes of isoenzyme I and II inhibitors and their mechanism of action. Kinetic and spectroscopic investigations on native and cobalt-substituted enzymes. *European Journal of Medicinal Chemistry* **31,** 1001–1010.

Brown, G. S., Navon, G., and Shulman, R.G. (1977) X-ray absorption studies of halide binding to carbonic anhydrase. *Proceedings of the National Academy of Sciences of the United States of America* **74,** 1794–1797.

Casini, A., Antel, J., Abbate, F., Scozzafava, A., David, S., Waldeck, H., Schafer, S., and Supuran, C.T. (2003) Carbonic anhydrase inhibitors: SAR and x-ray crystallographic study for the interaction of sugar sulfamates/sulfamides with isozymes I, II and IV. *Bioorganic and Medicinal Chemistry Letters* **13,** 837–840.

Chegwidden, W.R., and Carter, N.D. (2000) Introduction to the carbonic anhydrases. In *The Carbonic Anhydrases — New Horizons*, Chegwidden, W.R., Carter, N.D., and Edwards, Y.H. (Eds.), Birkhäuser, Basel, pp. 14–28.

Coleman, J.E. (1975) Chemical reactions of sulfonamides with carbonic anhydrase. *Annual Reviews in Pharmacology* **15,** 221–242.

Dodgson, S.J., Shank, R.P., and Maryanoff, B.E. (2000) Topiramate as an inhibitor of carbonic anhydrase isoenzymes. *Epilepsia* **41,** S35–S39.

Dodgson, S.J., Tashian, R.E., Gros, G., and Carter, N.D., (Eds.) (1991) *The Carbonic Anhydrases — Cellular Physiology and Molecular Genetics*, Plenum Press, New York.

Engberg, P., and Lindskog, S. (1984) Effects of pH and inhibitors on the absorption spectrum of cobalt(II)-substituted carbonic anhydrase III from bovine skeletal muscle. *FEBS Letters* **170,** 326–330.

Eriksson, A.E., Jones, T.A., and Liljas, A. (1988a) Refined structure of human carbonic anhydrase II at 2.0.ANG. resolution. *Proteins: Structure, Function, and Genetics* **4,** 274–282.

Eriksson, A.E., Kylsten, P.M., Jones, T.A., and Liljas, A. (1988b) Crystallographic studies of inhibitor binding sites in human carbonic anhydrase II: A pentacoordinated binding of the thiocyanate ion to the zinc at high pH. *Proteins: Structure, Function, and Genetics* **4,** 283–293.

Eriksson, A.E., and Liljas, A. (1991) X-ray crystallographic studies of carbonic anhydrase isozymes I, II, and III. In *Carbonic Anhydrases,* Dodgson, S.J. (Ed.), Plenum Press, New York, pp. 33–48.

Fitzgerald, J.J., and Chasteen, N.D. (1974) Electron paramagnetic resonance studies of the structure and metal ion exchange kinetics of vanadyl(IV) bovine carbonic anhydrase. *Biochemistry* **13,** 4338–4347.

Forster, R.E. (2000) Remarks on the discovery of carbonic anhydrase. In *The Carbonic Anhydrases — New Horizons*, Chegwidden, W.R., Carter, N.D., and Edwards, Y.H. (Eds.), Birkhauser, Basel, pp. 1–11.

Gruneberg, S., Stubbs, M.T., and Klebe, G. (2002) Successful virtual screening for novel inhibitors of human carbonic anhydrase: strategy and experimental confirmation. *Journal of Medicinal Chemistry* **45,** 3588–3602.

Guerri, A., Briganti, F., Scozzafava, A., Supuran, C.T., and Mangani, S. (2000) Mechanism of cyanamide hydration catalyzed by carbonic anhydrase II suggested by cryogenic x-ray diffraction. *Biochemistry* **39,** 12391–12397.

Haakansson, K., Briand, C., Zaitsev, V., Xue, Y., and Liljas, A. (1994) Wild-type and E106Q mutant carbonic anhydrase complexed with acetate. *Acta Crystallographica, Section D: Biological Crystallography* **D50,** 101–104.

Haakansson, K., Carlsson, M., Svensson, L.A., and Liljas, A. (1992) Structure of native and apo carbonic anhydrase II and structure of some of its anion-ligand complexes. *Journal of Molecular Biology* **227,** 1192–1204.

Haakansson, K., and Liljas, A. (1994) The structure of a complex between carbonic anhydrase II and a new inhibitor, trifluoromethane sulfonamide. *FEBS Letters* **350,** 319–322.

Haakansson, K., and Wehnert, A. (1992) Structure of cobalt carbonic anhydrase complexed with bicarbonate. *Journal of Molecular Biology* **228,** 1212–1218.

Heck, R.W., Tanhauser, S.M., Manda, R., Tu, C., Laipis, P.J., and Silverman, D.N. (1994) Catalytic properties of mouse carbonic anhydrase V. *Journal of Biological Chemistry* **269,** 24742–27446.

Jarvet, J., Olivson, A., Mets, U., Pooga, M., Aguraiuja, R., and Lippmaa, E. (1989) Carbon-13 and nitrogen-15 NMR and time-resolved fluorescence depolarization study of bovine carbonic anhydrase-4-methylbenzenesulfonamide complex. *European Journal of Biochemistry* **186,** 287–290.

Jin, Y., and Kim, D.H. (1998) Inhibition stereochemistry of hydroxamate inhibitors for thermolysin. *Bioorganic and Medicinal Chemistry Letters* **8,** 3515–3518.

Joensson, B.M., Haakansson, K., and Liljas, A. (1993) The structure of human carbonic anhydrase II in complex with bromide and azide. *FEBS Letters* **322,** 186–190.

Kannan, K.K., Petef, M., Fridborg, K., Cid Dresdner, H., and Lovgren, S. (1977) Structure and function of carbonic anhydrases: Imidazole binding to human carbonic anhydrase B and the mechanism of action of carbonic anhydrases. *FEBS Letters* **73,** 115–119.

Kernohan, J.C. (1964) The activity of bovine carbonic anhydrase in imidazole buffers. *Biochimica et Biophysica Acta* **81,** 346–356.

Kernohan, J.C. (1965) The pH–activity curve of bovine carbonic anhydrase and its relation to the inhibition of the enzyme by anions. *Biochimica et Biophysica Acta* **96,** 304–317.

Kernohan, J.C. (1966) A method for studying the kinetics of the inhibition of carbonic anhydrase by sulphonamides. *Biochimica et Biophysica Acta* **118,** 405–412.

Khalifah, R.G., Rogers, J.I., and Mukherjee, J. (1991) Interaction of carbon dioxide-competitive inhibitors with carbonic anhydrase. In *Carbonic Anhydrase*, Botre, F., Gros, G., and Storey, B.T. (Eds.), Proceedings of the International Workshop, VCH, Weinheim, pp. 65–74.

Krauss, M., and Garmer, D.R. (1991) Active site ionicity and the mechanism of carbonic anhydrase. *Journal of the American Chemical Society* **113,** 6426–6435.

Kumar, P.N.V.P., and Marynick, D.S. (1993) Ab initio comparison of the bonding modes of nitrate and bicarbonate in model carbonic anhydrase systems. *Inorganic Chemistry* **32,** 1857–1859.

Kumar, V., and Kannan, K.K. (1994) Enzyme–substrate interactions: Structure of human carbonic anhydrase I complexed with bicarbonate. *Journal of Molecular Biology* **241,** 226–232.

Lanir, A., Gradsztajn, S., and Navon, G. (1973) Proton magnetic relaxation in solutions of manganese-carbonic anhydrase. *FEBS Letters* **30,** 351–354.

Liang, J.Y., and Lipscomb, W.N. (1987) Hydration of carbon dioxide by carbonic anhydrase: Internal protein transfer of zinc(2+)-bound bicarbonate. *Biochemistry* **26,** 5293–5301.

Liang, J.Y., and Lipscomb, W.N. (1990) Binding of substrate carbon dioxide to the active site of human carbonic anhydrase II: A molecular dynamics study. *Proceedings of the National Academy of Sciences of the United States of America* **87,** 3675–3679.

Liljas, A., Haakansson, K., Jonsson, B.H., and Xue, Y. (1994) Inhibition and catalysis of carbonic anhydrase: Recent crystallographic analyses. *European Journal of Biochemistry* **219,** 1–10.

Lim, M., Hamm, P., and Hochstrasser, R.M. (1998) Protein fluctuations are sensed by stimulated infrared echoes of the vibrations of carbon monoxide and azide probes. *Proceedings of the National Academy of Sciences of the United States of America* **95,** 15315–15320.

Lindahl, M., Liljas, A., Habash, J., Harrop, S., and Helliwell, J.R. (1992) The sensitivity of the synchrotron Laue method to small structural changes: Binding studies of human carbonic anhydrase II (HCAII). *Acta Crystallographica, Section B: Structural Science* **B48,** 281–285.

Lindahl, M., Svensson, L.A., and Liljas, A. (1993) Metal poison inhibition of carbonic anhydrase. *Proteins: Structure, Function, and Genetics* **15,** 177–182.

Lindskog, S. (1963) Effects of pH and inhibitors on some properties related to metal binding in bovine carbonic anhydrase. *Journal of Biological Chemistry* **238,** 945–951.

Lindskog, S. (1966) Interaction of cobalt(II)–carbonic anhydrase with anions. *Biochemistry* **5,** 2641–2646.

Lindskog, S. (1982) Carbonic anhydrase. *Advanced Inorganic Biochemistry* **4,** 115–170.

Lindskog, S., and Coleman, J.E. (1973) The catalytic mechanism of carbonic anhydrase. *Proceedings of the National Academy of Sciences of the United States of America* **70,** 2505–2508.

Lindskog, S., Henderson, L.E., Kannan, K.K., Liljas, A., Nyman, P.O., and Strandberg, B. (1971) *Carbonic Anhydrase Enzymes*, 3rd ed., Boyer, P.D. (Ed.), Academic Press, New York, pp. 587–665.

Lindskog, S., and Liljas, A. (1993) Carbonic anhydrase and the role of orientation in catalysis. *Current Opinion in Structural Biology* **3,** 915–920.

Lindskog, S., and Silverman, D.N. (2000) The catalytic mechanism of mammalian carbonic anhydrases. In *The Carbonic Anhydrases — New Horizons*, Chegwidden, W.R., Carter, N.D., and Edwards, Y.H. (Eds.), Birkhäuser, Basel, pp. 175–195.

Lindskog, S., and Thorslund, A. (1968) On the interaction of bovine cobalt carbonic anhydrase with sulfonamides. *European Journal of Biochemistry* **3,** 453–460.

Lo, Y.S., Nolan, J.C., Maren, T.H., Welstead, W.J., Jr., Gripshover, D.F., and Shamblee, D.A. (1992) Synthesis and physiochemical properties of sulfamate derivatives as topical antiglaucoma agents. *Journal of Medicinal Chemistry* **35,** 4790–4794.

Luchinat, C., Monnanni, R., and Sola, M. (1990) Carbon-13 and proton NMR studies of imidazole binding to native and cobalt(II)-substituted human carbonic anhydrase I. *Inorganica Chimica Acta* **177,** 133–139.

Magid, E. (1967) The activity of carbonic anhydrases B and C from human erythrocytes and the inhibition of the enzymes by copper. *Scandinavian Journal of Haematology* **4,** 257–270.

Mangani, S., and Haakansson, K. (1992) Crystallographic studies of the binding of protonated and unprotonated inhibitors to carbonic anhydrase using hydrogen sulfide and nitrate anions. *European Journal of Biochemistry* **210,** 867–871.

Mangani, S., and Liljas, A. (1993) Crystal structure of the complex between human carbonic anhydrase II and the aromatic inhibitor 1,2,4-triazole. *Journal of Molecular Biology* **232,** 9–14.

Mansoor, U.F., Zhang, X.R., and Blackburn, G.M. (2000) The design of new carbonic anhydrase inhibitors. In *The Carbonic Anhydrases — New Horizons*, Chegwidden, W.R., Carter, N.D., and Edwards, Y.H. (Eds.), Birkhäuser, Basel, pp. 437–459.

Maren, T.H. (1967) Carbonic anhydrase: Chemistry, physiology, and inhibition. *Physiological Reviews* **47,** 595–781.

Maren, T.H. (2000) Carbonic anhydrase inhibition in ophthalmology: Aqueous humor secretion and the development of sulfonamide inhibitors. In *The Carbonic Anhydrases — New Horizons*, Chegwidden, W.R., Carter, N.D., and Edwards, Y.H. (Eds.), Birkhäuser, Basel, pp. 425–435.

Maren, T.H., and Conroy, C.W. (1993) A new class of carbonic anhydrase inhibitor. *Journal of Biological Chemistry* **268,** 26233–26239.

Maren, T.H., Rayburn, C.S., and Liddell, N.E. (1976) Inhibition by anions of human red cell carbonic anhydrase B: physiological and biochemical implications. *Science* **191,** 469–472.

Maren, T.H., and Sanyal, G. (1983) The activity of sulfonamides and anions against the carbonic anhydrases of animals, plants, and bacteria. *Annual Review of Pharmacology and Toxicology* **23,** 439–459.

Maryanoff, B.E., Costanzo, M.J., Nortey, S.O., Greco, M.N., Shank, R.P., Schupsky, J.J., Ortegon, M.P., and Vaught, J.L. (1998) Structure–activity studies on anticonvulsant sugar sulfamates related to topiramate. enhanced potency with cyclic sulfate derivatives. *Journal of Medicinal Chemistry* **41,** 1315–1343.

Maryanoff, B.E., Nortey, S.O., Gardocki, J.F., Shank, R.P., and Dodgson, S.P. (1987) Anticonvulsant *O*-alkyl sulfamates. 2,3:4,5-Bis-*O*-(1-methylethylidene)-beta-D-fructopyranose sulfamate and related compounds. *Journal of Medicinal Chemistry* **30,** 880–887.

Meirovitch, E., and Lanir, A. (1978) Electron spin resonance of manganese(II) complexes with proteins. *Chemical Physics Letters* **53,** 530–535.

Meldrum, N.U., and Roughton, F.J.W. (1933) Carbonic anhydrase: Its preparation and properties. *Journal of Physiology* **8,** 133-147.

Merz, K.M., Jr. (1990) Insights into the function of the zinc hydroxide-Thr199-Glu106 hydrogen bonding network in carbonic anhydrases. *Journal of Molecular Biology* **214,** 799–802.

Merz, K.M., Jr. (1991) Carbon dioxide binding to human carbonic anhydrase II. *Journal of the American Chemical Society* **113,** 406–411.

Merz, K.M., Jr., and Banci, L. (1996) Binding of azide to human carbonic anhydrase II: The role electrostatic complementarity plays in selecting the preferred resonance structure of azide. *Journal of Physical Chemistry* **100,** 17414–17420.

Moratal, J.M., Martinez-Ferrer, M.J., Donaire, A., and Aznar, L. (1992a) Thermodynamic parameters of the interaction between cobalt(II) bovine carbonic anhydrase and anionic inhibitors. *Journal of Inorganic Biochemistry* **45,** 65–71.

Moratal, J.M., Martinez-Ferrer, M.J., Jimenez, H.R., Donaire, A., Castells, J., and Salgado, J. (1992b) Proton NMR and UV-vis spectroscopic characterization of sulfonamide complexes of nickel(II)-carbonic anhydrase. Resonance assignments based on NOE effects. *Journal of Inorganic Biochemistry* **45,** 231–243.

Murakami, H., and Sly, W.S. (1987) Purification and characterization of human salivary carbonic anhydrase. *Journal of Biological Chemistry* **262,** 1382–1388.

Nair, S.K., Ludwig, P.A., and Christianson, D.W. (1994) Two-site binding of phenol in the active site of human carbonic anhydrase II: Structural implications for substrate association. *Journal of the American Chemical Society* **116,** 3659–3660.

Paranawithana, S.R., Tu, C., Laipis, P.J., and Silverman, D.N. (1990) Enhancement of the catalytic activity of carbonic anhydrase III by phosphates. *Journal of Biological Chemistry* **265,** 22270–22274.

Peng, Z., Merz, K.M., and Banci, L. (1993) Binding of cyanide, cyanate, and thiocyanate to human carbonic anhydrase II. *Proteins: Structure, Function, and Genetics* **17,** 203–216.

Pocker, Y., and Sarkanen, S. (1978) Carbonic anhydrase: Structure, catalytic versatility, and inhibition. *Advances in Enzymology and Related Areas of Molecular Biology* **47,** 149–274.

Pocker, Y., and Stone, J.T. (1965) The catalytic versatility of erythrocyte carbonic anhydrase. The enzyme-catalyzed hydrolysis of *p*-nitrophenyl acetate. *Journal of the American Chemical Society* **87,** 5497–5498.

Pocker, Y., and Stone, J.T. (1967) The catalytic versatility of erythrocyte carbonic anhydrase. 3. Kinetic studies of the enzyme-catalyzed hydrolysis of *p*-nitrophenyl acetate. *Biochemistry* **6,** 668–678.

Pocker, Y., and Stone, J.T. (1968) The catalytic versatility of erythrocyte carbonic anhydrase. VI. Kinetic studies of noncompetitive inhibition of enzyme-catalyzed hydrolysis of *p*-nitrophenyl acetate. *Biochemistry* **7,** 2936–2945.

Powers, J.C., and Harper, J.W. (1986) Inhibitors of metalloproteases. In *Research Monographs in Cell and Tissue Physiology,* Barett, A.J.S., and Salvensen, G. (Eds.), Elsevier, Amsterdam, pp. 219–298.

Pullman, A. (1981) Carbonic anhydrase: theoretical studies of different hypotheses. *Annals of the New York Academy of Sciences* **367,** 340–355.

Recacha, R., Costanzo, M. J., Maryanoff, B. E., and Chattopadhyay, D. (2002) Crystal structure of human carbonic anhydrase II complexed with an anti-convulsant sugar sulphamate. *Biochemical Journal* **361,** 437–441.

Roughton, F.J.W., and Booth, V.H. (1946) Effect of substrate concentration, pH, and other factors upon the activity of carbonic anhydrase. *Biochemical Journal* **40,** 319–330.

Rowlett, R.S., Gargiulo, N.J., III, Santoli, F.A., Jackson, J.M., and Corbett, A.H. (1991) Activation and inhibition of bovine carbonic anhydrase III by dianions. *Journal of Biological Chemistry* **266,** 933–941.

Sabers, A., and Gram, L. (2000) Newer anticonvulsants: Comparative review of drug interactions and adverse effects. *Drugs* **60,** 23–33.

Scolnick, L.R., Clements, A.M., Liao, J., Crenshaw, L., Hellberg, M., May, J., Dean, T.R., and Christianson, D.W. (1997) Novel binding mode of hydroxamate inhibitors to human carbonic anhydrase II. *Journal of the American Chemical Society* **119,** 850–851.

Scozzafava, A., Banciu, M.D., Popescu, A., and Supuran, C.T. (2000) Carbonic anhydrase inhibitors: Inhibition of isozymes I, II and IV by sulfamide and sulfamic acid derivatives: 85. *Journal of Enzyme Inhibition* **15,** 443–453.

Scozzafava, A., Saramet, I., Banciu, M.D., and Supuran, C.T. (2001) Carbonic anhydrase activity modulators: Synthesis of inhibitors and activators incorporating 2-substituted-thiazol-4-yl-methyl scaffolds. *Journal of Enzyme Inhibition* **16,** 351–358.

Scozzafava, A., and Supuran, C.T. (2000a) Carbonic anhydrase and matrix metalloproteinase inhibitors: sulfonylated amino acid hydroxamates with mmp inhibitory properties act as efficient inhibitors of ca isozymes I, II, and IV, and N-hydroxysulfonamides inhibit both these zinc enzymes. *Journal of Medicinal Chemistry* **43,** 3677–3687.

Scozzafava, A., and Supuran, C.T. (2000b) Protease inhibitors. Part 5. Alkyl/arylsulfonyl- and arylsulfonylureido-/arylureido- glycine hydroxamate inhibitors of *Clostridium histolyticum* collagenase. *European Journal of Medicinal Chemistry* **35,** 299–307.

Scozzafava, A., and Supuran, C.T. (2000c) Protease inhibitors: Synthesis of *Clostridium histolyticum* collagenase inhibitors incorporating sulfonyl-L-alanine hydroxamate moieties. *Bioorganic and Medicinal Chemistry Letters* **10,** 499–502.

Silverman, D.N., and Lindskog, S. (1988) The catalytic mechanism of carbonic anhydrase: Implications of a rate-limiting protolysis of water. *Accounts of Chemical Research* **21,** 30–36.

Simonsson, I., Jonsson, B.H., and Lindskog, S. (1982) Phenol, a competitive inhibitor of CO_2 hydration catalyzed by carbonic anhydrase. *Biochemical and Biophysical Research Communications* **108,** 1406–1412.

Supuran, C.T. (1992) Carbonic anhydrase activators. 4. A general mechanism of action for activators of isozymes I, II and III. *Revue Roumaine de Chimie* **37,** 411–421.

Supuran, C.T. (1994) Carbonic anhydrase inhibitors. In *Carbonic Anhydrase and Modulation of Physiologic and Pathologic Processes in the Organism*, Puscas, I. (Ed.), Helicon, Timisoara, pp. 29–111.

Supuran, C.T., Briganti, F., Mincione, G., and Scozzafava, A. (2000) Protease inhibitors: synthesis of L-alanine hydroxamate sulfonylated derivatives as inhibitors of *Clostridium histolyticum* collagenase. *Journal of Enzyme Inhibition* **15,** 111–128.

Supuran, C.T., Conroy, C.W., and Maren, T.H. (1994) Cyanate is a carbonic anhydrase substrate. *The European Bioinorganic Conference — EUROBIC II*, Florence, p. 306.

Supuran, C.T., Conroy, C.W., and Maren, T.H. (1995) Is cyanate a carbonic anhydrase substrate? *The 4th International Conference on Carbonic Anhydrases*, Oxford, U.K.

Supuran, C.T., Conroy, C.W., and Maren, T.H. (1997) Is cyanate a carbonic anhydrase substrate? *Proteins: Structure, Function, and Genetics* **27,** 272–278.

Supuran, C.T., and Manole, G. (1999) *The Carbonic Anhydrase Inhibitors: Syntheses, Reactions and Therapeutical Applications* (in Romanian), Romanian Academy Publishing House, Bucharest.

Supuran, C.T., and Scozzafava, A. (2000) Carbonic anhydrase inhibitors and their therapeutic potential. *Expert Opinion on Therapeutic Patents* **10,** 575-600.

Supuran, C.T., and Scozzafava, A. (2002a) Matrix metalloproteinase inhibitors. In *Proteinase and Peptidase Inhibition: Recent Potential Targets for Drug Development,* Smith, H.J., and Simons, C. (Eds.), Taylor & Francis, London, pp. 35–52.

Supuran, C.T., and Scozzafava, A. (2002b) Applications of carbonic anhydrase inhibitors and activators in therapy. *Expert Opinion on Therapeutic Patents* **12,** 217–242.

Taylor, J.S., and Coleman, J.E. (1971) Electron spin resonance of metallocarbonic anhydrases. *Journal of Biological Chemistry* **246,** 7058–7067.

Taylor, P.W., King, R.W., and Burgen, A.S.V. (1970) Influence of pH on kinetics of complex formation between aromatic sulfonamides and human carbonic anhydrase. *Biochemistry* **9,** 3894–3902.

Thorslund, A., and Lindskog, S. (1967) Studies of the esterase activity and the anion inhibition of bovine zinc and cobalt carbonic anhydrases. *European Journal of Biochemistry* **3,** 117–123.

Tibell, L., Forsman, C., Simonsson, I., and Lindskog, S. (1984) Anion inhibition of carbon dioxide hydration catalyzed by human carbonic anhydrase II. Mechanistic implications. *Biochimica et Biophysica Acta* **789,** 302–310.

Tibell, L., Forsman, C., Simonsson, I., and Lindskog, S. (1985) The inhibition of human carbonic anhydrase II by some organic compounds. *Biochimica et Biophysica Acta* **829,** 202–208.

Tilander, B., Strandeberg, B., and Fridborg, K. (1965) Crystal structure studies on human erythrocyte carbonic anhydrase C (II). *Journal of Molecular Biology* **12,** 740–760.

Tu, C., Wynns, G.C., and Silverman, D.N. (1981) Inhibition by cupric ions of ^{18}O exchange catalyzed by human carbonic anhydrase II. Relation to the interaction between carbonic anhydrase and hemoglobin. *Journal of Biological Chemistry* **256,** 9466–9470.

Vedani, A., Huhta, D.W., and Jacober, S.P. (1989) Metal-coordination, hydrogen-bond network formation, and protein-solvent interactions in native and complexed human carbonic anhydrase I: a molecular mechanics study. *Journal of the American Chemical Society* **111,** 4075–4081.

Vidgren, J., Liljas, A., and Walker, N.P.C. (1990) Refined structure of the acetazolamide complex of human carbonic anhydrase II at 1.9.ANG. *International Journal of Biological Macromolecules* **12,** 342–344.

Vidgren, J., Svensson, A., and Liljas, A. (1993) Refined structure of the aminobenzolamide complex of human carbonic anhydrase II at 1.9 A and sulphonamide modelling of bovine carbonic anhydrase III. *International Journal of Biological Macromolecules* **15,** 97–100.

Williams, T.J., and Henkens, R.W. (1985) Dynamic carbon-13 NMR investigations of substrate interaction and catalysis by cobalt(II) human carbonic anhydrase I. *Biochemistry* **24,** 2459–2462.

Wingo, T., Tu, C., Laipis, P.J., and Silverman, D.N. (2001) The catalytic properties of human carbonic anhydrase IX. *Biochemical and Biophysical Research Communications* **288,** 666–669.

Winum, J.-Y., Vullo, D., Casini A., Montero, J.-L., Scozzafava, A., and Supuran, C.T. (2003) Carbonic anhydrase inhibitors: Inhibition of cytosolic isozymes I and II and transmembrane, tumor-associated isozyme IX with sulfamates also acting as steroid sulfatase inhibitors. *Journal of Medicinal Chemistry,* **46,** 2187–2196.

Xue, Y., Vidgren, J., Svensson, L.A., Liljas, A., Jonsson, B.H., and Lindskog, S. (1993) Crystallographic analysis of Thr-200-His human carbonic anhydrase II and its complex with the substrate, bicarbonate ion. *Proteins: Structure, Function, and Genetics* **15,** 80–87.

Yachandra, V., Powers, L., and Spiro, T.G. (1983) X-ray absorption spectra and the coordination number of zinc and cobalt carbonic anhydrase as a function of pH and inhibitor binding. *Journal of the American Chemical Society* **105,** 6596–6604.

Zheng, Y.J., and Merz, K.M., Jr. (1992) Mechanism of the human carbonic anhydrase II-catalyzed hydration of carbon dioxide. *Journal of the American Chemical Society* **114,** 10498–104507.

8 Clinical Applications of Carbonic Anhydrase Inhibitors in Ophthalmology

Francesco Mincione, Luca Menabuoni and Claudiu T. Supuran

CONTENTS

8.1 Introduction 244
8.2 Sulfonamides in the Treatment of Glaucoma..244
8.3 Combination Antiglaucoma Therapy of CAIs with Other
    Pharmacological Agents..247
8.4 CAIs in Macular Edema, Macular Degeneration and Related Ocular
    Pathologies 251
References..252

Carbonic anhydrase inhibitors (CAIs), such as acetazolamide, methazolamide, ethoxzolamide and dichlorophenamide, were and still are widely used as systemic antiglaucoma drugs. Their mechanism of action involves inhibiting CA isozymes II and IV present in the ciliary processes of the eye, with the consequent reduction of bicarbonate and aqueous humor secretion, and of elevated intraocular pressure (IOP) characteristic of this disease. Because CAs II and IV are present in many other tissues and organs, generally systemic CAIs possess undesired side effects such as numbness and tingling of extremities, metallic taste, depression, fatigue, malaise, weight loss, decreased libido, gastrointestinal irritation, metabolic acidosis, renal calculi and transient myopia. To avoid these undesired side effects, topically effective CAIs have been recently developed. Two drugs are available clinically: dorzolamide (since 1995) and brinzolamide (since 1999). Both drugs are applied topically as aqueous solutions or suspensions, alone or in combination with other agents (such as β-blockers or prostaglandin derivatives) and produce a consistent and prolonged reduction of IOP. Furthermore, recent reports show both the systemically as well as topically acting sulfonamide CAIs to be effective in treating macular edema and other

0-415-30673-6/04/$0.00+$1.50
© 2004 by CRC Press LLC

macular degeneration diseases for which pharmacological treatment was unavailable until now. Much research is in progress in the search for even more effective topically acting CAIs, free of the inconveniences and side effects of the presently available drugs.

8.1 INTRODUCTION

The pioneering studies of Friedenwald (1949), Kinsey (1953) and Kinsey and Barany (1949) on the chemistry and dynamics of aqueous humor showed that the main constituent of this secretion is sodium bicarbonate. Next, Wistrand (1951) identified the CA in the anterior uvea of the eye and showed that this enzyme (present mainly in the ciliary processes) is responsible for bicarbonate secretion, as a consequence of the hydration reaction of carbon dioxide. Becker (1955) then showed that the sulfonamide CAI acetazolamide, **8.1**, produced a drop in the intraocular pressure (IOP) in experimental animals and humans, whereas Kinsey and Reddy (1959) proved that this phenomenon is due to a reduced bicarbonate secretion as a consequence of CA inhibition by the sulfonamide. This was the beginning of a novel treatment for glaucoma, a condition that is affecting an increasing number of people and is also the leading cause of blindness in Western countries (Sugrue 1989).

8.2 SULFONAMIDES IN THE TREATMENT OF GLAUCOMA

Glaucoma is a chronic, degenerative eye disease, characterized by high IOP, which causes irreversible damage to the optic nerve head, and, as a result, the progressive loss of visual function and eventually blindness (Bartlett and Jaanus 1989; Maren 1992, 1995). Elevated IOP (ocular hypertension) is generally indicative of an early stage of the disease (Bartlett and Jaanus 1989; Soltau and Zimmerman 2002; Hoyng and Kitazawa 2002). CAIs represent the most physiological treatment of glaucoma, because by inhibiting the ciliary process enzyme (the sulfonamide susceptible isozymes CA II and CA IV), a reduced rate of bicarbonate and aqueous humor secretion is achieved, which leads to a 25 to 30% decrease in IOP (Maren 1967, 1992, 1995). Acetazolamide **8.1**, methazolamide **8.2**, ethoxzolamide **8.3** or dichlorophenamide **8.4** were and are still extensively used as systemic drugs in the therapy of this disease (Maren 1967; Supuran and Scozzafava 2000, 2001, 2002; Supuran et al. 2003), as they all act as very efficient inhibitors of several CA isozymes (Table 8.1), principally of CA II and CA IV, the two isozymes involved in aqueous humor secretion (Maren 1967, 1992, 1995).

The best-studied drug is acetazolamide, which has been frequently administered for years because of its efficient reduction of IOP, very reduced toxicity and ideal pharmacokinetic properties (Wistrand and Lindqvist 1991). In long-term therapy, acetazolamide **8.1** is administered in doses of 250 mg every 6 h, whereas the more liposoluble, structurally related methazolamide **8.2** in doses of 25 to 100 mg thrice daily; an equal dosage of dichlorophenamide **8.4** is also useful in reducing ocular hypertension (Bartlett and Jaanus 1989). But as CAs are ubiquitous enzymes in vertebrates, administration of these systemic sulfonamides with such a high affinity

TABLE 8.1

Inhibition Data with Clinically Used Sulfonamides 8.1 to 8.6 Against Several α-CA Isozymes

Isozyme[a]	K_I (nM)					
	8.1	8.2	8.3	8.4	8.5	8.6
hCA I	200	10	1	350	50,000	nt[b]
hCA II	10	8	0.7	30	9	3
hCA III	3×10^5	1×10^5	5000	nt	8000	nt
hCA IV	66	56	13	120	45	45
mCA V	60	nt	5	nt	nt	nt
hCA VI	1100	560	nt	nt	nt	nt
mCA VII	16	nt	0.5	nt	nt	nt

[a] h: human; m: murine isozyme.

[b] nt: not tested (no data available).

Source: From Supuran, C.T., and Scozzafava, A. (2000) *Expert Opinion on Therapeutic Patents* **10**, 575–600. With permission.

for the enzyme (Table 8.1) leads to CA inhibition in organs other than the target one (i.e., the eye), and, as a result, to undesired side effects of these drugs. The most frequent ones are numbness and tingling of extremities, metallic taste, depression, fatigue, malaise, weight loss, decreased libido, gastrointestinal irritation, metabolic acidosis, renal calculi and transient myopia (Bartlett and Jaanus 1989; Maren 1992, 1995; Sugrue 1989). Although producing all these unpleasant side effects, systemic sulfonamide CAIs of the type mentioned previously (compounds **8.1** to **8.4**) are particularly useful to manage glaucoma resistant to other antiglaucoma therapies and in the control of acute glaucoma attacks (Kaur et al. 2002). Even if the launch of the topically acting compounds **8.5** and **8.6** (see later) initially seemed to have produced a marked loss of interest in acetazolamide and other systemic sulfona- mides, there are several reasons why acetazolamide and the related drugs **8.2** – **8.4** still find a place in the clinical armamentarium of antiglaucoma drugs: (1) dorzola- mide **8.5** reduces IOP by up to 23% as monotherapy and an extra 15% when combined with timolol, whereas acetazolamide **8.1** alone reduces IOP by 30% (Maren 1967; Sugrue 2000); (2) dorzolamide inhibits aqueous flow by 17%, whereas acetazolamide inhibits this parameter by 30% (Maus et al. 1997); (3) ocular burning and stinging was reported in 12 to 19% of patients undergoing topical dorzolamide treatment, whereas the other topically acting sulfonamide available, brinzolamide (**8.6**), is much less effective than acetazolamide and dorzolamide in reducing high IOP [IOP reduction of 20% when brinzolamide was administered thrice a day as a 1% suspension (Silver 2000)]. Thus, systemically acting sulfonamide CAIs are on many occasions preferred by ophthalmologists to manage glaucoma, mainly because of their well-known pharmacological properties, low toxicity and much lower cost than that of the newer drugs dorzolamide and brinzolamide.

The idea to administer topically, directly into the eye, a sulfonamide CAI was initially addressed by Becker (1955). This and other studies involving the clinically used compounds **8.1** to **8.4** gave negative results only, and it was concluded that sulfonamide CAIs are effective as antiglaucoma drugs only via the systemic route (Bartlett and Jaanus 1989; Maren 1992). The lack of efficiency of sulfonamides **8.1** to **8.4** via the topical route was attributed to the drug's inability to arrive at the ciliary processes where CA is present (Maren 1995). Inadequate drug penetration through the cornea was due to the fact that sulfonamides **8.1** to **8.4** possess inappropriate physicochemical properties for such a route of administration. In 1983, in a seminal paper Maren et al. (1983) postulated that a water-soluble sulfonamide possessing a relatively balanced lipid solubility (to be able to penetrate through the cornea) as well as sufficiently strong CA inhibitory properties would be an effective IOP-lowering drug via the topical route. However, at that time, no inhibitors possessing such physicochemical properties existed, as the bioorganic chemistry of this class of compounds had remained relatively unexplored (Supuran and Scozzafava 2001; 2001). Water-soluble sulfonamide CAIs started to be developed in several laboratories soon thereafter, and by 1995 the first such pharmacological agent, dorzolamide **8.5**, was launched by Merck & Co. (Sugrue 2000). A second compound, brinzolamide **8.6**, developed by Alcon Laboratories, structurally very similar to dorzolamide, was approved for the topical treatment of glaucoma (Silver 2000). The detailed history of the development of the topically active sulfonamide CAIs is presented in Chapter 4 of this book.

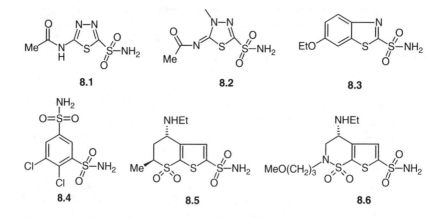

Dorzolamide **8.5** and brinzolamide **8.6** are nanomolar CA II and CA IV inhibitors (Table 8.1), possess a good water solubility, are sufficiently liposoluble to penetrate through the cornea and can be administered topically, directly into the eye, as a 2% aqueous solution (of the dorzolamide hydrochloride salt) or as a 1% suspension (as the brinzolamide hydrochloride salt) twice or thrice a day (Sugrue 2000; Silver 2000). The two drugs are effective in reducing IOP and show fewer side effects as compared with the systemically applied drugs. The observed side effects include stinging, burning or reddening of the eye, blurred vision, pruritus and a bitter taste (Sugrue 2000; Silver 2000). All but the last are probably experienced because dorzolamide (the best studied topical CAI) is the salt of a weak base with a very strong acid, so that the pH of the drug solution is rather acidic (generally ca. 5.5). The bitter taste is probably due to drug-laden lachrymal fluid draining into the oropharynx and inhibition of CA present in the saliva (CA VI) and the taste buds (CA II and CA VI), with the consequent accumulation of bicarbonate, and has been seen both with systemic as well as topical CAIs (Maren 1967, Sugrue 2000). Less is known at present about the side effects of brinzolamide, but it seems that this drug produces less stinging but more blurred vision as than dorzolamide (Sugrue 2000; Silver 2000). Unfortunately, dorzolamide has already shown some more serious side effects, such as contact allergy (Aalto-Korte 1998), nephrolithiasis (Carlsen et al. 1999), anorexia, depression and dementia (Thoe Schwartzenberg and Trope 1999) and irreversible corneal decompensation in patients who already presented corneal problems (Konowal et al. 1999). Thus, even if dorzolamide and brinzolamide represent major progress in the fight against glaucoma with therapies based on CAIs, novel types of topically effective inhibitors belonging to this class of pharmacological agents are urgently needed and sought for.

One approach recently reported for obtaining novel types of such agents was to attach water-solubilizing moieties to the molecules of aromatic/heterocyclic sulfonamides (Supuran and Scozzafava 2001; Supuran et al. 2003). Such moieties include pyridine-carboximido, carboxypyridine-carboxamido, quinolinesulfonamido, picolinoyl, isonicotinoyl, perfluoroalkyl/arylsulfonyl-, as well as amino acyl groups, whereas ring systems derivatized by using these moieties include 2-, 3- or 4-aminobenzenesulfonamides, 4-(ω-aminoalkyl)-benzenesulfonamides, 3-halogeno-substituted-sulfanilamides, 1,3-benzene-disulfonamides, 1,3,4-thiadiazole-2-sulfonamides, benzothiazole-2-sulfonamides and thienothiopyran-2-sulfonamides, and were chosen in such a way as to demonstrate that the proposed approach is a general one (Borras et al. 1999; Casini et al. 2000; Menabuoni et al. 1999; Mincione et al. 1999; Scozzafava et al. 1999a, 1999b, 2000, 2002; Supuran et al. 1999). Compounds such as **8.7** to **8.12** showed two to three times more effective topical IOP-lowering effects in rabbits than dorzolamide did (Borras et al. 1999; Casini et al. 2000; Menabuoni et al. 1999; Mincione et al. 1999; Scozzafava et al. 1999a, 1999b, 2000, 2002; Supuran et al. 1999). They possessed good water solubility (as hydrochlorides, triflates or trifloroacetates), inhibition potency in the low nanomolar range against hCA II (the figures after the number characterizing the compound represent the inhibition constant against hCA II), good penetrability through the cornea and very good IOP-lowering properties in both normotensive and glaucomatous rabbits (Table 8.2). More importantly, this effect lasted for a prolonged period of time as compared with the similar effect of dorzolamide.

8.3 COMBINATION ANTIGLAUCOMA THERAPY OF CAIs WITH OTHER PHARMACOLOGICAL AGENTS

Another approach for improving topically administered CAIs, mainly dorzolamide or brinzolamide, consisted in formulating the eye drops containing the two sulfonamides in combination with other topically or systemically acting antiglaucoma

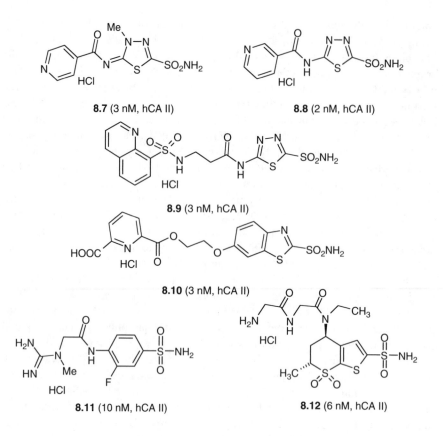

8.7 (3 nM, hCA II)   **8.8** (2 nM, hCA II)

8.9 (3 nM, hCA II)

8.10 (3 nM, hCA II)

8.11 (10 nM, hCA II)   **8.12** (6 nM, hCA II)

TABLE 8.2
**Fall in IOP of Normotensive Rabbits after Treatment with
One Drop (50 μl) Solution (2%) of CAIs of Types 8.7 to 8.12
Directly into the Eye 30, 60 and 90 min after Administration**

Inhibitor	pH[a]	t = 0	ΔIOP (mmHg)[b]		
			t = 30 min	t = 60 min	t = 90 min
Dorzolamide	5.5	0	2.2 ± 0.1	4.1 ± 0.15	2.7 ± 0.1
8.7	6.5	0	5.9 ± 0.2	11.2 ± 0.5	13.1 ± 0.3
8.8	6.5	0	5.4 ± 0.1	10.9 ± 0.4	12.5 ± 0.3
8.9	5.5	0	4.8 ± 0.1	8.2 ± 0.1	7.0 ± 0.2
8.10	7.0	0	4.0 ± 0.2	7.2 ± 0.1	9.0 ± 0.2
8.11	7.5	0	3.0 ± 0.1	7.2 ± 0.2	5.5 ± 0.3
8.12	5.5	0	4.9 ± 0.1	8.7 ± 0.3	6.9 ± 0.4

[a] The pH of the ophthalmic solution used in the experiments.
[b] ΔIOP = IOP$_{\text{control eye}}$ − IOP$_{\text{treated eye}}$; mean ± average spread ($n = 3$).

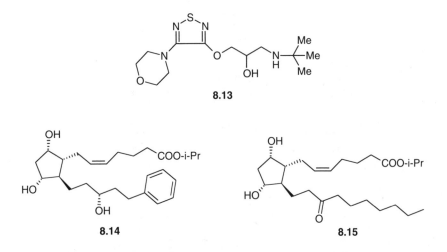

8.13

8.14   **8.15**

drugs such as β-blockers, prostaglandins or acetazolamide (Wayman et al. 1997; Alm et al. 1998; Larsson and Alm 1998; Vanlandingham and Brubaker 1998; Higginbotham et al. 2002). Xanthan gum (0.5%), which contains a high-molecular-weight polysaccharide possessing pseudoplastic properties and therefore an unusually high ocular penetration associated to dorzolamide hydrochloride (2% aqueous solution) has also been used in association with dorzolamide, which is more efficient as an IOP-lowering agent in this combination (Sugrue 1999). A 2% aqueous solution of dorzolamide hydrochloride (or brinzolamide hydrochloride, 1% suspension) was shown to effectively lower IOP when associated with the β-blocker timolol **8.13** (0.5%), with the prostaglandin derivatives latanoprost **8.14** (0.005–0.1%) or unoprostone **8.15** (0.005–0.1%) or with systemically administered acetazolamide (Waymann et al. 1997; Alm et al. 1998; Larsson and Alm 1998; Vanlandingham and Brubaker 1998; Higginbotham et al. 2002). In all these cases, an additive effect of ca. 15% on IOP lowering of the two drugs has been observed, whereas side effects seemed to be reduced or of diminished intensity as compared with those of each drug alone.

Complexes of several 1,3,4-thiadiazole-2-sulfonamide derivatives of types **8.16** to **8.19**, possessing strong CA inhibitory properties with hydroxypropyl-β-cyclodextrin (HPβCD) have also recently been investigated for IOP-lowering properties, it being known that the association of ophthalmologic drugs with cyclodextrins can lead to enhanced penetration of the drug within the eye (Maestrelli et al. 2002). Although the investigated CAIs **8.16** to **8.19** possessed very powerful inhibitory properties against the two CA isozymes involved in aqueous humor production within the eye, i.e., CA II and CA IV (in the low nanomolar range), these compounds were topically ineffective as IOP-lowering agents in normotensive or hypertensive rabbits, because of their very low water solubility and lack of penetratation through the cornea and consequent nonarrival at the ciliary processes enzymes. On the contrary, the cyclodextrin–sulfonamide complexes proved to be effective and long-lasting IOP-lowering agents in the two animal models of glaucoma, leading to strong IOP lowering, as shown in Figure 8.1, in which data for the standard drug dorzola-

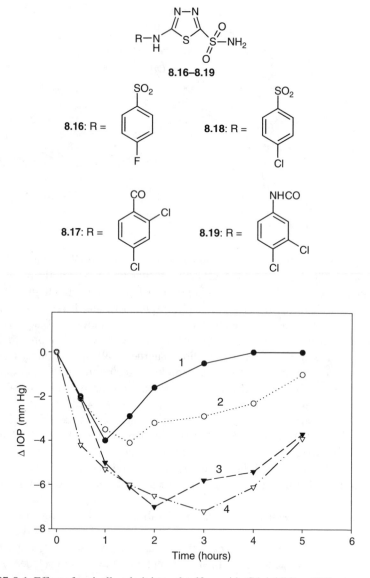

FIGURE 8.1 Effect of topically administered sulfonamide CA inhibitors (2% aqueous solutions) on the IOP of normotensive albino rabbits. Curve 1: dorzolamide (hydrochloride salt, pH 5.5); curve 2: complex **8.16**–HPβCD (pH 7.4); curve 3: complex **8.17**–HPβCD (pH 7.4); curve 4: complex **8.18**–HPβCD (pH 7.4).

mide are also provided. It is obvious that association of sulfonamide CAIs with cyclodextrins leads to a very efficient IOP lowering in this glaucoma animal model, constituting a thought-provoking approach for developing antiglaucoma medications for human use.

8.4 CAIs IN MACULAR EDEMA, MACULAR DEGENERATION AND RELATED OCULAR PATHOLOGIES

Optic nerve blood flow is diminished in the eyes of primary open-angle glaucoma suspects and patients (Pilts-Seymour et al. 2001). Because sulfonamides with CAI properties act as vasodilators (Supuran and Scozzafava 2000), this might explain the use of such drugs for treating retinal edema and age-related macular degeneration. In consequence, these pharmacological agents represent a new approach for improving visual function. Retinal edema (also referred to as cystoid macular edema) consists of a swelling process within the critically important central visual zone, and might develop in association with a variety of ocular conditions, such as diabetic retinopathy, ischemic retinopathies, intraocular surgery (such as cataract procedures) or laser photocoagulation (Cox et al. 1988; Grover et al. 1997; Sponsel et al. 1997; Barnes et al. 2000). It is also common in patients affected by retinitis pigmentosa, a hereditary disorder leading to total blindness (Orzalesi et al. 1993). The precise mechanism by which the swelling process is triggered is uncertain, but natural metabolic toxins might play a role in this disease (Sponsel et al. 1997). Macular degeneration is characterized by fluid accumulation in the outer retina, accompanied by lipofuscin (a metabolic waste product) accumulation between photoreceptors and the villi of the retina pigment epithelium. These catabolic waste products (also called drusen) are generally cleared by the blood in the healthy retina, but with aging they tend to accumulate and coalesce, so that vast areas of the retinal photoreceptors become disengaged from their neighboring retinal pigment epithelial villi (Cox et al. 1988; Grover et al. 1997; Sponsel et al. 1997; Barnes et al. 2000). As a consequence of drusen confluencing in the foveal area, the affected sections of the retina become blind, which can trigger a macular degenerative disease with dramatic loss of the visual function. No satisfactory therapy for this condition is currently known (Sponsel et al. 1997; Barnes et al. 2000).

The use of CAIs to treat macular edema is based on the important observation of Cox et al. (1988) that acetazolamide **8.1** (as sodium salt) is effective in treating this condition when administered systemically. A similar efficiency has also recently been reported for dorzolamide (Grover et al. 1997; Sponsel et al. 1997) after topical administration (without the side effects of the systemic inhibitor **8.1**). It is generally assumed that the disappearance of the edema and the improvement of visual function are independent of the hypotensive activity of the sulfonamide, being due to direct effects of the drug on the circulation in the retina (Sponsel et al. 1997). Practically, acetazolamide, dorzolamide or brinzolamide act as local vasodilators (Supuran and Scozzafava 2000) and improve blood flow in this organ, thereby clearing metabolic waste products, drusen, etc. The improvement of visual function after such a treatment (in early phases of the disease) seems to be very good (Sponsel et al. 1997).

It has also been reported that acetazolamide (375 mg/d) is effective in treating serous retinal detachment of various etiologies (this disease is also generally not amenable to treatment) (Gonzalez 1992). Additional studies are needed to establish whether the topically acting sulfonamide CAIs might be also used in this serious medical problem.

REFERENCES

Aalto-Korte, K. (1998) Contact allergy to dorzolamide eyedrops. *Contact Dermatitis* **39**, 206.

Alm, A. (1998) Prostaglandin derivates as ocular hypotensive agents. *Progress in Retinal and Eye Research* **17**, 291–312.

Barnes, G.E., Li, B., Dean, T., and Chandler, M.L. (2000) Increased optic nerve head blood flow after 1 week of twice daily brinzolamide treatment in Dutch-belted rabbits. *Survey in Ophthalmology* **44** (Suppl. 2), 131–140.

Bartlett, J.D., and Jaanus, S.D. (1989) *Clinical Ocular Pharmacology*, Butterworth, Boston, pp. 254–263.

Becker, B. (1955) The mechanism of the fall in intraocular pressure by the carbonic anhydrase inhibitor Diamox. *American Journal of Ophthalmology* **39**, 177–183.

Borras, J, Scozzafava, A, Menabuoni, L, Mincione, F, Briganti, F, Mincione, G., and Supuran, C.T. (1999) Carbonic anhydrase inhibitors: Synthesis of water-soluble, topically effective intraocular pressure lowering aromatic/heterocyclic sulfonamides containing 8-quinoline-sulfonyl moieties: is the tail more important than the ring? *Bioorganic and Medicinal Chemistry* **7**, 2397–2406.

Carlsen, J., Durcan, J., Swartz, M., and Crandall, A. (1999) Nephrolithiasis with dorzolamide. *Archives of Ophthalmology* **117**, 1087–1088.

Casini, A., Scozzafava, A., Mincione, F., Menabuoni, L., Ilies, M.A., and Supuran, C.T. (2000) Carbonic anhydrase inhibitors: Water soluble 4-sulfamoylphenylthioureas as topical intraocular pressure lowering agents with long lasting effects. *Journal of Medicinal Chemistry* **43**, 4884–4892.

Cox, S.N., Hay, E., and Bird, A.C. (1988) Treatment of chronic macular edema with acetazolamide. *Archives of Ophthalmology* **106**, 1190–1195.

Friedenwald, J.S. (1949) The formation of the intraocular fluid. *American Journal of Ophthalmology* **32**, 9–27.

Gonzalez, C. (1992) Serous retinal detachment: Value of acetazolamide. *Journal Francais d' Ophthalmologie* **15**, 529–536.

Grover, S., Fishman, G.A., Fiscella, R.G., and Adelman, A.E. (1997) Efficacy of dorzolamide hydrochloride in the management of chronic cystoid macular edema in patients with retinitis pigmentosa. *Retina* **17**, 222–231.

Higginbotham, E.J., Diestelhorst, M., Pfeiffer, N., Rouland, J.F., and Alm, A. (2002) The efficacy and safety of unfixed and fixed combinations of latanoprost and other anti-glaucoma medications. *Survey of Ophthalmology* **47**, S133–S140.

Hoyng, P.F.J., and Kitazawa, Y. (2002) Medical treatment of normal tension glaucoma. *Survey of Ophthalmology* **47**, S116–S124.

Kaur, I.P., Smitha, R., Aggarwal, D., and Kapil, M. (2002) Acetazolamide: Future perspectives in topical glaucoma therapeutics. *International Journal of Pharmaceutics* **248**, 1–14.

Kinsey, V.E. (1953) Comparative chemistry of of aqueous humor in posterior and anterior chambers of rabbit eye. *Archives of Ophthalmology* **50**, 401–417.

Kinsey, V.E., and Barany, E. (1949) The rate flow of aqueous humor. II. Derivation of rate of flow and its physiologic significance. *American Journal of Ophthalmology* **32**, 189–202.

Kinsey, V.E., and Reddy, D.V.N. (1959) Turnover of total carbon dioxide in aqueous humors and the effect thereon of acetazolamide. *Archives of Ophthalmology* **62**, 78–83.

Konowal, A., Morrison, J.C., Brown, S.V., Cooke, D.L., Maguire, L.J., Verdier, D.V., Fraunfelder, F.T., Dennis, R.F., and Epstein, R.J. (1999) Irreversible corneal decompensation in patients treated with topical dorzolamide. *American Journal of Ophthalmology* **127**, 403–406.

Larsson, L.I., and Alm, A. (1998) Aqueous humor flow in human eyes treated with dorzola-mide and different doses of acetazolamide. *Archives of Ophthalmology* **116**, 19–24.

Maestrelli, F., Mura, P., Casini, A., Mincione, F., Scozzafava, A., and Supuran, C.T. (2002) Cyclodextrin complexes of sulfonamide carbonic anhydrase inhibitors as long lasting topically acting antiglaucoma agents. *Journal of Pharmaceutical Sciences* **91**, 2211–2219.

Maren, T.H. (1967) Carbonic anhydrase: chemistry, physiology and inhibition. *Physiological Reviews* **47**, 595–781.

Maren, T.H. (1992) Role of carbonic anhydrase in aqueous humor and cerebrospinal fluid formation. In *Barriers and Fluids of the Eye and Brain*, Segal, M.B. (Ed.), MacMillan Press, London, 1992, pp. 37–48.

Maren, T.H. (1995) The development of topical carbonic anhydrase inhibitors. *Journal of Glaucoma* **4**, 49–62.

Maren, T.H., Jankowska, L., Sanyal, G., and Edelhauser, H.F. (1983) The transcorneal per-meability of sulfonamide carbonic anhydrase innhibitors and their effect on aqueous humor secretion. *Experimental Eye Research* **36**, 457–480.

Maus, T.L., Larsson, L.I., McLaren, J.W., and Brubaker, R.F. (1997) Comparison of dorzola-mide and acetazolamide as suppresors of aqueous humor flow in humans. *Archives of Ophthalmology* **115**, 45–49.

Menabuoni, L., Scozzafava, A., Mincione, F., Briganti, F., Mincione, G., and Supuran, C.T. (1999) Carbonic anhydrase inhibitors: Water-soluble, topically effective intraocular pressure lowering agents derived from isonicotinic acid and aromatic/heterocyclic sulfonamides — Is the tail more important than the ring? *Journal of Enzyme Inhibition* **14**, 457–474.

Mincione, G., Menabuoni, L., Briganti, F., Scozzafava, A., Mincione, F., and Supuran, C.T. (1999) Carbonic anhydrase inhibitors. Part 79. Synthesis and ocular pharmacology of topically acting sulfonamides incorporating GABA moieties in their molecule, with long-lasting intraocular pressure lowering properties. *European Journal of Phar-maceutical Sciences* **9**, 185–199.

Orzalesi, N., Pierrottet, C., Porta, A., and Aschero, M. (1993) Long-term treatment of retinitis pigmentosa with acetazolamide: A pilot study. *Graefes Archives of Clinical and Experimental Ophthalmology* **231**, 254–256.

Piltz-Seymour, J.R., Grunwald, J.E., Hariprasad, S.M., and DuPont, J. (2001) Optic nerve blood flow is diminished in eyes of primary open-angle glaucoma suspects. *American Journal of Ophthalmology* **132**, 63–69.

Scozzafava, A., Briganti, F., Mincione, G., Menabuoni, L., Mincione, F., and Supuran, C.T. (1999a) Carbonic anhydrase inhibitors: Synthesis of water-soluble, amino acyl/dipep-tidyl sulfonamides possessing long-lasting intraocular pressure-lowering properties via the topical route. *Journal of Medicinal Chemistry* **42**, 3690–3700.

Scozzafava, A., Menabuoni, L., Mincione, F., Briganti, F., Mincione, G., and Supuran, C.T. (1999b) Carbonic anhydrase inhibitors: Synthesis of water-soluble, topically effective intraocular pressure lowering aromatic/heterocyclic sulfonamides containing cationic or anionic moieties: is the tail more important than the ring? *Journal of Medicinal Chemistry* **42**, 2641–2650.

Scozzafava, A., Menabuoni, L., Mincione, F., Briganti, F., Mincione, G., and Supuran, C.T. (2000) Carbonic anhydrase inhibitors: Perfluoroalkyl/aryl-substituted derivatives of aromatic/heterocyclic sulfonamides as topical intraocular pressure lowering agents with prolonged duration of action. *Journal of Medicinal Chemistry* **43**, 4542–4551.

Scozzafava, A., Menabuoni, L., Mincione, F., and Supuran, C.T. (2002) Carbonic anhydrase inhibitors: A general approach for the preparation of water soluble sulfonamides incorporating polyamino-polycarboxylate tails and of their metal complexes possessing long lasting, topical intraocular pressure lowering properties. *Journal of Medicinal Chemistry* **45**, 1466–1476.

Silver, L.H. (2000) Dose-response evaluation of the ocular hypotensive effect of brinzolamide ophthalmic suspension (Azopt). Brinzolamide dose-response study group. *Survey in Ophthalmology* **44** (Suppl. 2), 147–153.

Soltau, J.B., and Zimmerman, T.J. (2002) Changing paradigms in the medical treatment of glaucoma. *Survey of Ophthalmology* **47**, S2–S5.

Sponsel, W.E., Harrison, J., Elliott, W.R., Trigo, Y., Kavanagh, J., and Harris, A. (1997) Dorzolamide hydrochloride and visual function in normal eyes. *American Journal of Ophthalmology* **123**, 759–766.

Sugrue, M.F. (1989) The pharmacology of antiglaucoma drugs. *Pharmacology and Therapy* **43**, 91–138.

Sugrue, M.F. (2000) Pharmacological and ocular hypotensive properties of topical carbonic anhydrase inhibitors. *Progress in Retinal and Eye Research* **19**, 87–112.

Supuran, C.T., and Scozzafava, A. (2000) Carbonic anhydrase inhibitors and their therapeutic potential. *Expert Opinion on Therapeutic Patents,* **10**, 575–600.

Supuran, C.T., and Scozzafava, A. (2001) Carbonic anhydrase inhibitors. *Current Medicinal Chemistry — Immunologic, Endocrine and Metabolic Agents* **1**, 61–97.

Supuran, C.T., and Scozzafava, A. (2002) Applications of carbonic anhydrase inhibitors and activators in therapy. *Expert Opinion on Therapeutic Patents* **12**, 217–242.

Supuran, C.T., Scozzafava, A., and Casini, A. (2003) Carbonic anhydrase inhibitors. *Medicinal Research Reviews* **23**, 146–189.

Supuran, C.T., Scozzafava, A., Menabuoni, L., Mincione, F., Briganti, F., and Mincione, G. (1999) Carbonic anhydrase inhibitors. Part 70. Synthesis and ocular pharmacology of a new class of water-soluble, topically effective intraocular pressure lowering agents derived from nicotinic acid and aromatic/heterocyclic sulfonamides. *European Journal of Medicinal Chemistry* **34**, 799–808.

Thoe Schwartzenberg, G.W., and Trope, G.E. (1999) Anorexia, depression and dementia induced by dorzolamide eyedrops (Trusopt). *Canadian Journal of Ophthalmology* **34**, 93–94.

Vanlandingham, B.D., and Brubaker, R.F. (1998) Combined effect of dorzolamide and latanoprost on the rate of aqueous humor flow. *American Journal of Ophthalmology* **126**, 191–196.

Wayman, L., Larsson, L.I., Maus, T., Alm, A., and Brubaker, R. (1997) Comparison of dorzolamide and timolol as suppressors of aqueous humor flow in humans. *Archives of Ophthalmology* **115**, 1368–1371.

Wistrand, P.J. (1951) Carbonic anhydrase in the anterior uvea of the rabbit. *Acta Physiologica Scandinavica* **24**, 144–148.

Wistrand, P.J., and Lindqvist, A. (1991) Design of carbonic anhydrase inhibitors and the relationship between the pharmacodynamics and pharmacokinetics of acetazolamide. In *Carbonic Anhydrase — From Biochemistry and Genetics to Physiology and Clinical Medicine,* Botrè F., Gros G., and Storey B.T. (Eds.), VCH, Weinheim, pp. 352–378.

9 Cancer-Related Carbonic Anhydrase Isozymes and Their Inhibition

Silvia Pastoreková and Jaromír Pastorek

CONTENTS

9.1 Introduction 256
9.2 Identification, Structure and Distribution of CA IX 257
9.3 Regulation of CA IX Expression...259
9.4 CA IX as a Marker of Tumor Hypoxia...260
9.5 Identification, Distribution and Regulation of CA XII 263
9.6 Catalytic Properties and Proposed Roles of CA IX and CA XII..............265
9.7 Influence of Extracellular pH on Tumor Progression and Possible
      Implication of CA Activity 270
9.8 CA Inhibition as an Approach to Anticancer Therapy...............................270
9.9 Inhibition Profile of CA IX 272
9.10 Conclusions 274
Acknowledgments...274
References..274

Changes in tumor metabolism and microenvironment connected with adaptation of cells to hypoxia are important components of tumor progression. Transmembrane carbonic anhydrase (CA) isozymes CA IX and CA XII containing extracellular enzyme active sites appear to participate in this process via their ability to catalyze hydration of CO_2 to bicarbonate and a proton and regulate intratumoral pH. In addition, CA IX, possessing a unique N-terminal domain, can perturb E-cadherin-mediated cell–cell adhesion via interaction with β-catenin and potentially contribute to tumor invasion. CA IX shows restricted expression in normal tissues but is tightly associated with different types of tumors, mostly because of its strong induction by tumor hypoxia that involves HIF-1 binding to a hypoxia response element in the *CA9* promoter. CA IX has been proposed to serve as a marker of tumor hypoxia and its predictive and prognostic potential has been demonstrated in numerous clinical studies. CA XII is present in many normal tissues and overexpressed in some tumors. It is also induced by hypoxia, but the underlying molecular mechanism remains undetermined. Both CA IX and CA XII are negatively regulated by von Hippel-Lindau tumor suppressor protein and their expression in renal cell carcinomas is related to inactivating mutations of the *VHL* gene. High catalytic activity of these two CA isoforms supports their role in acidification of the tumor microenvironment, which facilitates acquisition of the metastatic phenotype. Therefore, modulating extracellular tumor pH by inhibiting CA activity is a promising approach to anti-cancer therapy. Sulfonamide CAIs have been shown to compromise tumor cell proliferation and invasion *in vitro* and improve the effect of conventional chemo-therapy *in vivo*; however, their precise targets are not known. Because the tumor-associated CA IX is a sulfonamide-avid enzyme, it is possible that at least some of the observed anticancer effects of the sulfonamide inhibitors are mediated by inhi-bition of this CA isoform. Moreover, the inhibition profile of CA IX differs from those of CA I, II and IV, indicating that the design and synthesis of CA IX-specific inhibitors with therapeutic properties could in principle be feasible.

9.1 INTRODUCTION

In the course of cancer development and progression, cells undergo multiple genetic and epigenetic alterations that lead to accumulation of aberrant phenotypic features, which significantly affect tumor metabolism and microenvironment. The hallmark of cancer is deregulated proliferation and survival; however, to achieve this, malig-nant cells not only have to turn on positive regulatory pathways and turn off negative control genes but also adapt metabolic performance to assure higher availability of basic substrates for synthesis of their components (Evan and Vousden 2001; Hanahan and Weinberg 2000). To allow sustained expansion, tumor cells need to overcome additional restrictions, e.g., those connected with insufficient oxygen supply (i.e., hypoxia) to areas distant from existing blood vessels. They do all this by dramatic changes in the gene expression profile, which are accompanied by a switch in energy metabolism from oxidative respiration to anaerobic glycolysis (Dang and Semenza 1998). These changes further influence tumor phenotype, triggering neoangiogene-sis, clonal selection of more aggressive cells and acidification of extracellular milieu, which facilitates invasion (Semenza 2001). At the same time, to prevent apoptosis and protect different intracellular processes, intracellular pH is maintained within a narrow range by the modified activity of numerous ion transporters, pumps, exchang-ers and channels (Stubbs et al. 2000). As an overall consequence, tumor physiology only partially reflects the physiology of normal tissues, because many processes are abnormally shifted or amplified to compensate for the acquired imbalance.

This rather fragmented picture is far from the real complexity of the tumor phenotype, but illustrates several situations that might require CA catalytic activity in the reversible hydration of carbon dioxide to bicarbonate and a proton. As explained in Chapter 1, via the catalysis of this simple but fundamental reaction, CAs participate in myriad physiological processes in normal tissues. These almost ubiquitous enzymes exist in mammals in at least 14 isoforms, of which 11 are active catalysts (CAs I–VII, CAs IX, XII–XIV) and 3 contain a conserved but inactive CA domain (CA-RPs VIII, X and XI). In addition, there are two CA-related receptor protein tyrosine phosphatases (RPTPβ and γ) whose CA domains are also inactive and possess different cellular functions. The active isozymes are major players in mammalian physiology, acting in ion transport and acid–base balance and assisting in respiration, formation of body fluids, renal acidification, etc. (Chegwidden et al. 2000). It is conceivable that CA activity might also be well exploited in tumors. Improved production of bicarbonate ions might be needed for the synthesis of nucleotide and lipid precursors in highly proliferating tumor cells. In addition, tumors often display a reversed pH gradient across the plasma membrane when compared with normal tissues, and CA activity can significantly contribute to this altered pH regulation by providing protons for acidification of the microenvironment and bicarbonate ions to maintain a neutral intracellular milieu (Chegwidden et al. 2000; Stubbs et al. 2000).

Despite the apparent demand for a CA function in malignant transformation, efforts to find a clear-cut relationship between expression of CAs and tumor phenotype had remained generally unsuccessful until the discovery of the tumor-associated isozymes CA IX and CA XII.

9.2 IDENTIFICATION, STRUCTURE AND DISTRIBUTION OF CA IX

CA IX was originally detected in a human carcinoma cell line HeLa as a cell density-regulated membrane antigen named MN (Pastoreková et al. 1992). Shortly after, it was recognized that expression of the MN antigen correlates with tumorigenic phenotype of somatic cell hybrids of HeLa and normal human fibroblasts (Závada et al. 1993). The same study revealed an association of MN with tumor cell lines and surgical tumor specimens and its absence from the corresponding normal tissues. This made MN antigen an interesting subject for evaluation as a tumor marker, and the first thorough immunohistochemical examination demonstrated its diagnostic potential for carcinomas of the cervix uteri (Liao et al. 1994).

Later, isolation and sequence analysis of MN cDNA and the corresponding gene disclosed the presence of a large CA domain located in the extracellular part of the encoded protein (Pastorek et al. 1994; Opavský et al. 1996). This CA domain showed 40.8 and 35.8% identity with secreted isozyme CA VI and cytosolic isozyme CA II, respectively, and contained a perfectly conserved and active enzyme site. Moreover, exon–intron distribution in the genomic region coding for the CA domain of MN antigen was similar to those of other α-CA genes. At that time, this was the ninth mammalian CA identified, and thus, based on the suggestion of Hewett-Emmett and Tashian (1996), the MN protein was renamed CA IX.

Sequencing data provided information that allowed deduction of the overall CA IX composition. In addition to a central CA domain, CA IX contains a transmembrane anchor followed by a short C-terminal cytoplasmic tail. The N-terminal side of the CA IX molecule is extended with a so-called PG-like region that is homologous to the keratan sulfate attachment domain of a large proteoglycan aggrecan (Opavský et al. 1996). This PG-like region is absent from the other CA isozymes

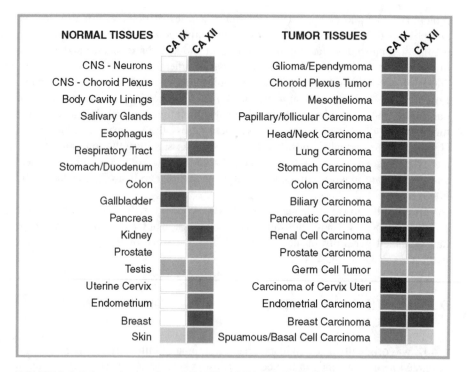

FIGURE 9.1 Schematic overview of CA IX and CA XII distribution in normal human tissues and derived tumors. Intensity of the gray tone in rectangles corresponds to both level and frequency of expression according to literary data cited in the text.

known at present and thus represents a unique feature that appears to endow CA IX with the capacity to act in cell adhesion (Zavada et al. 2000; Svastova et al. 2003).

Further studies have confirmed the earlier observations that CA IX exhibits a particular expression pattern (Figure 9.1). In fact, only a few normal tissues have been found to express CA IX. These belong predominantly to the gastrointestinal tract (Pastoreková et al. 1997). The most abundant expression is detected in all major cell types of the gastric mucosa, including the surface pit cells, parietal cells and glandular chief cells. CA IX is also present in the epithelia of the small intestine and proximal colon, but its level decreases toward the rectum. Intestinal CA IX is confined to the epithelial cells in the cryptal area possessing greatest proliferative activity (Saarnio et al. 1998b). CA IX is also abundant in the gallbladder mucosa, whereas pancreatic ducts show a weak expression. In all gastrointestinal epithelia, CA IX is present in the basolateral membranes, suggesting its possible involvement in intercellular communication and maintenance of tissue integrity. This assumption agrees with a recently described phenotype of CA IX-deficient mice constructed by targeted *Car9* gene disruption (Ortova Gut et al. 2002). The knockout mice displayed gastric hyperplasia with aberrant cell lineage development, resulting in an increased number of surface pit cells and a decreased number of glandular chief cells. However, they did not show any significant change in gastric pH, hydrochloric acid production and systemic electrolyte status. Although it is possible that other gastric CA isozymes complement the enzyme activity of CA IX, the hyperplastic phenotype supports the role of CA IX in gastric morphogenesis and control of cell proliferation and differentiation. Variable degrees of CA IX expression were later detected in the lining cells of the body cavity, rete ovarii, rete testis and ductular efferens, in vetricular linings of CNS and in the choroid plexus, but most normal tissues remained negative (Ivanov et al. 2001; Karhumaa et al. 2001).

In contrast to the relatively limited presence of CA IX in normal tissues, the spectrum of cancers expressing CA IX has successively expanded to various types of benign and malignant tumors derived from the kidney, esophagus, colon, lung, pancreas, liver, endometrium, ovary, brain, skin and breast (Liao et al. 1997; McKiernan et al. 1997; Turner et al. 1997; Saarnio et al. 1998a; Vermylen et al. 1999; Kivela et al. 2000; Saarnio et al. 2001; Ivanov et al. 2001; Bartosova et al. 2002). Tumors originating from tissues with high natural expression, such as hepatobiliary epithelial tumors, revealed decreasing levels of CA IX with increasing grades of dysplasia and carcinoma (Saarnio et al. 2001). A similar phenomenon was observed in gastric carcinomas (Pastoreková et al. 1997; Leppilampi et al. 2003), in accord with the proposed involvement of CA IX in differentiation of gastrointestinal epithelia. However, tumors derived from CA IX-negative tissues showed its ectopic activation (Figure 9.1). The most striking observation was the very intense and homogeneous expression of CA IX in a high percentage of renal clear cell carcinomas (Liao et al. 1997; McKiernan et al. 1997). In nonsmall cell lung carcinomas, CA IX was absent from the preneoplastic lesions but present in 80% of malignant tumors (Vermylen et al. 1999). Aberrant expression of CA IX was detected in colorectal tumors, wherein it correlated with proliferation evaluated according to the Ki-67 index, on which basis it has been proposed that it serves as a marker of increased proliferation in the colorectal mucosa (Saarnio et al. 1998a).

9.3 REGULATION OF CA IX EXPRESSION

Broad carcinoma-related distribution indicated that expression of CA IX might represent a more general attribute of tumor tissues or that it might be regulated by a mechanism or pathway common to many tumor types. A sequence comparison between *CA9* cDNA derived from HeLa carcinoma cells and that from the normal human stomach did not show any mutation in the coding region, suggesting that mutations do not play any role in the differential expression of CA IX. Rather, an involvement of other regulatory events, e.g., tumor-specific transcription factors, has been proposed (Pastoreková et al. 1997). Initial efforts to characterize the *CA9* promoter and analyze its transcriptional regulation led to a description of important regulatory elements that bind transcription activators AP-1, AP-2 and SP-1 and identification of a silencer region implicated in repression of *CA9* transcription in nontumorigenic HeLa × fibroblast hybrid cells (Kaluz et al. 1999). Further study has shown that synergistic cooperation between SP and AP-1 transcription factors is necessary for *CA9* transcriptional activity (Kaluzova et al. 2001).

However, the first clue to a major pathway involved in CA IX control was given by a demonstration of its downregulation by the wild-type von Hippel–Lindau (VHL) tumor suppressor protein (Ivanov et al. 1998). In this study, CA IX expression was not suppressed by the mutated variants of pVHL that lacked the elongin-binding domain, which is required for the interaction of pVHL with elongin C and integration within a ubiquitin ligase complex (Iwai et al. 1999; Lisztwan et al. 1999). This important finding explained overexpression of CA IX in the majority of renal cell carcinomas (RCCs) that frequently carry an inactivating mutation in the *VHL* gene (Gnarra et al. 1994). In some RCC cell lines with mutated *VHL*, expression of CA IX was also regulated by promoter methylation, an epigenetic mechanism involved in control of many cancer-related genes (Ashida et al. 2002). Recently, expression of CA IX was demonstrated as a very early sign of premalignant lesions in VHL patients (Mandriota et al. 2002).

In tumors other than RCC, pVHL plays a critical role as an upstream negative regulator of an α-subunit of the hypoxia-inducible transcription factor (HIF; Maxwell et al. 1999). In normoxic tumor cells with adequate supply of oxygen, pVHL binds an HIF-α subunit modified by prolyl hydroxylases, targets it to proteasomal degradation and thus abrogates its functioning in transcriptional activation of downstream genes (Jaakkola et al. 2001; Ivan et al. 2001). In hypoxia, which frequently occurs in tumors as a result of aberrant vasculature, pVHL fails to recognize HIF-α because of the absence of prolyl hydroxylation (Figure 9.2). This leads to HIF-α stabilization, translocation to nucleus, heterodimerization with a constitutive HIF-β subunit, formation of the HIF transcription complex and activation of a large spectrum of genes involved in the response of tumor cells to hypoxia, including those triggering neo-angiogenesis (VEGF), erythropoesis (EPO-1), glucose transport (GLUT-1, GLUT-3), glycolysis (LDHA) and other adaptive processes (Semenza 2001; Maxwell 2001).

Based on the relationship between pVHL and CA IX on the one hand and between pVHL and hypoxia on the other, it was finally construed and proven that hypoxia is the major factor underlying the widespread expression of CA IX in tumors (Wykoff et al. 2000). Hypoxic regulation of *CA9* gene is accomplished via HIF-1 binding to the hypoxia response element (HRE), which was defined in the minimal *CA9* promoter at position −10/−3 with respect to the transcription initiation site (Wykoff et al. 2000). The tight regulation by the HIF/pVHL system was reflected in the pattern of CA IX expression within tumors. In RCC tumors defective for VHL, CA IX upregulation was associated with generalized activation of the hypoxia-related pathway, whereas in non-VHL tumors, CA IX exhibited focal perinecrotic pattern linked to hypoxia (Wykoff et al. 2000).

9.4 CA IX AS A MARKER OF TUMOR HYPOXIA

Because hypoxia is a clinically important tumor parameter that has significant impact on the treatment outcome and disease progression (Höckel and Vaupel 2001), the discovery that CA IX is a HIF target has opened a new and exciting era in CA IX research. During the subsequent period, many studies of CA IX expression in hypoxic tumors were performed in a hope that it might potentially serve as an intrinsic marker of tumor hypoxia and possibly also as a therapeutic target. CA IX distribution was often examined in relation to the extent of necrosis as an indicator

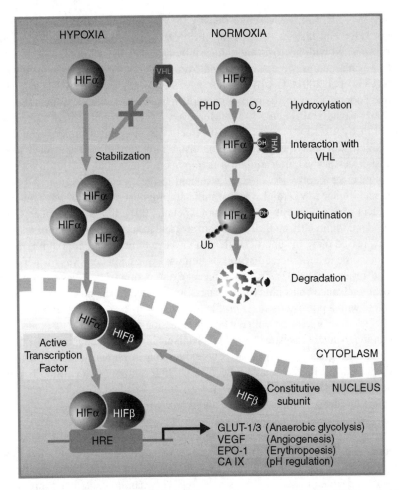

FIGURE 9.2 Mechanism of hypoxia-induced gene expression mediated by HIF transcription factor. At normal oxygen levels (normoxia), prolyl-4-hydroxylases (PHDs) hydroxylate two conserved proline residues of HIF-α. The von Hippel–Lindau protein (VHL) binds hydroxylated HIF-α and targets it for degradation by the ubiquitin-proteasome system. Under hypoxia, HIF-α is not hydroxylated, because PHDs are inactive in absence of dioxygen. Nonhydroxylated HIF-α is not recognized by VHL protein; it is stabilized and accumulates. After translocation to nucleus, HIF-α dimerizes with the HIF-β constitutive subunit to form the active transcription factor. HIF transcription factor then binds to the hypoxia response element (HRE) in target genes and activates their transcription. Target genes include glucose transporters (GLUT-1 and GLUT-3), which participate in glucose metabolism; vascular endothelial growth factor (VEGF), which triggers neoangiogenesis; erythropoietin (EPO-1), which is involved in erythropoiesis; CA IX, which is proposed to contribute to pH regulation; and additional genes with functions in cell survival, proliferation, metabolism and other processes (reviewed in Semenza, G.L. (2001) *Trends in Molecular Medicine* **7**, 345–350).

of severe hypoxia, to microvascular density (MVD) as a measure of angiogenesis, to tumor stage and disease progression. In breast carcinomas, CA IX was significantly associated with necrosis and a high tumor grade (Wykoff et al. 2001) as well as with a worse relapse-free survival and overall survival in patients with invasive tumors (Chia et al. 2001). CA IX was also associated with necrosis, MVD, advanced stage and poor response to chemoradiotherapy in head and neck carcinomas (Beasley et al. 2001; Koukourakis et al. 2001). Moreover, CA IX expression correlated with the level of hypoxia measured by needle electrodes in cervical tumors, wherein it was a significant and independent prognostic indicator of overall survival and metastasis-free survival after radiation therapy (Loncaster et al. 2001). In nonsmall cells lung cancer, expression of CA IX related to the extent of necrosis, advanced T stage and higher MVD, and was a significant prognostic factor of poor outcome, independent of angiogenesis (Giatromanolaki et al. 2001). In bladder cancer, CA IX was found predominantly on the luminal surface and in surrounding areas of necrosis. It was expressed more in superficial than in invasive tumors, and although it did not predict outcome in superficial disease, its luminal expression deserves further investigation (Turner et al. 2002). The most recent examination of biopsies from patients with locally advanced nasopharyngeal carcinoma treated by chemoradiation showed that tumors with a positive hypoxic profile (defined as high expression of both CA IX and HIF-1α) were associated with a worse progression-free survival (Hui et al. 2002).

In an independent investigation of mechanisms involved in adaptation of tumors to hypoxia, using serial analysis of gene expression in human glioblastoma cells, CA9 displayed the highest magnitude of induction among 32 identified hypoxia-responsive genes (Lal et al. 2001). Of the 12 genes selected, CA9 was induced in the highest number of tumor cell lines and was the most consistently induced gene in human solid tumors.

Analysis of spheroids and tumor xenografts generated from human cervical carcinoma and glioma cells confirmed that CA IX-expressing cells are clonogenic, more likely to be resistant to killing by ionizing radiation and bind significantly more pimonidazole, a chemical marker of hypoxia, than do cells that express little or no CA IX (Olive et al. 2001).

All these studies clearly demonstrate the potential clinical utility of CA IX as an intrinsic marker of hypoxia in a large spectrum of tumors and justify its further investigation as a prognostic indicator and a therapeutic target. However, almost all papers describing the CA IX expression pattern in comparison with other hypoxia markers showed some discrepancies in their intratumoral distribution. CA IX was more widespread than VEGF in bladder cancer (Turner et al. 2002). The extent of GLUT-1 was higher than that of CA IX in breast carcinoma, and the proportion of breast tumors positive for CA IX was much higher than for HIF-1α (Tomes et al. 2003). There was a lack of association between CA IX and VEGF and only borderline association between CA IX and HIF-1α in nonsmall cell lung carcinoma (Giatromanolaki et al. 2001), and CA IX staining extended beyond the regions binding pimonidazole in cervical carcinoma (Olive et al. 2001). These differences suggest supplementary or alternative modes of CA IX regulation.

It has been repeatedly shown that CA IX expression is induced by high cell density (Pastoreková et al. 1992; Závada et al. 1993; Lieskovska et al. 1999). Recent investigations of density-regulated *CA9* transcription revealed that the mechanism involves lowered pericellular oxygen without stabilization of HIF-1α, and requires cooperation of HRE with the juxtaposed promoter element through the separate but interdependent pathways of PI3 kinase activation and a subhypoxic level of HIF-1α (Kaluz et al. 2002). Moreover, in some tumors, CA IX is coexpressed with proteins involved in angiogenesis, apoptosis inhibition and cell–cell adhesion disruption, including oncoproteins EGFR and c-ErbB2, and thus it is plausible that it might be also regulated by the oncogenic pathways (Giatromanolaki et al. 2001; Bartosova et al. 2002). Finally, the specific expression pattern of CA IX might be related to its high posttranslational stability in reoxygenated cells, which was proposed in tumor studies as a reason for lack of full overlap with other hypoxic markers that are either short-lived (e.g., HIF-1α) or secreted (e.g., VEGF; Turner et al. 2002). Pulse-chase analysis determined the CA IX protein half-life to be ca. 38 h and showed that it is independent of the duration of hypoxia (Rafajova et al. 2004). This high protein stability allows for the long persistence of CA IX in reoxygenated tumor areas and might contribute to adaptation of tumor cells to reoxygenation.

Despite hypoxia remaining the major factor underlying CA IX expression in non-VHL human tumors, these adverse regulatory pathways can considerably affect its overall distribution pattern and their understanding can have important implications for clinical interpretation of immunohistochemical data as well as for the use of CA IX as a therapeutic target.

9.5 IDENTIFICATION, DISTRIBUTION AND REGULATION OF CA XII

It is probably not a coincidence that the second tumor-associated isozyme CA XII is also regulated by hypoxia (Wykoff et al. 2000). Its cDNA sequence was published in 1998 by two independent groups and allowed the classification of CA XII as a CA IX/CA VI-related transmembrane protein with an extracellularly exposed enzyme domain containing all three histidines needed for catalytic activity (Türeci et al. 1998; Ivanov et al. 1998). Türeci et al. (1998) identified CA XII in a human renal cell carcinoma by serological expression screening with autologous antibodies. They cloned and sequenced the corresponding cDNA and proved that its mRNA is overexpressed in ca. 10% of RCC patients. Ivanov et al. (1998) cloned CA XII as a novel pVHL target by using RNA differential display. They showed that the expression of *CA12* mRNA is strongly inhibited by the wild-type pVHL in RCC cell lines, suggesting that it is subjected to similar regulation as *CA9*. However, suppression of *CA12* requires both the central pVHL domain involved in the HIF-1α binding and the C-terminal elongin-binding domain, whereas only the latter is needed for the negative regulation of *CA9*. In addition to this important finding, the same study attempted to determine chromosomal positions of *CA9* and *CA12* genes by fluorescence *in situ* hybridization with the corresponding cDNAs (Ivanov et al. 1998). The *CA12* locus was placed to 15q22 in accordance with the position determined by Türeci et al. (1998) as well as with the genome sequencing data (International Human Genome Sequencing Consortium 2001). The *CA9* locus was misplaced to position 17q21.2–21.3 because the genome sequencing data revealed that it is local-ized at 9p13 (International Human Genome Sequencing Consortium 2001).

Türeci et al. (1998) and Ivanov et al. (1998) have also shown that *CA12* mRNA is expressed in adult kidney, pancreas, colon, prostate, ovary, testis, lung and brain (Figure 9.1). The CA XII protein was found in normal endometrium (Karhumaa et al. 2000), colon (Kivela et al. 2000) and kidney (Parkkila et al. 2000a), suggesting an important physiological role for this enzyme in ion transport and fluid concen-tration. In the comparative immunohistochemical study, expression of CA XII and CA IX in normal adult tissues was detected in highly specialized cells and for most tissues did not overlap (Ivanov et al. 2001).

In contrast, some tumors displayed coexpression of CA XII and CA IX (Figure 9.1) mostly in the perinecrotic areas, in accordance with the fact that both isozymes are induced by hypoxia (Wykoff et al. 2000; Ivanov et al. 2001). However, *CA12* gene does not contain any HRE element that would correspond to the HRE of *CA9* present in the promoter region close to the transcription initiation site (and found also in the mouse *Car9* gene). In fact, there are additional candidate core HRE sequences in the *CA9* 5′ upstream region that are more distant from the transcription start (Table 9.1). The upstream region of the *CA12* gene possesses several putative

TABLE 9.1

Consensus Sequences of Hypoxia Response Elements (HREs) in the 5′ Regions of the Human *CA9* Gene, Its Mouse *Car9* Counterpart and Human *CA12* Gene

	TGCACGTA	CACGY	RCGTG
CA9	−10/−3	−8/−4	−1410/−1406
			−1711/−1707
			−2129/−2125
Car9	−1/+7	+1/+6	−181/−177
		−1349/−1345	−879/−875
			−1533/−1529
CA12		−515/−511	−1046/−1042
		−245/−241	−417/−413
		+28/+32	

Note: Positions of HRE are numbered with respect to the transcription initiation site. Sequence of the functional *CA9* HRE that is localized on the negative strand close to the transcription start site is written above the first column. It is found also in the mouse *Car9* gene. Putative core HRE sequences are present more upstream on both strands of all three genes (second and third column).

HRE elements with a core HIF-binding sequence, but their functionality has not been examined so far. According to tissue distribution and *in vitro* experiments, CA XII does not appear to be so tightly regulated by the hypoxia/pVHL pathway and so strongly linked to cancer as CA IX (Wykoff et al. 2000; Ivanov et al. 2001). In the breast, CA IX expression is rare in normal epithelium and in benign lesions occurring primarily in preinvasive DCIS and invasive breast carcinomas. It is significantly associated with necrosis and a high tumor grade. However, CA XII is frequently expressed in normal breast tissue as well as in benign, preinvasive and invasive breast lesions. In DCIS, expression of CA XII is associated with the absence of necrosis and low-grade lesions. This indicates that hypoxia might be a dominant factor in the regulation of CA IX, and that factors related to differentiation dominate the regulation of CA XII (Wykoff et al. 2001). In the colon, CA XII is normally expressed by the differentiated surface epithelium and increased basal/deep mucosal expression is associated with increasing dysplasia and an invasive tumor stage (Kivela et al. 2000). CA IX is normally expressed in the crypts and is abnormally induced in the superficial area in adenomas and carcinomas (Saarnio et al. 1998a). These differences imply the existence of factors or pathways that differentially regulate expression of *CA9* and *CA12* and modulate their response to the pVHL/hypoxia pathway involved in control of both genes.

9.6 CATALYTIC PROPERTIES AND PROPOSED ROLES OF CA IX AND CA XII

The finding that CA IX and CA XII are overexpressed in tumors because of the absence of functional pVHL led to the assumption that these CA isozymes might be involved in sensing and maintaining the acidic tumor environment, a fundamental property of solid tumors (Ivanov et al. 1998). This assumption has become even more meaningful when hypoxia entered the scene as an important positive regulator of *CA9* and *CA12* (Wykoff et al. 2000), because metabolic changes triggered by hypoxia significantly contribute to the acidic extracellular milieu of tumor cells (Stubbs et al. 2000). This idea is quite conceivable, because both CA IX and CA XII contain well-conserved enzyme active sites and exert considerable enzyme activity (Wingo et al. 2001; Ulmasov et al. 2001). Wingo et al. (2001) have shown that the CA IX catalytic domain produced in bacteria displays ca. 55% of the hydration activity of the most efficient isozyme CA II, whereas its proton transfer rate is a little higher than that of CA II. Similarly, the recombinant catalytic domain of CA XII exhibits ca. 23% of the CA II activity but has a slightly lower proton transfer rate (Ulmasov et al. 2001). These data demonstrate that CA IX and CA XII are highly active enzymes; however, the activity values obtained with single enzyme domains expressed in a prokaryotic system might not fully correspond to the situation in eukaryotic cells and tissues carrying full-length native proteins. Differences in sequences, domain composition and folding can significantly influence the enzyme activities (Figure 9.3).

Crystallographic data show that CA XII is a bitopic protein whose short intracellular C-termini are placed in an opposite position to the enzyme domains with

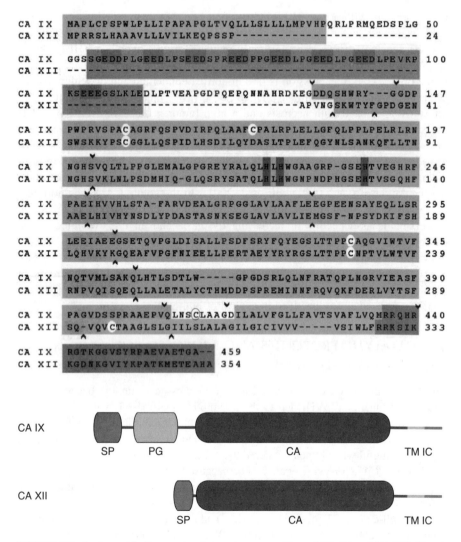

FIGURE 9.3 Amino acid alignment and domain composition of CA IX and CA XII. Signal peptides (SP), proteoglycan-related region (PG), carbonic anhydrase domains (CA), transmembrane regions (TM) and intracellular tails (IC) are designated by different background colors. Acidic tripeptides present in the PG region of CA IX are enclosed in medium gray rectangles and cysteine residues are encircled. Exon borders are indicated by arrowheads. Schemes of the proteins below the alignment illustrate the domain composition of their monomeric forms.

active-site clefts facing the extracellular space (Whittington et al. 2001). CA domains associate to form an isologous dimer that is presumably further stabilized by two putative amino acid signature GXXXG and GXXXS motifs in the transmembrane portion of the protein. CA XII contains two extracellular glycosylation sites (N52 and N136), and a single disulfide linkage is present between C23 and C203 [numbered according to Whittington et al. (2001)]. An additional cysteine residue is localized close to the transmembrane region. The intracellular domain contains potential phosphorylation sites with a possible role in regulating structure, enzyme activity and signal transduction.

CA IX differs from CA XII in that it forms trimers linked by disulfidic bonds, as revealed by mobility in nonreducing PAGE (Pastoreková et al. 1992). Putative transmembrane dimerization motifs of CA XII are absent from the CA IX sequence. There is only one predicted extracellular glycosylation site (N346), but the CA domain contains four cysteine residues and at least some of them appear to be required for intermolecular oligomerization. The cytoplasmic tail of CA IX shows high homology to that of CA XII and also contains putative phosphorylation sites. Deletion of this region results in loss of proper transport and plasma membrane localization of CA IX despite the presence of the transmembrane anchor (Svastova et al. submitted), suggesting involvement of this region in protein–protein interactions at the cytoplasmic side of the plasma membrane.

However, the major difference between the two cancer-related CA isozymes is in the presence of a unique PG-like region located at the N-terminus of CA IX but not of CA XII (Figure 9.3). This domain contains an imperfect repetitive sequence of six amino acids and has been implicated in cell adhesion. Zavada et al. (2000) have shown that it is required for CA IX-mediated cell adhesion to a solid support, suggesting its possible role in cell–matrix interactions. This adhesion was abrogated by the monoclonal antibody M75, which binds to the epitope within the repetitive region and by the peptides derived from this region. The PG domain also appears to be important for the capacity of hCA IX to reduce cell–cell adhesion of MDCK cells. Svastova and coworkers (2003) have recently shown that in these cells, CA IX colocalizes with a key cell–cell adhesion molecule E-cadherin, responds by internalization to same external stimuli (i.e., calcium depletion and hypoxia) as E-cadherin and weakens intercellular adhesion by destabilization of E-cadherin links to the cytoskeleton via a mechanism that involves direct binding of CA IX to β-catenin. The effect of CA IX is pronounced in hypoxia, wherein it increases dissociation of MDCK cells. This observation conforms to a recent proposal that hypoxia can initiate tumor invasion via decreased E-cadherin-mediated cell–cell adhesion and offers a possibility that CA IX is functionally implicated in this process (Beavon 1999).

Involvement of CA IX in cell adhesion represents a novel aspect of CA function, the concept related RPTPβ, expressed in the form of a proteoglycan, regulates cellular processes linked to cell adhesion and differentiation. However, this function of RPTPβ is executed by the inactive CA domain via binding of the neuronal cell recognition molecule contactin. It cannot be excluded that the CA domain of CA IX is used in a manner analogous to that for RPTPβ and that an interplay between PG and CA domains is needed for its full biological activity.

Cooperation between PG and CA domains might be relevant also for the enzyme activity of CA IX. PG domain contains many acidic amino acid residues arranged in tripeptide clusters (Figure 9.3). Similar peptides, externally added to CA II isozyme, were shown to substantially enhance CA II catalytic performance (Scozzafava and Supuran 2002). Thus, it is possible that the acidic PG domain might work as an intrinsic modulator of CA IX, cross-talking with the catalytic domain.

To add more complexity to this view, additional data have to be taken into account, despite not being directly obtained with CA IX and CA XII. During the last few years, there has been remarkable progress in understanding how CAs function to ensure efficient ion flux and pH regulation. Sterling and coworkers (2001, 2002a, 2002b) have shown that CAs are directly involved in the transport metabolon, an operative protein complex consisting of a membrane-bound CA IV that interacts with an extracellular part of transmembrane anion exchangers (AE1, AE2, AE3 and DRA), which in turn bind to cytosolic CA II at the intracellular side. The interaction between CA II and AE1 requires the acidic binding motif LDADD in the C-terminal region of AE1, which also seems to be responsible for the stimulation of CA II catalytic activity (Vince and Reithmeier 2000). The external peptides enhancing CA II activity were derived from this motif (Scozzafava and Supuran 2002). These data show that CAs are closely associated functionally with bicarbonate transport proteins, including anion exchangers and natrium/bicarbonate cotransporters (Sterling et al. 2002b; Li et al. 2002). They potentiate the bicarbonate transport activity by a push–pull mechanism, in which the CA-mediated production of bicarbonate on one side of the membrane provides push for its movement across the membrane and the CA-mediated conversion to CO_2 on the other side provides pull by bicarbonate minimization (Sterling et al. 2002a).

There is no reason to deny that CA IX and CA XII might participate in the analogous metabolon in tumor cells. Actually, many tumor cells maintain the expression of CA II, which might serve as an intracellular pull component of the metabolon (Parkkila et al. 1995a, 1995b; Bekku et al. 2000). The performance of such a tumor metabolon might then be regulated by availability of either ion transporter or CA IX/CA XII. In support of the possible functional cooperation between CA IX/CA XII and ion transporters in tumor cells, Karumanchi and coworkers (2001) identified AE2 as a candidate pVHL target whose mRNA and protein levels were upregulated by wild-type pVHL, but its transport activity was reduced to 50%. They proposed that downregulation of the AE2 exchange activity might represent secondary adaptation to pVHL-mediated downregulation of CA IX/CA XII, leading then to compensatory upregulation of the AE2 level. This proposal is in accordance with the observation of Sterling and coworkers (2002a) that inhibition of CA IV abrogated activity of AEs in the transport metabolon.

Translation of all this data into a simplified model relevant for the tumor phenotype suggests that loss of pVHL negative regulatory function either due to inactivating mutations or due to hypoxia might lead to upregulation of CA IX/CA XII expression or activity, or both; assembly of the putative transport metabolon via incorporation of transmembrane ion transporters and CA II; and finally to activation of ion transport. In this putative metabolon, extracellular CA IX/CA XII might preferentially catalyze hydration of CO_2, producing bicarbonate, which is efficiently transported to the intracellular space and dehydrated back to CO_2, which then leaves the cells by diffusion. The dehydration reaction relies on consumption of intracellular protons and might help neutralize the cell interior, whose pH is regulated by a very complex interplay of proton extrusion paths (Figure 9.4). On the other hand, outside CO_2 hydration generates not only bicarbonate but also protons that remain in extracellular milieu and can contribute to its acidification.

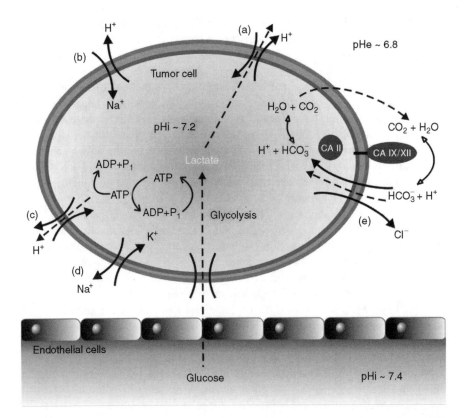

FIGURE 9.4 Mechanisms of pH regulation and ion transport in tumors. Lactate ion and H^+ produced by glycolysis are extruded to the extracellular space by the monocarboxylate carrier (a) and acidify extracellular pH (pH_e). The Na^+/H^+ antiport (b) activated in transformed cells exports H^+ and imports Na^+. Vacuolar H^+ pump (c) and ATP-dependent Na^+/K^+ antiport (d) further contribute to maintenance of neutral intracellular pH (pH_i). According to a recent view, HCO_3^- generated from interstitial CO_2 via catalysis by extracellular CAs enters the cell through the HCO_3^-/Cl^- anion exchanger (e). In the cytoplasm, CA II catalyzes conversion of HCO_3^- to CO_2, consuming intracellular H^+. CO_2 leaves the cell by diffusion and can be used for further hydration reactions catalyzed by extracellular CAs. H^+ produced from this reaction appears to contribute to acidification of pH_e. (Adapted from Stubbs, M. et al. (2000) *Molecular Medicine Today* **6**, 15–19 and modified to include CA activity.)

In accord with this model, human CA IX protein ectopically expressed in MDCK cells elicited acidification of the culture medium. This acidification occurred only under hypoxia and was not observed in mock-transfected control cells. Its inhibition by the membrane impermeant sulfonamide supported involvement of CA IX enzyme activity (Svastova et al. submitted). Because CA IX was expressed from the constitutive promoter, this finding indicates that hypoxia can regulate also the catalytic performance of CA IX.

9.7 INFLUENCE OF EXTRACELLULAR PH ON TUMOR PROGRESSION AND POSSIBLE IMPLICATION OF CA ACTIVITY

It had been believed for long that acidification of the tumor microenvironment is caused by accumulation of lactic acid produced during aerobic and anaerobic glycolysis, coupled with the poor removal of this catabolite by the inadequate tumor vasculature (Stubbs et al. 2000). However, experiments with glycolysis-deficient cells indicated that the production of lactic acid via glycolysis is not the only mechanism leading to tumor acidity. The deficient cells did not produce lactic acid and did not acidify culture medium when exposed to either aerobic or hypoxic conditions, but formed acidic tumors *in vivo* (Newell et al. 1993). Moreover, a recently published comparison of the metabolic profiles of glycolysis-impaired and parental cells in both *in vitro* and *in vivo* models revealed that CO_2, in addition to lactic acid, was a significant source of acidity in tumors (Helmlinger et al. 2002). These results strongly imply the contribution of CAs that is now emerging as a new paradigm.

Low extracellular pH has been associated with tumor progression via multiple effects, including upregulation of growth factors (e.g., IL-8, VEGF, bFGF) and proteases (e.g., MMPs, cathepsin B), degradation of the interstitial matrix, loss of intercellular adhesion, increased migration and metastasis, increased rate of mutations and reduced nucleotide excision repair (Xu and Fidler 2000; Fukumura et al. 2001; D'Arcangelo et al. 1999; Kato et al. 1992; Martinez-Zaguilan et al. 1996; Reynolds et al. 1996). It is a prominent feature of the reaction-diffusion model of cancer invasion that conveys a selective advantage to tumor cells invading normal tissues (Gatenby and Gawlinski 1996). Acidic pH can also influence the uptake of anticancer drugs exhibiting weakly electrolytic properties and impose varying effects on the response of tumor cells to conventional anticancer therapy. Manipulation of the extracellular pH was found to improve the cytotoxic efficacy of the weakly basic drugs mitoxantrone, paclitaxel and topotecan (Vukovic and Tannock 1997). Raghunand and coworkers. (1999) have shown that the tumor growth delay induced by the weak base doxorubicin was enhanced by increasing the extracellular pH of tumors following chronic ingestion of a sodium bicarbonate solution. This short summary suggests that modulation of interstitial tumor pH might have therapeutic value.

9.8 CA INHIBITION AS AN APPROACH TO ANTICANCER THERAPY

Although there are no complete data that would link the phenotypic or therapeutic effects of CA inhibition as a means of tumor pH manipulation to perturbed activity of particular CA isozymes, the literature available so far clearly indicates that this is an exploitable avenue toward treating cancer. Teicher and co-workers (1993) showed that acetalozamide, a prototype carbonic anhydrase inhibitor (CAI) of several CA isozymes, reduced *in vivo* growth of tumors when given alone and produced additive tumor growth delays when administered in combination with various chemotherapeutic agents. Parkkila and co-workers (2000b) investigated the effect of acetazolamide on the invasive capacity of renal carcinoma cell lines and found that a 10-μM concentration in the culture medium inhibited relative cell invasion rate through the matrigel membrane by 18 to 74%. Based on the levels of CA isoenzymes, this effect was attributed to the inhibition of CA II or CA XII, or both. The other two examined isoforms were either absent (CA IV) or present in only in one cell line (CA IX). However, the possible role of CA IX in this process warrants further investigation, especially in light of the current knowledge that its level and activity are regulated by hypoxia, whereas the previous experiments were performed under normoxic conditions.

There is also extensive literature showing *in vitro* antiproliferative activities of CAIs in a broad range of human tumor cell lines. Inhibition of human cancer cell proliferation by classical sulfonamide CAIs was reported by Chegwidden and Spencer (1995), who demonstrated that methazolamide (0.4 mM) and ethoxzolamide (10 μM) inhibited growth of a human lymphoma cell line U937. Interestingly, no or only weak inhibition was observed in cells cultured in a medium containing the nucleotide precursors hypoxanthine and thymidine, assuming that sulfonamides inhibited synthesis of nucleotides. This explanation was deduced from involvement of CA activity in production of bicarbonate that is required by carbamoyl phosphate synthetase I for synthesis of pyrimidines, but other mechanisms have not been excluded (Chegwidden et al. 2000).

Supuran and collaborators, motivated by the emerging role of CAs in cancer and by the possibility of using them as therapeutic targets, synthesized and tested several hundreds of potent sulfonamide CAIs containing the aromatic or heterocyclic moiety, or both (Supuran and Scozzafava 2000a, 2000b, 2000c; Scozzafava and Supuran 2000; Supuran et al. 2001). These compounds were subjected to screening for their ability to inhibit the growth of tumor cells *in vitro* by using a panel of 60 cancer cells lines. The screening performed at the National Cancer Institute in the U.S. led to the identification of lead compounds that exhibited considerably higher inhibitory properties (in the low micromolar range) than did classical sulfonamides (Supuran and Scozzafava 2000a, 2000b, 2000c). These leads were used to design novel classes of derivatives with enhanced antitumor activities by using the tail approach, in which new tails were attached to precursor sulfonamides (Supuran and Scozzafava 2000c, 2001; Casini et al. 2002). The active compounds showed GI$_{50}$ values (i.e., 50% inhibition of tumor cell growth after 48 h of exposure) in micromolar to nanomolar concentrations. Tumor cell lines used in the study were derived from different types of cancers, including leukemia, nonsmall cell lung carcinoma; melanoma; and carcinoma of the colon, ovary, kidney, prostate and breast (Supuran et al. 2001; Casini et al. 2002). Many, but not all, of these cell lines are known to express either CA IX or CA XII, or both; therefore, it is not impossible that at least in some cases, the observed growth-restricting effect of sulfonamides was mediated by inhibition of these cancer-related isozymes. However, CA IX and CA XII appear to be functionally linked with the tumor metabolism and microenvironment *in vivo* rather than with growth dynamics in a two-dimensional culture *in vitro*; future experiments designed to address these aspects of tumor phenotype might reveal whether they are targets of these anticancer sulfonamides. Tumor cell growth-inhibiting effects of the examined drugs might be related to other isozymes, such as cytosolic CA II and mitochondrial CA V, which donate bicarbonate to carboxy-lation reactions in biosynthetic processes and might be required to uphold increased tumor cell growth.

In addition to drugs developed for CA inhibition and then tested for antitumor effects, a new and very potent anticancer sulfonamide E-7070 (indisulam) has been discovered by elaborate preclinical screening (Owa et al. 1999, Owa and Nagasu 2000). Although it was selected regardless of CA-inhibitory capacity, it has been recently shown to act as a nanomolar CA inhibitor (Abbate et al. 2004). Its anticancer effects were shown to involve decrease in the S-phase fraction along with cell cycle perturbations in G1 or G2, or both; downregulation of the cyclins E, A, B1, H, CDK2 and CDC2; reduction of CDK2 activity; inhibition of pRb phosphorylation; and differential expression of many additional molecules known to participate in metab-olism, the immune response, signaling and cell adhesion (Fukuoka et al. 2001; Yokoi et al. 2002). In animal models *in vivo*, intraperitoneally or intravenously administered E7070 exhibited significant antitumor effect against human tumor xenografts gen-erated from various colon and lung carcinoma cell lines (Ozawa et al. 2001). Based on the high antitumor potency in preclinical studies, E7070 has progressed to Phase I and Phase II clinical trials for treating solid tumors and holds promise as a novel therapeutic strategy. For the CA research field, it will be of great interest to examine whether cancer-related or other CA isozymes belong to molecular targets of E7070 in tumor cells.

9.9 INHIBITION PROFILE OF CA IX

To learn more about the role of cancer-related CAs and consequences of their inhibition, it is also important to obtain information on their inhibition profile with different sulfonamides, particularly with those exerting an anticancer effect. For CA XII, such comprehensive information is missing. However, recent analyses of CA IX inhibition profiles by using several dozens of aromatic and heterocyclic sulfonamides reveal very interesting data (Vullo et al. 2003; Ilies et al. 2003). These studies report inhibition of recombinant CA IX catalytic domain by six classical sulfonamides that have been used as either systemic or topical antiglaucoma drugs (acetazolamide, methazolamide, ethoxzolamide, dichlorophenamide, dorzolamide and brinzolamide) in parallel with the newly synthesized compounds. In addition, inhibition of CA I, CA II and CA IV by the same collection of sulfonamides was determined for comparing isozyme-related inhibition patterns. CA I, CAII and CA IV isozymes showed a general preference for the heterocyclic compounds. The overall range of CA I and CA IV inhibition constants differed by three to four orders of magnitude. In contrast, CA IX behaved as a sulfonamide-avid enzyme, with a relatively narrow range of inhibition constants (14 to 305 nM). In that respect it was similar to CA II, considered up to now to be responsible for the majority of the pharmacological effects of sulfonamides. In addition, CA IX was effectively inhib-ited by all the examined sulfonamides, including the aromatic compounds. Most importantly, these studies identified several strong inhibitors of CA IX that are weak or medium inhibitors of CA I, CA II and CA IV, including orthanilamide derivatives, 1,3-benzene-disulfonamide derivatives and dihalogenated sulfanilamide derivatives

INHIBITOR	K_I (nM)	ISOZYME
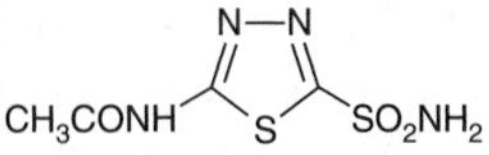	250	hCA I[a]
	12	hCA II[a]
CH_3CONH—[thiadiazole N—N, S]—SO_2NH_2	70	bCA IV[b]
	25	hCA IX[c]
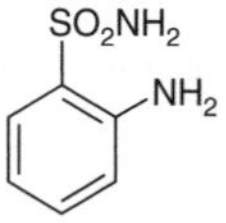 SO_2NH_2 / NH_2 (benzene ring)	45 400	hCA I
	295	hCA II
	1 310	bCA IV
	33	hCA IX
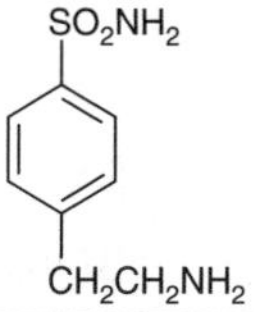 SO_2NH_2 (benzene ring) $CH_2CH_2NH_2$	21 000	hCA I
	160	hCA II
	2 450	bCA IV
	33	hCA IX
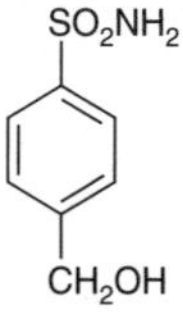 SO_2NH_2 (benzene ring) CH_2OH	24 000	hCA I
	125	hCA II
	560	bCA IV
	21	hCA IX
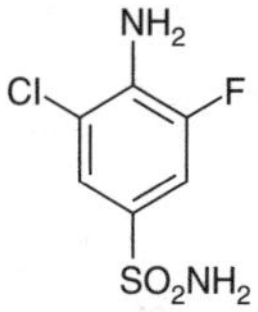 NH_2, Cl, F (benzene ring) SO_2NH_2	3 800	hCA I
	32	hCA II
	95	bCA IV
	12	hCA IX

FIGURE 9.5 CA IX inhibition constant (K_i) values obtained with the prototype CA inhibitor acetazolamide (upper formula) and four newly synthesized aromatic sulfonamides selected from the collections tested by Vullo, D. et al. (2003) *Bioorganic and Medicinal Chemistry Letters*, and Ilies et al. (2003) *Journal of Medicinal Chemistry*. The new compounds efficiently inhibit CA IX while acting as weak or medium inhibitors of CA I, CA II and CA IV isozymes. [a] Human cloned isozyme, analyzed by the esterase method. [b] Bovine isozyme isolated from lung microsomes, esterase method. [c] Catalytic domain of human cloned isozyme analyzed by the CO_2 hydration method.

(Figure 9.5, Vullo et al. (2003); Ilies et al. 2003). The best CA IX inhibitor was the very simple dihalogenated compound 3-fluoro-5-chloro-4-aminobenzenesulfona-mide ($K_I = 12$ nM), which was twice as active as acetazolamide ($K_I = 25$ nM; Ilies et al. 2003). These findings indicate the feasibility of using such differentially acting potent CA IX inhibitors as lead compounds to design CA IX-specific sulfonamides that can be potentially used as novel therapeutic tools against cancer. This strategy can be further refined by another approach that has led to the generation of membrane-impermeable sulfonamides (Supuran et al. 2000; Scozzafava et al. 2000), thus combining enzyme specificity with spatial selectivity.

9.10 CONCLUSIONS

Contributions from several streams of investigation are required to achieve the major goal of exploiting therapeutically CA activity, which is presumably involved in tumor development and progression. The principal and interdependent tasks include a better understanding of the functional relevance of CAs for malignant phenotype and metabolism by characterizing inhibition profiles of these CAs, developing isozyme-specific inhibitors and determining *in vitro* and *in vivo* consequences of the selective isozyme inhibition. This chapter allows us to foresee that the research rapidly progressing in all these directions will soon approach the point of convergence and answer the question of whether inhibitory targeting of CA isozymes can benefit cancer patients.

ACKNOWLEDGMENTS

We thank Dr. Juraj Kopacek (Institute of Virology, Bratislava) for help with graphical illustrations. Our research is supported by the Bayer Corporation and by APVT (51-005802) and VEGA (2/2025,2/3055) Slovak Grant Agencies.

REFERENCES

Abbate, F., Casini, A., Owa, T., Scozzafava, A., Supuran, C.T. (2004) Carbonic anhydrase inhibitors: E7070, a sulfonamide anticancer agent, potently inhibits cytosolic isoenzymes I and II, and transmembrane, tumor-associated isozyme IX. *Bioorganic and Medicinal Chemistry Letters* **14**, 217–223.

Ashida, S., Nishimori, I., Tanimura, M., Onishi, S., and Shuin, T. (2002) Effects of von Hippel-Lindau gene mutation and methylation status on expression of transmembrane carbonic anhydrases in renal cell carcinoma. *Journal of Cancer Research and Clinical Oncology* **128**, 561–568.

Bartosova, M., Parkkila, S., Pohlodek, K., Karttunen, T.J., Galbavy, S., Mucha, V., Harris, A.L., Pastorek, J., and Pastorekova, S. (2002) Expression of carbonic anhydrase IX in breast is associated with malignant tissues and related to overexpression of c-erbB2. *Journal of Pathology* **197**, 1–8.

Beasley, N.J.P., Wykoff, C.C., Watson, P.H., Leek, R., Turley, H., Gatter, K., Pastorek, J., Cox, G.J., Ratcliffe, P., and Harris, A.L. (2001) Carbonic anhydrase IX, an endogenous hypoxia marker, expression in head and neck squamous cell carcinoma and its relationship to hypoxia, necrosis and microvessel density. *Cancer Research* **61**, 5262–5267.

Beavon, I.R.G. (1999) Regulation of E-cadherin: does hypoxia initiate the metastatic cascade? *Journal of Clinical and Molecular Pathology* **52**, 179–188.

Bekku, S., Mochizuki, H., Yamamoto, T., Ueno, H., Takayama, E., and Tadakuma, T. (2000) Expression of carbonic anhydrase I or II and correlation to clinical aspects of colorectal cancer. *Hepatogastroenterology* **47**, 998–1001.

Casini, A., Scozzafava, A., Mastrolorenzo, A., and Supuran, C.T. (2002) Sulfonamides and sulfonylated derivatives as anticancer agents. *Current Cancer Drug Targets* **2**, 55–75.

Chegwidden, W.R., Carter, N., and Edwards, Y. (Eds.) (2000) *The Carbonic Anhydrases: New Horizons*. Birkhäuser Verlag, Basel.

Chegwidden, W.R., and Spencer, I.M. (1995) Sulphonamide inhibitors of carbonic anhydrase inhibit the growth of human lymphoma cells in culture. *Inflammopharmacology* **3**, 231–239.

Chia, S.K., Wykoff, C.C., Watson, P.H., Han, C., Leek, R.D., Pastorek, J., Gatter, K.C., Ratcliffe, P., and Harris, A.L. (2001) Prognostic significance of a novel hypoxia-regulated marker, carbonic anhydrase IX, in invasive breast carcinoma. *Journal of Clinical Oncology* **19**, 3660–3668.

Dang, C.V., and Semenza, G.L. (1999) Oncogenic alterations of metabolism. *Trends in Biochemical Sciences* **24**, 68–72.

D'Arcangelo, D., Facchiano, F., Barlucchi, L.M., Melillo, G., Illi, B., Testolin, L., Gaetano, C., and Capogrossi, M.C. (1999) Acidosis inhibits endothelial cell apoptosis and function and induces basic fibroblast growth factor and vascular endothelial growth factor expression. *Circulation Research* **86**, 312–318.

Evan, G.I., and Vousden, K.H. (2001) Proliferation, cell cycle and apoptosis in cancer. *Nature* **411**, 342–348.

Fukumura, D., Xu, L., Chen, Y., Gohongi, T., Seed, B., and Jain, R.K. (2001) Hypoxia and acidosis independently up-regulate vascular endothelial growth factor transcription in brain tumors *in vivo*. *Cancer Research* **61**, 6020–6024.

Fukuoka, K., Usuda, J., Iwamoto, Y., Fukumoto, H., Nakamura, T., Yoneda, T., Narita, N., Saijo, N., and Nishio, K. (2001) Mechanism of action of the novel sulfonamide anticancer agent E7070 on cell cycle progression in human non-small cell lung cancer cells. *Investigation of New Drugs* **19**, 219–227.

Gatenby, R.A. and Gawlinski, E.T. (1996) A reaction-diffusion model of cancer invasion. *Cancer Research* **56**, 5745–5753.

Giatromanolaki, A., Koukourakis, M.I., Sivridis, E., Pastorek, J., Wykoff, C.C., Gatter, K.C., and Harris, A.L. (2001) Expression of hypoxia-inducible carbonic anhydrase-9 relates to angiogenic pathways and independently to poor outcome in non-small cell lung cancer. *Cancer Research* **61**, 7992–7998.

Gnarra, J.R., Tory, K., Weng, Y., Schmidt, L., Wei, M.H., Li, H., Latif, F., Liu, S., Chen, F., Duh, F.M., Lubensky, I., Duan, D.R., Florence, C., Pozzatti, R., Walther, M.M., Bander, N.H., Grossmann, H.B., Branch, H., Pomer, S., Brooks, J.D., Isaacs, W.B., Lerman, M.I., Zbar, B., and Lineham, W.M. (1994). Mutations of the VHL tumour suppressor gene in renal carcinoma. *Nature Genetics* **7**, 85–90.

Hanahan, D., and Weinberg, R.A. (2000) The hallmarks of cancer. *Cell* **100**, 57–70.

Helmlinger, G., Sckell, A., Dellian, M., Forbes, N.S., and Jain, R.K. (2002) Acid production in glycolysis-impaired tumors provides new insights into tumor metabolism. *Clinical Cancer Research* **8**, 1284–1291.

Hewett-Emmett, D., and Tashian, R.E. (1996) Functional diversity, conservation and convergence in the evolution of the alpha, beta and gamma carbonic anhydrase gene families. *Molecular Phylogenesis and Evolution* **5**, 50–77.

Höckel, M., and Vaupel, P. (2001) Tumor hypoxia: Definitions and current clinical, biologic and molecular aspects. *Journal of National Cancer Institute* **93**, 266–276.

Hui, E.P., Chan, A.T., Pezzella, F., Turley, H., To, K.F., Poon, T.C., Zee, B., Mo, F., Teo, P.M., Huang, D.P., Gatter, K.C., Johnson, P.J., and Harris, A.L. (2002) Coexpression of hypoxia-inducible factors 1alpha and 2alpha, carbonic anhydrase IX, and vascular endothelial growth factor in nasopharyngeal carcinoma and relationship to survival. *Clinical Cancer Research* **8**, 2595–2604.

Ilies, M.A., Vullo, D., Pastorek, J., Scozzafava, A., Ilies, M., Caproiu, T., Pastorekova, S., and Supuran, C. (2003) Carbonic anhydrase inhibitors: Inhibition of tumor-associated isozyme IX by halogeno-sulfanilamide and halogeno-aminobenzolamide derivatives. *Journal of Medicinal Chemistry*, **46**, 2187-2196.

International Human Genome Sequencing Consortium (2001) Initial sequencing and analysis of the human genome. *Nature* **409**, 860–921.

Ivan, M., Kondo, K., Yang, H., Kim, W., Valiando, J., Ohh, M., Salic, A., Asara, J.M., Lane, W.S., and Kaelin, W.G. (2001) HIFα targeted for VHL-mediated destruction by proline hydroxylation: implications for O_2 sensing. *Science* **292**, 464–468.

Ivanov, S.V., Kuzmin, I., Wei, M.H., Pack, S., Geil, L., Johnson, B.E., Stanbridge, E.J., and Lerman, M.I. (1998) Down-regulation of transmembrane carbonic anhydrases in renal cell carcinoma cell lines by wild-type von Hippel-Lindau transgenes. *Proceedings of the National Academy of Science of the USA* **95**,12596–12601.

Ivanov, S., Lia, S.Y., Ivanova, A., Danilkovich-Miagkova, A., Tarasova, N., Weirich, G., Merrill, M.J., Proescholdt, M.A., Oldfield, E.H., Lee, J., Zavada, J., Waheed, A., Sly, W., Lerman, M.I., and Stanbridge, E.J. (2001) Expression of hypoxia-inducible cell-surface transmembrane carbonic anhydrases in human cancer. *American Journal of Pathology* **158**, 905–919.

Iwai, K., Yamanaka, K., Kamura, T., Minato, N., Conaway, R.C., Conaway, J.W., Klausner, R.D., and Pause, A. (1999) Identification of the von Hippel-Lindau tumor-suppressor protein as a part of an active E3 ubiquitin ligase complex. *Proceedings of the National Academy of Science of the USA* **96**, 12436–12441.

Jaakkola, P., Mole, D.R., Tian, Y.M., Wilson, M.I., Gielbert, J., Gaskell, S.J., von Kriegsheim, A., Heberstreist, H.F., Mukherji, M., Schofield, C.J., Maxwell, P.H., Pugh, C.W., and Ratcliffe, P.J. (2001) Targeting of HIFα to the von Hippel Lindau ubiquitilation complex by O_2-regulated prolyl hydroxylation. *Science* **292**, 468–472.

Kaluz, S., Kaluzova, M., Chrastina, A., Olive, P.L., Pastorekova, S., Pastorek, J., Lerman, M.I., and Stanbridge, E.J. (2002) Lowered oxygen tension induces expression of the hypoxia marker MN/carbonic anhydrase IX in the absence of hypoxia-inducible factor 1α stabilization: a role for phosphatidylinositol 3′-kinase. *Cancer Research* **62**, 4469–4477.

Kaluz, S., Kaluzova, M., Opavsky, R., Pastorekova, S., Gibadulinova, A., Dequiedt, F., Kettmann, R., and Pastorek, J. (1999) Transcriptional regulation of the MN/CA9 gene coding for the tumor-associated carbonic anhydrase IX: Identification and characterization of a proximal silencer element. *Journal of Biological Chemistry* **274**, 32588–32595.

Kaluzova, M., Pastorekova, S., Svastova, E., Pastorek, J., Stanbridge, E.J., and Kaluz, S. (2001) Characterization of the MN/CA9 proximal region: a role for SP and AP1 factors. *Biochemical Journal* **359**, 669–677.

Karhumaa, P., Kaunisto, K., Parkkila, S., Waheed, A., Pastorekova, S., Pastorek, J., Sly, W.S., and Rajaniemi, H. (2001) Expression of the transmembrane carbonic anhydrases, CA IX and CA XII, in the human male excurrent ducts. *Molecular Human Reproduction* **7**, 611–616.

Karumanchi, S.A., Jiang, L., Knebelmann, B., Stuart-Tilley, A.K., Alper, S.L., and Sukhatme, V.P. (2001) VHL tumor suppressor regulates Cl^-/HCO_3^+ exchange and Na^+/H^+ exchange activities in renal carcinoma cells. *Physiological Genomics* **5**, 119–128.

Kato, Y., Nakayama, Y., Umeda, M., and Miyazaki, K. (1992) Induction of 103-kDa gelatinase/type IV collagenase by acidic culture conditions in mouse metastatic meloma cell lines. *Journal of Biological Chemistry* **267**, 11424–11430.

Kivela, A.J., Parkkila, S., Saarnio, J., Karttunen, T.J., Kivela, J., Parkkila, A.K., Pastorekova, S., Pastorek, J., Waheed, A., Sly, W.S., and Rajaniemi, H. (2000a) Expression of transmembrane carbonic anhydrase isoenzymes IX and XII in normal human pancreas and pancreatic tumours. *Histochemistry and Cell Biology* **114**, 197–204

Kivela, A., Parkkila, S., Saarnio, J., Karttunen, T.J., Kivela, J., Parkkila, A.K., Waheed, A., Sly, W.S., Grubb, J.H., Shah, G., Tureci, O., and Rajaniemi H. (2000b) Expression of a novel transmembrane carbonic anhydrase XII in normal human gut and colorectal tumors. *American Journal of Pathology* **156**, 577–584.

Kivela, A.J., Saarnio, J., Karttunen, T.J., Kivela, J., Parkkila, A.K., Pastorekova, S., Pastorek, J., Waheed, A., Sly, W.S., Parkkila, S., and Rajaniemi, H. (2001) Differential expression of cytoplasmic carbonic anhydrases, CA I and II, and membrane-associated isozymes, CA IX and XII, in normal mucosa of large intestine and in colorectal tumors. *Digestion Disease Science* **46**, 2179–2186.

Koukourakis, M.I., Giatromanolaki, A., Sivridis, E., Simopoulos, K., Pastorek, J., Wykoff, C.C., Gatter, K.C., and Harris, A.L. (2001) Hypoxia-regulated carbonic anhydrase-9 (CA9) relates to poor vascularization and resistance of squamous cell head and neck cancer to chemoradiotherapy. *Clinical Cancer Research* **7**, 3399–3403.

Lal, A., Peters, H., St. Croix, B., Haroon, Z.A., Dewhirst, M.W., Strausberg, R.L., Kaanders, J.H., van der Kogel, A.J., and Riggins, G.J. (2001) Transcriptional response to hypoxia in human tumors. *Journal of the National Cancer Institute* **93**, 1337–1343.

Leppilampi, M., Saarnio, J., Karttunen, T.J., Kivela, J., Pastorekova, S., Pastorek, J., Waheed, A., Sly, W.S., and Parkkila, S. (2003) Carbonic anhydrase isozymes IX and XII in gastric tumors. *World Journal of Gastroenterology* **9**, 1398–1403.

Li, X., Alvarez, B., Casey, J.R., Reithmeier, R.A., Fliegel, L. (2002) Carbonic anhydrase II binds to and enhances activity of the Na+/H+ exchanger. *Journal of Biological Chemistry* **277**, 36085–36091.

Liao, S.Y., Aurelio, O.N., Jan, K., Zavada, J., and Stanbridge, E.J. (1997) Identification of the MN/CA9 protein as a reliable diagnostic biomarker of clear cell carcinoma of the kidney. *Cancer Research* **57**, 2827–2831.

Liao, S.Y., Brewer, C., Zavada, J., Pastorek, J., Pastorekova, S., Manetta, A., Berman, M.L., DiSaia, P.J., and Stanbridge, E.J. (1994) Identification of the MN antigen as a diagnostic biomarker of cervical intraepithelial neoplasia and cervical carcinoma. *American Journal of Pathology* **145**, 598–609.

Lieskovska, J., Opavsky, R., Zacikova, L., Glasova, M., Pastorek, J., and Pastorekova, S. (1999) Study of *in vitro* conditions modulating expression of MN/CA IX protein in human cell lines derived from cervical carcinoma. *Neoplasma* **46**, 17–24.

Lisztwan, J., Imbert, G., Wirbelauer, C., Gstaiger, M., and Krek, W. (1999) The von Hippel-Lindau tumor suppressor is a component of a E3 ubiquitin-protein ligase activity. *Genes and Development* **13**, 1822–1833.

Loncaster, J.A., Harris, A.L., Davidson, S.E., Logue, J.P., Hunter, R.D., Wykoff, C.C., Pastorek, J., Ratcliffe, P., Stratford, I.J., and West, C.M.L. (2001) Carbonic anhydrase IX expression, a potential new intrinsic marker of hypoxia: correlations with tumour oxygen measurements and prognosis in locally advanced carcinoma of the cervix. *Cancer Research* **61**, 6394–6399.

Mandriota, S.J., Turner, K.J., Davies, D.R., Murray, P.G., Morgan, N.V., Sowter, H.M., Wykoff, C.C., Maher, E.R., Harris, A.L., Ratcliffe, P.J., and Maxwell P.H. (2002) HIF activation identifies early lesions in VHL kidneys: evidence for site-specific tumor suppressor function in the nephron. *Cancer Cell* **1**, 459–468.

Martinez-Zaguilan, R., Seftor, E.A., Seftor, R.E., Chu, Y.W., Gillies, R.J., and Hendrix, M.J. (1992) Acidic pH enhances the invasive behavior of human melanoma cells. *Clinical and Experimental Metastasis* **14**, 176–186.

Maxwell, P.H., Pugh, C.W., and Ratcliffe, P.J. (2001) Activation of HIF pathway in cancer. *Current Opinion in Genetics and Development* **11**, 293–299.

Maxwell, P.H., Wiesener, M.S., Chang, G.W., Clifford, S.C., Vaux, E.C., Cockman, M.E., Wykof, C.C., Pugh, C.W., Maher, E.R., and Ratcliffe, P.J. (1999) The tumour suppressor protein VHL targets hypoxia-inducible factors for oxygen-dependent proteolysis. *Nature* **399**, 271–275.

McKiernan, J.M., Buttyan, R., Bander, N.H., Stifelman, M.D., Katz, A.E., Olsson, C.A., and Sawczuk, I.S. (1997) Expression of the tumour-associated gene MN: A potential biomarker for human renal cell carcinoma. *Cancer Research* **57**, 2362–2365.

Newell, K., Franchi, A., Pouyssegur, J., and Tannock, I. (1993) Studies with glycolysis-deficient cells suggest that production of lactic acid is not the only cause of tumor acidity. *Proceedings of the National Academy of Science of the USA* **90**, 1127–1131.

Olive, P.L., Aquino-Parsons, C., MacPhail, S.H., Laio, S.Y., Raleigh, J.A., Lerman, M.I., and Stanbridge, E.J. (2001) Carbonic anhydrase 9 as an endogenous marker for hypoxic cells in cervical cancer. *Cancer Research* **61**, 8924–8929.

Opavský, R., Pastoreková, S., Zelník, V., Gibadulinová, A., Stanbridge, E.J., Závada, J., Kettmann, R., and Pastorek, J. (1996) Human MN/CA9 gene, a novel member of the carbonic anhydrase family: structure and exon to protein domain relationship. *Genomics* **33**, 480–487.

Ortova Gut, M., Parkkila, S., Vernerová, Z., Rohde, E., Závada, J., Höcker, M., Pastorek, J., Karttunen, T., Zavadová, Z., Knobeloch, K.P., Wiedenmann, B., Svoboda, J., Horak, I., and Pastoreková, S. (2002) Gastric hyperplasia in mice with targeted disruption of the carbonic anhydrase gene *Car9*. *Gastroenterology* **123**, 1889–1903.

Owa, T., and Nagasu, T. (2000) Novel sulphonamide derivatives for the treatment of cancer. *Expert Opinion on Therapeutic Patents* **10**, 1725–1740.

Owa, T., Yoshino, H., Okauchi, T., Yoshimatsu, K., Ozawa, Y., Sugi, N.H., Nagasu, T., Koyanagi, N., and Kitoh, K. (1999) Discovery of novel antitumor sulfonamides targeting G1 phase of the cell cycle. *Journal of Medicinal Chemistry* **42**, 3789–3799.

Ozawa, Y., Sugi, N.H., Nagasu, T., Owa, T., Watanabe, T., Koyanagi, N., Yoshino, H., Kitoh, K., and Yoshimatsu, K. (2001) E7070, a novel sulfonamide agent with potent antitumor activity *in vitro* and *in vivo*. *European Journal of Cancer* **37**, 2275–2282.

Parkkila, A.K., Herva, R., Parkkila, S., and Rajaniemi, H. (1995a) Immunohistochemical demonstration of human carbonic anhydrase isoenzyme II in brain tumors. *Histochemical Journal* **27**, 974–982.

Parkkila, S., Parkkila, A.K., Juvonen, T., Lehto, V.P., and Rajaniemi, H. (1995b) Immunohistochemical demonstration of the carbonic anhydrase isoenzymes I and II in pancreatic tumors. *Histochemical Journal* **27**, 133–138.

Parkkila, S., Parkkila, A.K., Saarnio, J., Kivela, J., Karttunen, T., Kaunisto, K., Waheed, A., Sly, W.S., Tureci, O., Virtanen, I., and Rajaniemi, H. (2000a) Expression of the membrane-associated carbonic anhydrase XII in the human kidney and renal tumors. *Journal of Histochemistry and Cytochemistry* **48**, 1601–1608.

Parkkila, S., Rajaniemi, H., Parkkila, A.K., Kivela, J., Waheed, A., Pastorekova, S., Pastorek, J., and Sly, W.S. (2000b) Carbonic anhydrase inhibitor suppresses invasion of renal cancer cells in vitro. *Proceedings of the National Academy of Science of the USA*, **97**, 2220–2224.

Pastorek, J., Pastorekova, S., Callebaut, I., Mornon, J.P., Zelnik, V., Opavský, R., Zatovicova, M., Liao, S., Portetelle, D., Stanbridge, E.J., Zavada, J., Burny, A., and Kettmann, R. (1994) Cloning and characterization of MN, a human tumor-associated protein with a domain domologous to carbonic anhydrase and a putative helix-loop-helix DNA binding segment. *Oncogene* **9**, 2788–2888.

Pastoreková, S., Parkkila, S., Parkkila, A.K., Opavský, R., Zelník, V., Saarnio, J., and Pastorek, J. (1997) Carbonic anhydrase IX, MN/CA IX: Analysis of stomach complementary DNA sequence and expression in human and rat alimentary tracts. *Gastroenterology* **112**, 398–408.

Pastoreková, S., Závadová, Z., Košál, M., Babušíková, O., and Závada, J. (1992) A novel quasi-viral agent, MaTu, is a two-component system. *Virology* **187**, 620–626.

Peles, E., Nativ, M., Campbell, P.L., Sakurai, T., Martinez, R., Lev, S., Clary, D., Schilling, J., Barnea, G., Plowman, G.D., Grumet, M., and Schlessinger, J. (1995) The carbonic anhydrase domain of receptor tyrosine phosphatase β is a functional ligand for the axonal cell recognition molecule contactin. *Cell* **82**, 251–260.

Rafajova, M., Zatovicova, M., Kettmann, R., Pastorek, J., and Pastorekova, S. (2004) Induction by hypoxia combined with low glucose or low bicarbonate and high posttranslational stability upon reoxygenation contribute to carbonic anhydrase IX expression in cancer cells. *International Journal of Oncology* **24**, 995–1004.

Raghunand, N., He, X., van Sluis, R., Mahoney, B., Baggett, B., Taylor, C.W., Paine-Murrieta, G., Roe, D., Bhujwalla, Z.M., and Gillies, R.J. (1999) Enhancement of chemotherapy by manipulation of tumour pH. *British Journal of Cancer* **80**, 1005–1011.

Reynolds, T.Y., Rockwell, S., and Glazer, P.M. (1996) Genetic instability induced by the tumor microenvironment. *Cancer Research* **56**, 5754–5757.

Saarnio, J., Parkkila, S., Parkkila, A.K., Haukipuro, K., Pastorekova, S., Pastorek, J., Kairaluoma, M.I., and Karttunen, T.J. (1998a) Immunohistochemical study of colorectal tumors for expression of a novel transmembrane carbonic anhydrase, MN/CA IX, with potential value as a marker of cell proliferation. *American Journal of Pathology* **153**, 279–285.

Saarnio, J., Parkkila, S., Parkkila, A.K., Pastorekova, S., Haukipuro, K., Pastorek, J., Juvonen, T., and Karttunen, T. (2001) Transmembrane carbonic anhydrase, MN/CA IX, is a potential biomarker for biliary tumors. *Journal of Hepatology* **35**, 643–649.

Saarnio, J., Parkkila, S., Parkkila, A.K., Waheed, A., Casey, M.C., Zhou, X.Y., Pastorekova, S., Pastorek, J., Karttunen, T., Haukipuro, K., Kairaluoma, M.I., and Sly, W.S. (1998b) Immunohistochemistry of carbonic anhydrase isozyme IX (MN/CA IX) in human gut reveals polarized expression in the epithelial cells with the highest proliferative capacity. *Journal of Histochemistry and Cytochemistry* **46**, 497–504.

Scozzafava, A., Briganti, F., Ilies, M.A., and Supuran, C.T. (2000) Carbonic anhydrase inhibitors: synthesis of membrane-impermeant low molecular weight sulfonamides possessing *in vivo* selectivity for the membrane-bound versus cytosolic isozymes. *Journal of Medicinal Chemistry* **43**, 292–300.

Scozzafava, A., and Supuran, C.T. (2000) Carbonic anhydrase inhibitors: synthesis of N-morpholyl-thiocarbonylsulfenylamino aromatic/heterocyclic sulfonamides and their interaction with isozymes I, II and IV. *Bioorganic and Medicinal Chemistry Letters* **10**, 1117–1120.

Scozzafava, A., and Supuran, C.T. (2002) Carbonic anhydrase activators: human isozyme II is strongly activated by oligopeptides incorporating the carboxyterminal sequence of the bicarbonate anion exchanger AE1. *Bioorganic and Medicinal Chemistry Letters* **12**, 1177–1180.

Semenza, G.L. (2000) Hypoxia, clonal selection, and the role of HIF-1 in tumor progression. *Critical Reviews in Biochemistry and Molecular Biology* **35**, 71–103.

Semenza, G.L. (2001) Hypoxia-inducible factor 1: Oxygen homeostasis and disease pathophysiology. *Trends in Molecular Medicine* **7**, 345–350.

Sterling, D., Alvarez, B.V., and Casey, J.R. (2002) The extracellular component of a transport metabolon: Extracellular loop 4 of the human AE1 Cl⁻/HCO₃⁺ exchanger binds carbonic anhydrase IV. *Journal of Biological Chemistry* **277**, 25239–25246.

Sterling, D., Brown, N.J., Supuran, C.T., and Casey, J.R. (2002) The functional and physical relationship between the DRA bicarbonate transporter and carbonic anhydrase II. *American Journal of Physiology: Cell Physiology* **283**, C1522–1529.

Sterling, D., Reithmeier, R.A.F., and Casey, J.R. (2001) A transport metabolon. Functional interaction of carbonic anhydrase II and chloride/bicarbonate exchangers. *Journal of Biological Chemistry* **276**, 47886–47894.

Stubbs, M., McSheehy, P.M.J., Griffiths, J.R., and Beshford, C.L. (2000) Causes and consequences of tumour acidity and implications for treatment. *Molecular Medicine Today* **6**, 15–19.

Supuran, C.T., Briganti, F., Tilli, S., Chegwidden, W.R., and Scozzafava, A. (2001) Carbonic anhydrase inhibitors: sulfonamides as antitumor agents? *Bioorganic and Medicinal Chemistry* **9**, 703–714.

Supuran, C.T., and Scozzafava, A. (2000a) Carbonic anhydrase inhibitors. Part 94. 1,3,4-thiadiazole-2-sulfonamide derivatives as antitumor agents? *European Journal of Medicinal Chemistry* **35**, 867–874.

Supuran, C.T., and Scozzafava, A. (2000b) Carbonic anhydrase inhibitors: Aromatic sulfonamides and disulfonamides act as efficient tumor growth inhibitors. *Journal of Enzyme Inhibition* **15**, 597–610.

Supuran, C.T., and Scozzafava, A. (2000c) Carbonic anhydrase inhibitors and their therapeutic potential. *Expert Opinion on Therapeutic Patents* **10**, 575–600.

Supuran, C.T., and Scozzafava, A. (2001) Carbonic anhydrase inhibitors. *Current Medicinal Chemistry — Immunologic, Endocrine and Metabolic Agents* **1**, 61–97.

Supuran, C.T., Scozzafava, A., Ilies, M.A., and Briganti, F. (2000) Carbonic anhydrase inhibitors: Synthesis of sulfonamides incorporating 2,4,6-trisubstituted-pyridinium-ethycarboxamido moieties possessing membrane impermeability and *in vivo* selectivity for the membrane-bound (CA IV) versus the cytosolic (CA I and CA II) isozymes. *Journal of Enzyme Inhibition* **15**, 381-401.

Svastova, E., Zilka, N., Zatovicova, M., Gibadulinova, A., Ciampor, F., Pastorek, J., and Pastorekova, S. (2003) Carbonic anhydrase IX reduces E-cadherin-mediated adhesion of MDCK cells via interaction with β-catenin. *Experimental Cell Research* **290**, 332–345.

Teicher, B.A., Liu, S.D., Liu, J.T., Holden, S.A., and, Herman, T.S. (1993) A carbonic anhydrase inhibitor is a potential modulator of cancer therapies. *Anticancer Research* **13**, 1549–1556.

Tomes, L., Emberley, E., Niu, Y.L., Troup, S., Pastorek, J., Strange, K., Harris, A., and Watson, P.H. (2003) Necrosis and hypoxia in invasive breast carcinoma. *Breast Cancer Research and Treatment* **31**, 61–69.

Turner, J.R., Odze, R.D., Crum, C.P., and Resnick, M.B. (1997) MN antigen expression in normal, preneoplastic and neoplastic esophagus: a clinicopathological study of a new cancer-associated biomarker. *Human Pathology* **28**, 740–744.

Turner, K.J., Crew, J.P., Wykoff, C.C., Watson, P.H., Poulsom, R., Pastorek, J., Ratcliffe, P.J., Cranston, D., and Harris, A.L. (2002) The hypoxia-inducible genes VEGF and CA9 are differentially regulated in superficial vs invasive bladder cancer. *British Journal of Cancer* **86**, 1276–1282.

Türeci, O., Sahin, U., Vollmar, E., Siemer, S., Göttert, E., Seitz, G., Parkkila, A.K., Shah, G.N., Grubb, J.H., Pfreundschuh, M., and Sly, W.S. (1998) Human carbonic anhydrase XII: cDNA cloning, expression and chromosomal localization of a carbonic anhydrase gene that is overexpressed in some renal cancers. *Proceedings of the National Academy of Science of the USA* **95**, 7608–7613.

Ulmasov, B., Waheed, A., Shah, G.N., Grubb, J.H., Sly, W.S., Tu, C., and Silverman, D.N. (2000) Purification and kinetic analysis of recombinant CA XII, a membrane carbonic anhydrase overexpressed in certain cancers. *Proceedings of the National Academy of Science of the USA* **97**, 14212–14217.

Vermylen, P., Roufosse, C., Burny, A., Verhest, A., Bossehaerts, T., Pastorekova, S., Ninane, V., and Sculier, J.P. (1999) Carbonic anhydrase IX antigen differentiates between preneoplastic and malignant lesions in non-small cell lung carcinomas. *European Respiratory Journal* **14**, 806–811.

Vince, J.W. and Reithmeier, R.A. (2000) Identification of the carbonic anhydrase II binding site in the Cl⁻/HCO₃⁺ anion exchanger AE1. *Biochemistry*, **39**, 5527–5533.

Vullo, D., Franchi, M., Gallori, E., Pastorek, J., Scozzafava, A., Pastorekova, S., and Supuran, C.T. (2003) Carbonic anhydrase inhibitors: Inhibition of the tumor-associated isozyme IX with aromatic and heterocyclic sulfonamides. *Bioorganic and Medicinal Chemistry Letters*, **13**, 1005–1009.

Whittington, D.A., Waheed, A., Ulmasov, B., Shah, G.N., Grubb, J.H., Sly, W.S., and Christianson, D.W. (2001) Crystal structure of the dimeric extracellular domain of human carbonic anhydrase XII, a bitopic membrane protein overexpressed in certain cancer cells. *Proceedings of the National Academy of Science of the USA* **98**, 9545–9550.

Wingo, T., Tu, C., Laipis, P.J., and Silverman, D.N. (2001) The catalytic properties of human carbonic anhydrase IX. *Biochemical and Biophysical Research Communications* **288**, 666–669.

Wykoff, C.C., Beasley, N., Watson, P.H., Campo, L., Chia, S.K., English, R., Pastorek, J., Sly, W.S., Ratcliffe, P., and Harris, A.L. (2001) Expression of the hypoxia-inducible and tumor-associated carbonic anhydrases in ductal carcinoma *in situ* of the breast. *American Journal of Pathology* **158**, 1011–1019.

Wykoff, C., Beasley, N., Watson, P., Turner, L., Pastorek, J., Wilson, G., Turley, H., Maxwell, P., Pugh, C., Ratcliffe, P., and Harris, A. (2000) Hypoxia-inducible regulation of tumor-associated carbonic anhydrases. *Cancer Research* **60**, 7075–7083.

Xu, L., and Fidler, I.J. (2000) Acidic pH-induced elevation in interleukin 8 expression by human ovarian carcinoma cells. *Cancer Research* **60**, 4610–4616.

Yokoi, A., Kuromitsu, J., Kawai, T., Nagasu, T., Sugi, N.H., Yoshimatsu, K., Yoshino, H., and Owa, T. (2002) Profiling novel sulfonamide antitumor agents with cell-based phenotypic screens and array-based gene expression analysis. *Molecular Cancer Therapy* **1**, 275–286.

Závada, J., Závadová, Z., Pastoreková, S., Ciampor, F., Pastorek, J., and Zelník, V. (1993) Expression of MaTu-MN protein in human tumor cultures and in clinical specimens. *International Journal of Cancer* **54**, 268–274.

Závada, J., Závadová, Z., Pastorek, J., Biesová, Z., Jeûek, K., and Velek, J. (2000) Human tumour-associated cell adhesion protein MN/CA IX: identification of M75 epitope and of the region mediating cell adhesion. *British Journal of Cancer* **82**, 1808–1813.

10 Role of Carbonic Anhydrase and Its Inhibitors in Gastroenterology, Neurology and Nephrology

Seppo Parkkila, Anna-Kaisa Parkkila and Jyrki Kivelä

CONTENTS

10.1 Role of CAs in the Alimentary Tract 284
    10.1.1 Upper Alimentary Tract 284
    10.1.2 Colon 285
    10.1.3 Pancreas, Liver and Gallbladder 286
10.2 CAIs in the Alimentary Tract 287
10.3 Role of CAs in the Kidney 287
10.4 CAIs in the Kidney 289
10.5 Role of CAs in the Nervous System 289
10.6 CAIs in the Nervous System 291
Acknowledgments 293
References 293

Research on carbonic anhydrases (CAs) has become increasingly attractive in the past 20 years, especially because of discoveries of multiple new isozymes in this gene family. The diverse family of mammalian CAs includes at least 11 enzymatically active CA isoforms and 1 nonclassical form expressed specifically in the nucleus (Karhumaa et al. 2000). At the same time, CA research has gathered information that has linked these enzymes to very fundamental physiological phenomena, such as cell growth, metabolism, apoptosis and development of cancer (Sly and Hu 1995; Henkin et al. 1999; Räisänen et al. 1999; Parkkila et al. 2000b; Ortova Gut et al. 2002; Supuran and Scozzafava 2002). Availability of CA inhibitors (CAIs) and specific antibodies for different isozymes, development of immunocytochemical techniques and application of molecular biology techniques have considerably helped increase our understanding of the distribution and role of these exciting

283

enzymes. This chapter discusses the distribution and potential roles of CA isozymes in the alimentary tract (Table 10.1), kidney and nervous system. It also touches on the clinical aspects of CAIs in treating gastrointestinal, renal and neurological diseases.

10.1 ROLE OF CAs IN THE ALIMENTARY TRACT

10.1.1 UPPER ALIMENTARY TRACT

The most prominent isozyme of the salivary glands is CA VI, which is heretofore the only known secretory isozyme of the CA gene family. Fernley et al. (1979) first described CA VI in the ovine parotid gland and saliva. The enzyme was later purified and characterized from rat and human saliva (Feldstein and Silverman 1984; Murakami and Sly 1987; Kadoya et al. 1987). Fernley et al. (1988) reported the amino acid sequence of sheep CA VI, and Aldred et al. (1991) determined the cDNA sequence of human CA VI. CA VI has two cysteine residues that form an intramolecular disulfide bond, which probably stabilizes the structure of the enzyme in the harsh environment of the alimentary canal (Fernley et al. 1988; Aldred et al. 1991; Parkkila et al. 1997). The CA domain of CA VI is highly homologous to those of four other CAs (CA IV, CA IX, CA XII and CA XIV) and they together form a group of extracellular CAs (Fujikawa-Adachi et al. 1999b; Mori et al. 1999).

CA VI is produced by the serous acinar cells of the parotid and submandibular glands (Kadoya et al. 1987; Parkkila et al. 1990), and its secretion is controlled by the autonomic nervous system (Fernley 1991). Functionally, CA VI probably accelerates neutralization of the protons produced by cariogenic bacteria, thus protecting teeth from caries (Kivelä et al. 1999). CA VI has also been linked to the neutralization processes in the upper gastrointestinal tract (Parkkila et al. 1997) and pancreas (Fujikawa-Adachi et al. 1999a). A novel role for CA VI was also suggested when Thatcher et al. (1998) linked CA VI to taste function. Henkin et al. (1999) proposed that CA VI might function as a trophic factor for the taste bud stem cells. Recently, Karhumaa et al. (2001) reported the presence of CA VI in milk. Interestingly, the CA VI concentration in human colostral milk was ca. eight times higher than in mature milk and saliva. CA VI concentrations in newborn saliva are much lower than in adults. Therefore, high concentrations of CA VI in colostrum might compensate the low concentrations of salivary CA VI and thus play an important functional role in the newborn gastrointestinal canal during the early postnatal period.

The gastric mucosa was the first epithelial tissue found to contain CA (Davenport and Fisher 1938; Davenport 1939), the activity of which was detected by histochemical methods in the parietal cells and surface epithelial cells (O´Brien et al. 1977; Sato et al. 1980). Immunohistochemical techniques have further revealed that the major isozyme present in the gastric mucosa is CA II (Parkkila et al. 1994), its main physiological function being to regulate the acidity of the gastric juice. Both the surface epithelial and parietal cells in the stomach contain CA II, which is involved in the production of bicarbonate in the former cells and of gastric acid in the latter. Gastroduodenal bicarbonate secreted by the surface epithelial cells neutralizes the gastric acid (Parkkila and Parkkila 1996). Although some bicarbonate reaches the

TABLE 10.1
Summary of the Distribution of CA Isozymes in the Human Alimentary Tract

Isozyme	Site of Expression	References
CA I	Epithelial cells of colon, subepithelial capillary endothelium, α-cells of Langerhans islets	Lönnerholm et al. 1985; Parkkila et al. 1994
CA II	Epithelial cells of salivary glands, esophagus, stomach, small intestine, colon, bile duct, gallbladder, pancreatic duct, hepatocytes	Lönnerholm et al. 1985; Parkkila et al. 1994
CA III	Hepatocytes	Jeffery et al. 1980
CA IV	Epithelial cells of duodenum, colon, bile duct, gallbladder, subepithelial capillary endothelium, pancreas	Fleming et al. 1995; Parkkila et al. 1996; Fujikawa-Adachi et al. 1999a
CA V	Hepatocytes, β-cells of Langerhans islets	Nagao et al. 1993; Parkkila et al. 1998
CA VI	Acinar cells of salivary glands, pancreas	Parkkila et al. 1990; Fujikawa-Adachi et al. 1999a
CA IX	Epithelial cells of stomach, small intestine, colon, esophagus, pancreas, gallbladder, bile duct	Pastorek et al. 1994; Pastoreková et al. 1997; Turner et al. 1997; Saarnio et al. 1998; Kivelä et al. 2000a
CA XII	Salivary glands, epithelial cells of colon, pancreas	Türeci et al. 1998; Ivanov et al. 1998; Fujikawa-Adachi et al. 1999a; Kivelä et al. 2000a, 2000b
CA XIV	Liver, small intestine, colon	Fujikawa-Adachi et al. 1999b

lumen, much of what is secreted remains below or within the mucous layer covering the epithelium. Thus, the mucosal surface is continuously in contact with a bicarbonate-rich fluid of a high pH relative to the lumen of the stomach (Flemström and Isenberg 2001). Under normal conditions, the bicarbonate neutralizes protons as they diffuse through the mucous gel layer, thus establishing a pH gradient between the lumen and the surface epithelial cells. It has been proposed that CA VI might play a role in this physiological neutralization process (Parkkila et al. 1997).

CA IX is another important CA isozyme of the gastric epithelium (Pastorek et al. 1994), and Pastoreková et al. (1997) first defined CA IX-positive cell types in the gastric mucosa. A recent study on CA IX-deficient mice produced by targeted mutagenesis revealed that the enzyme deficiency results in a marked hyperplasia of mucus-producing cells in the gastric mucosa (Ortova Gut et al. 2002). In addition, these mice frequently developed cystic changes in the gastric mucosa. Although the exact mechanisms still remained unclear, the results suggested that CA IX is functionally implicated in gastric morphogenesis via the control of cell proliferation and differentiation.

10.1.2 COLON

CAs play an important role in ion and water transport in the intestines. Turnberg et al. (1970a, 1970b) observed that acetazolamide markedly inhibited both Na^+ and Cl^- absorption in human intestines and proposed that most of the absorption is mediated by electroneutral Na^+–H^+ and Cl^- $HCO3^-$ exchange processes. Because water absorption follows ion movements, CAs are also implicated in water absorption. In these processes, CA I and II are probably key players, because both are abundantly expressed in the nongoblet epithelial cells of the mammalian colon (Lönnerholm et al. 1985; Parkkila et al. 1994). The colon alkalizes its luminal content by bicarbonate secretion, which is dependent on apical Cl^- $HCO3^-$ exchange (Feldman et al. 1990). It also acidifies the luminal content by active proton secretion (Suzuki and Kaneko 1987). Although the exact physiological significance of colonic proton secretion is uncertain, it is possible that it facilitates nonionic fatty acid uptake by promoting apical Na^+–H^+ exchange (Sellin and DeSoignie 1990) or a proton ATPase pump (Gustin and Goodman 1981).

In addition to cytosolic enzymes, intestinal enterocytes express four membrane-associated isozymes: CAs IV, IX, XII and XIV. The location of CA IV in the apical brush border of the colonic epithelium suggests that this isozyme plays a role in the regulation of colonic ion homeostasis (Fleming et al. 1995). Saarnio et al. (1998), Kivelä et al. (2000b) and Parkkila et al. (2002) have reported the presence of CAs IX, XII and XIV in the colonic enterocytes. The existence of several CA isozymes in the colonic epithelium reflects the complex regulation of intestinal physiology in which the role of each isozyme has remained unclear.

10.1.3 PANCREAS, LIVER AND GALLBLADDER

The role of CA II in secreting bicarbonate into the pancreatic juice by the epithelial duct cells is well documented (Parkkila and Parkkila 1996). CA I is expressed in the α-cells of the endocrine Langerhans islets, where its role has remained unclear (Parkkila et al. 1994). CA V is the second isozyme described in the endocrine pancreas, and its expression is solely confined to the β-cells (Parkkila et al. 1998). It is proposed that mitochondrial CA regulates insulin secretion probably by providing bicarbonate ions for the pyruvate–malate shuttle operating in these cells.

In the hepatobiliary tract, CA II is expressed in the epithelial cells of the bile ducts and gallbladder (Parkkila et al. 1994). In the hepatic bile ducts, CA II facilitates alkalization of bile (Parkkila and Parkkila 1996). In contrast, the same enzyme is one of the key factors involved in bile acidification in the gallbladder (Juvonen et al. 1994). This dual function of CA II is based on its enzymatic reaction producing both protons and bicarbonate. The ionic content of the final secretion depends on the transport machinery present at the plasma membrane in each cell type.

The mammalian liver expresses high levels of mitochondrial CA V. Physiologically, CA V has been implicated in two metabolic processes in the mitochondria of hepatocytes — urea synthesis and gluconeogenesis – supplying bicarbonate for carbamyl phosphate synthetase I, the first urea cycle enzyme in urea synthesis, and for pyruvate carboxylase in gluconeogenesis (Dodgson 1991). CAIs have been observed to retard both these processes in the livers of guinea pigs and rats (Dodgson et al. 1983; Metcalfe et al. 1985; Dodgson 1991).

The low activity isozyme, CA III, is expressed in hepatocytes (Jeffery et al. 1980). Cabiscol and Levine (1995) have demonstrated that it functions in an oxidizing environment and is the most oxidatively modified protein in the liver known so far.

These and other recent results suggest that CA III provides protection from oxidative damage and might serve as a useful marker protein to investigate *in vivo* the mechanisms that contribute to oxidative damage in the liver (Parkkila et al. 1999). Lower levels of free radicals in cells overexpressing CA III might affect growth-signaling pathways (Räisänen et al. 1999).

Earlier studies have indicated that both CA IV and CA IX are expressed in the biliary epithelial cells, whereas hepatocytes are devoid of immunoreactivity for these isozymes (Parkkila et al. 1996; Pastoreková et al. 1997). Interestingly, a recent study (Parkkila et al. 2002) has shown that CA XIV is abundantly expressed at the plasma membrane of murine hepatocytes. In this strategic location, CA XIV might regulate the pH and ion homeostasis between the bile canaliculi and hepatic sinusoids.

10.2 CAIs IN THE ALIMENTARY TRACT

Regulation of the acid–base balance in the alimentary tract is a physiological process involving a number of proteins, such as ion transport proteins, plasma membrane receptors and their ligands and CAs. Even though, or maybe because, several CA isozymes are involved in this process, applicability of CAIs has been so far limited in the gastrointestinal tract. Most attractive have been speculations that CAIs can be useful for the therapy of peptic ulcers. This was an early approach to attack the machinery of the acid-producing cell by acetazolamide (Baron 2000). In 1939, Davenport suggested that CA might be essential for acid production, and, consequently, an inhibitor of this enzyme would inhibit gastric acid secretion. A few years later, Davenport (1946) retracted this theory, because early inhibitors such as sulfanilamide failed to inhibit gastric acid secretion. Janowitz et al. (1952, 1957) demonstrated marked but very brief acid inhibition by acetazolamide and concluded that its action was too brief to be therapeutically useful. Despite this pessimism, a few later studies showed that acetazolamide might be effective in treating gastric ulcers (Puscas et al. 1989; Erdei et al. 1990). However, acetazolamide was never generally approved for treating gastric ulcer because of its many unfavorable side effects. This discussion was aborted more than 10 years ago when many other effective and safer drugs, such as histamine 2 receptor antagonists and $H^+K^+ATPase$ inhibitors, were introduced into clinical practice.

10.3 ROLE OF CAs IN THE KIDNEY

The histochemical localization of CA activity has been extensively studied in the kidney (Lönnerholm 1973, 1983; Lönnerholm and Ridderstråle 1974, 1980; Lönnerholm and Wistrand 1984). The first isozyme-specific antibodies enabled researchers to compare distributions of CAs I, II, and III in the kidney and various other sites (Spicer et al. 1979, 1982, 1990). Of the 11 known active CA isozymes, at least 4 (CAs II, IV, XII and XIV) are expressed in the epithelial cells of renal tubules.

Based on immunohistochemical studies, the intercalated cells of the late distal tubule, the collecting tubule and the collecting duct have been reported to express high levels of CA II (Sato and Spicer 1982; Brown et al. 1983; Lönnerholm and Wistrand 1984; Brown and Kumpulainen 1985). In addition, some immunoreaction has been observed in the loop of Henle (Spicer et al. 1982; Lönnerholm et al. 1986), the proximal tubules (Spicer et al. 1982, 1990; Lönnerholm et al. 1986) and the principal cells of the collecting ducts (Holthöfer et al. 1987). In a recent review, Schwartz (2002) summarized the CA II-positive segments along the mouse nephron and collecting ducts, which include the proximal convoluted (S1 segment) and straight tubules (S2-S3 segments), the descending thin limb of Henle, the thick ascending limb of Henle, the distal convoluted tubule and the intercalated and principal cells of the collecting ducts. Based on the distribution pattern described previously, it is clear that CA II is widely expressed in the kidney and thus plays a key role in renal functions. The fundamental role of CA II in renal physiology was documented when Sly et al. (1983) reported renal tubular acidosis in patients with CA II-deficiency syndrome. CA II-deficient mice produced by chemical mutagenesis also had impaired renal acidification (Lewis et al. 1988).

More than 95% of renal CA activity is cytosolic and corresponds to CA II, whereas 3 to 5% is membrane associated (McKinley and Whitney 1976; Wistrand and Kinne 1977). Nonetheless, functional studies have shown that an impermeant CAI blocked up to 90% of bicarbonate absorption in an isolated, perfused kidney and *in vivo* micropuncture (Lucci et al. 1980, 1983; DuBose and Lucci 1983; Frommer et al. 1984), suggesting a pivotal role for a luminal membrane-bound CA that was initially identified as CA IV, a glycosylphosphatidylinositol (GPI)-linked membrane-associated protein (Zhu and Sly 1990). Immunohistochemical staining of CA IV gives somewhat contradictory results, which partly can be attributable to species differences, which are common for various CA isozymes in other tissues. Brown et al. (1990) studied the localization of CA IV in rat kidney and found positive staining in the plasma membranes of the proximal tubule and thick ascending limb of Henle. Apical membrane staining was weak in the S1 segment, strong in the S2 segment and absent from the S3 segment of the proximal tubule. There was basolateral staining in the S1 and S2 proximal tubules and thick ascending limbs. Studies from Schwartz's laboratory have indicated that in the rabbit kidney, the acidifying nephron segments express CA IV. It is present in the S1 and S2 segments of the proximal tubules, medullary collecting ducts and α-intercalated cells of the cortical collecting duct (Schwartz 2002).

It was accepted for many years that the membrane-associated CA is CA IV. Interestingly, this prediction became questionable when Zhu and Sly (1990) used PI-PLC to release CA IV from the membranes of the human kidney and found that only 54% of CA activity was released. The finding that some membrane-associated CA activity might be insensitive to PI-PLC digestion suggested that there was some alternatively spliced form of CA IV that is expressed as a membrane-spanning protein or there were other membrane-bound isozymes without a GPI linkage.

At the end of the 1990s, two novel membrane-associated CA isozymes were discovered, CA XII and CA XIV, both expressed in the kidney (Türeci et al. 1998; Ivanov et al. 1998; Mori et al. 1999; Fujikawa-Adachi et al. 1999b). The first immunohistochemical study on CA XII in the human kidney revealed that the enzyme is expressed at the basolateral plasma membrane of the epithelial cells in the thick ascending limb of Henle and distal convoluted tubules and in the principal cells of the collecting ducts (Parkkila et al. 2000a). A weak positive reaction was observed in the epithelium of the proximal convoluted tubules. Recently, Schwartz et al. (2003) showed by *in situ* hybridization in the rabbit kidney that CA XII mRNA was expressed in the proximal convoluted and straight tubules, cortical and medullary collecting ducts and papillary epithelium. RT-PCR showed a positive signal in the proximal tubules (none in S3 segment), cortical and medullary collecting ducts and the thick ascending limb of Henle.

A recent study based on immunofluorescence staining, RT-PCR and western blotting indicated that CA XIV is highly expressed in some segments of rodent nephron (Kaunisto et al. 2002). A strong signal was found in apical plasma membranes of the S1 and S2 segments of the proximal tubules and weaker staining in the basolateral membranes. Also, strong staining was reported in the initial portion of the thin descending limb of Henle. The positive staining in the proximal tubules suggested that CA XIV might also function in urinary acidification.

10.4 CAIs IN THE KIDNEY

A number of studies that use CAIs have provided the functional basis on the role of CA activity in renal acidification. Despite their promising past, CAIs have not been very useful in the therapy of renal diseases (Supuran and Scozzafava 2000). Acetazolamide, methazolamide, ethoxzolamide and dichlorophenamide can be used to treat edema induced by drugs or congestive heart failure (Supuran and Scozzafava 2000). These inhibitors can, however, cause several undesired side effects, such as metabolic acidosis, nephrolithiasis, CNS symptoms and allergic reactions, which have limited their exploitation in therapy (Tawil et al. 1993; Supuran and Scozzafava 2001). The acute response to CA inhibition is an increase in the excretion of bicarbonate, sodium and potassium and an increase in urinary flow and titratable acid (Bagnis et al. 2001). The loss of bicarbonate and sodium is considered to be self-limited on continued administration of the inhibitor, probably because the initial acidosis resulting from bicarbonate loss activates bicarbonate reabsorption via CA-independent mechanisms. Chronic CA inhibition stimulates morphologic changes in the collecting ducts (Bagnis et al. 2001). Therefore, CA activity appears to play an important role in determining the differentiated phenotype of the renal epithelium.

10.5 ROLE OF CAs IN THE NERVOUS SYSTEM

It has been known for 60 years that CA occurs in the mammalian brain (Ashby 1943), where it has a number of physiological roles, e.g., in fluid and ion compartmentation (Bourke and Kimelberg 1975), the formation of cerebrospinal fluid (CSF; Maren 1967), seizure activity (Anderson et al. 1984), the respiratory response to carbon dioxide (Ridderstråle and Hanson 1985) and the generation of bicarbonate for biosynthetic reactions (Tansey et al. 1988; Cammer 1991).

The earliest reports on CA activity in glial cells were based on the dissection of neurons and glial cell clumps from the rat brainstem. Glial cells had six times the CA activity per cell as that in the neurons, and it was suggested that the glia might be the main site of CA expression in the central nervous system (CNS; Giacobini 1961, 1962). Similarly, CA activity was shown in an early histochemical

TABLE 10.2
Summary of the Distribution of CA Isozymes in the Brain

Isozyme	Site of Expression	References
CA II	Oligodendrocytes, astrocytes, myelin, choroid plexus, neurons	Ghandour et al. 1979, 1980; Roussel et al. 1979; Langley et al. 1980; Kumpulainen and Korhonen 1982; Kumpulainen et al. 1983; Kimelberg et al. 1982; Snyder et al. 1983; Cammer and Tansey 1988; Cammer 1991; Cammer and Zhang 1992; Jeffrey et al. 1991; Neubauer 1991
CA III	Choroid plexus, microglial cells	Nogradi 1993; Nogradi et al. 1993
CA IV	Endothelial cells	Ghandour et al. 1992
CA V	Astrocytes, neurons	Ghandour et al. 2000
CA XIV	Neurons	Parkkila et al. 2001

study to be highest in areas of the mouse brain that were rich in myelinated fibers and glial cells (Korhonen et al. 1964). When glial cells and neurons were separated by bulk isolation, the glia were found to have higher CA activity, although the various glial cell types were not distinguished (Sinha and Rose 1971; Nagata et al. 1974). After methods had been devised for purifying myelin and each type of glial cell from rat brain, the definitive biochemical localization of CNS enzymes became more feasible. Significant CA activity was observed in myelin isolated from rat, mouse, monkey, cat and rabbit brains (Cammer et al. 1976, 1977; Sapirstein and Lees 1978; Ridderstråle and Hanson 1985). When CA was assayed in bulk-isolated cells, the highest activities were observed in the oligodendrocytes, and the specific activities in the neurons and astrocytes were low but measurable (Snyder et al. 1983).

The presence of the various isozymes in the brain is summarized in Table 10.2. When sections from rodent and human brains and spinal cords were immunostained for CA II, several investigators concluded that the mammalian brain only contained CA in the oligodendrocytes (Roussel et al. 1979; Ghandour et al. 1979, 1980; Langley et al. 1980; Kumpulainen and Korhonen 1982; Kumpulainen et al. 1983). Myelin sheaths were also found to contain CA II (Roussel et al. 1979; Kumpulainen and Korhonen 1982). The presence of CA II in the astrocytes remained a controversial matter, as findings in cultured cells (Kimelberg et al. 1982), one study of bulk isolated cells (Snyder et al. 1983) and one immunohistochemical study (Roussel et al. 1979) suggested that they might contain CA at relatively low levels. *In situ* hybridization with CA II-specific probes showed that the expression of CA II mRNA is limited to oligodendrocytes in culture (Ghandour and Skoff 1991). It is also proposed that CA II is located in the astrocytes in the normal gray matter of the brain and also in the reactive astrocytes in severely gliotic white matter of the jimpy mutant mouse and of rats with experimental autoimmune encephalomyelitis (Cammer and Tansey 1988; Cammer 1991; Cammer and Zhang 1992). In addition, the reactive astrocytes, oligodendrocytes and neurons surrounding brain tumors and various types of neoplastic cells in these tumors contain variable amounts of CA II (Nakagawa et al. 1986, 1987; Parkkila et al. 1995).

Neurons have been considered to lack CA, but there seems to be a growing list of exceptions to this widely held notion, including subpopulations of neurons in the retina, peripheral sensory ganglia and CNS (Neubauer 1991; Ghandour et al. 2000; Parkkila et al. 2001). One of the neuronal CAs is the mitochondrial isoform, which is expressed in astrocytes and neurons but not in oligodendrocytes (Ghandour et al. 2000). In the astrocytes, mitochondrial CA activity might play an important role in gluconeogenesis by providing bicarbonate ions for the pyruvate carboxylase, which is known to be expressed in those cells (Yu et al. 1983). Because pyruvate carboxylase expression was not previously detected in neurons, two novel roles for neuronal CA V were speculated that were not linked to the gluconeogenesis: (1) regulation of intramitochondrial calcium levels and (2) regulation of neuronal transmission by facilitating the bicarbonate-ion-induced GABA responses (Ghandour et al. 2000). Interestingly, recent studies have found evidence that carboxylation of pyruvate to malate occurs in neurons and that it supports formation of the transmitter glutamate (Hassel 2001). This finding suggests that the previous predictions were not the final word in the neuronal metabolism of pyruvate, or, perhaps, CA V can serve a number of different physiological processes in neurons.

A recent study by Parkkila et al. (2001) demonstrated that a novel membrane-bound isozyme, CA XIV, is highly expressed in some neurons, where it might have an important role in producing the alkaline shift linked to neuronal signal transduction. The highest expression was reported on large neuronal bodies and axons in the anterolateral part of the pons and medulla oblongata. Other CA XIV-positive sites included the hippocampus, corpus callosum, cerebellar white matter and peduncles, pyramidal tract and choroid plexus.

CA activity plays an important role in the production of CSF and in the regulation of its pH and ionic constituents (Maren 1967). The presence of CA II and CA III has been demonstrated in the cytoplasm and microvilli of the epithelial cells of the choroid plexus (Kumpulainen and Korhonen 1982; Nogradi et al. 1993). The distribution of CA IV in the CNS of adult rats and CA II-deficient mice is limited to the luminal surface of the capillary endothelial cells (Ghandour et al. 1992). The unique location of CA IV in the luminal face of the cerebral capillaries suggests that it has an important role at the blood–brain barrier, regulating carbon dioxide and bicarbonate homeostasis in the brain.

10.6 CAIs IN THE NERVOUS SYSTEM

CAIs have profound effects on the function of the CNS. For example, acetazolamide can reduce CSF production by ca. 50% (Maren 1972; McCarthy and Reed 1974). However, it has been known for long that acetazolamide dilates intracranial vessels (Maren 1967; Hauge et al. 1983). This vasodilatation increases cerebral blood volume (CBV), which is considered to reflect an intrinsic, direct volume load to the intracranial cavity. In subjects with normal CSF circulation and absorption, the CBV increase does not significantly elevate the intracranial pressure through a reduction in the intracranial CSF volume. In this context, the so-called Diamox® challenge test has been used to determine indications for CSF shunting in patients with normal pressure hydrocephalus (Miyake et al. 1999). An increase of intracranial pressure of more than 10 mmHg in response to 1000 mg of Diamox® indicates considerable impairment of the CSF circulation.

Pseudotumor cerebri is a syndrome for which acetazolamide is considered a drug of choice (Shin and Balcer 2002). This disease is defined clinically by four criteria: (1) elevated intracranial pressure, (2) normal cerebral anatomy, (3) normal CSF fluid composition, and (4) signs and symptoms of intracranial pressure, including papilledema (http://www.revoptom.com/handbook/SECT53a.HTM). Administration of acetazolamide to these patients typically results in long-lasting control of transient visual obscuration, headache, and diplopia, all of which are manifestations of intra-cranial hypertension.

High-altitude cerebral edema is a severe form of acute mountain sickness occur-ring at heights above 4500 m. The clinical features are of headache, impairment of consciousness and a variety of neurological signs (Clarke 1988). Clinically and pathophysiologically, high-altitude cerebral edema is the end stage of acute mountain disease. Recent evidence has suggested that on ascent to high altitudes, all people have swelling of the brain (Hackett and Roach 2001). It has been speculated based on preliminary experimental data that the disease might be related to a person's ability to compensate for the swelling of the brain. Those with a higher ratio of cranial CSF to brain volume can better compensate the swelling through the dis-placement of CSF and might therefore be less likely to have acute mountain sickness. For preventing high-altitude illness, the best strategy is a gradual ascent to promote acclimatization. Medical prophylaxis by acetazolamide has been recommended for those who plan to ascend from sea level to over 3000 m in one day and for those with a history of acute mountain disease. Acetazolamide is an important drug for managing high-altitude illness, especially in patients with mild or moderate clinical presentation. The management strategy can also include administration of dexa-methasone and oxygen as well as use of a portable hyperbaric chamber (Hackett and Roach 2001).

For several decades, acetazolamide has been used to treat epilepsy (Reiss and Oles 1996). It is primarily used in combination therapy with other antiepileptic medications in both children and adults and should be especially considered in refractory epilepsy (Reiss and Oles 1996; Katayama et al. 2002). Although aceta-zolamide might be useful in partial, myoclonic, absence and primary generalized tonic-clonic seizures uncontrolled by other marketed agents, it has been inadequately studied by current standards and its use has been limited. It should also be noted that one might develop partial tolerance to the antiepileptic activity (Reiss and Oles 1996).

Topiramate is a sulfamate fructopyranose derivative currently available for treat-ing partial onset and generalized epileptic seizures in adults and children (Bialer et al. 1999). It is structurally distinct from the other commonly used antiepileptic drugs (Perucca 1997). Interestingly, topiramate shares with other sulfamate or sul-fonamide derivatives the ability to inhibit CA activity. *In vitro* studies have indicated that topiramate is more potent as an inhibitor of CA II and CA IV than of CA I, CA III, and CA VI (Dodgson et al. 2000). Studies from Supuran's laboratory have indicated that it is a strong CAI — similar to acetazolamide — of human CA II isozyme (Supuran and Scozzafava 2000). Recent CA inhibition data from the same laboratory clearly confirmed that topiramate is a potent inhibitor of CA activity, with inhibition constants 5 nM, 54 nM and 250 nM against human CA II, bovine CA IV, and human CA I, respectively (Casini et al. 2003). Detailed structural analysis of the CA II–topiramate complex has revealed a strong network of seven hydrogen bonds, which bind the inhibitor tightly to the active site of the enzyme. In addition, the sulfamate moiety present in topiramate binds to the catalytical Zn(II) ion. From this point of view, it is of considerable interest that the sulfamate moiety might be responsible for the anticonvulsant activity of topiramate. However, there is some evidence that inhibition of CA is not the main mechanism responsible for its therapeutic effect (Perucca 1997; Stringer 2000). It has been suggested that the anticonvulsant actions of topiramate involve several mechanisms, such as enhancement of GABAergic transmission and inhibitory action on neuronal sodium currents. As a third mechanism, topiramate inhibits excitatory transmission by antagonizing some types of glutamate receptors.

ACKNOWLEDGMENTS

This work was supported by grants from the Sigrid Juselius Foundation and Academy of Finland (SP) and the Finnish Cultural Foundation and the Finnish Dental Society (JK).

REFERENCES

Aldred, P., Fu, P., Barret, G., Penschow, J.D., Wright, R.D., Coghlan, J.P., and Fernley, R.T. (1991) Human secreted carbonic anhydrase: cDNA cloning, nucleotide sequence, and hybridization histochemistry. *Biochemistry* **30**, 569–575.

Anderson, R.E., Engström, F.L., and Woodbury, D.M. (1984) Localization of carbonic anhydrase in the cerebrum and cerebellum of normal and audiogenic seizure mice. *Annals of the New York Academy of Sciences* **429**, 502–524.

Ashby, W. (1943) Carbonic anhydrase in mammalian tissue. *Journal of Biological Chemistry* **151**, 521–527.

Bagnis, C., Marshansky, V., Breton, S., and Brown, D. (2001) Remodeling the cellular profile of collecting ducts by chronic carbonic anhydrase inhibition. *American Journal of Physiology* **280**, F437–F448.

Baron, J.H. (2000) Treatments of peptic ulcer. *Mount Sinai Journal of Medicine* **67**, 63–67.

Bialer, M., Johannessen, S.I., Kupferberg, H.J., Levy, R.H., Loiseau, P., and Perucca, E. (1999) Progress report on new antiepileptic drugs: a summary of the fourth Eilat conference (EILAT IV). *Epilepsy Research* **34**, 1–41.

Bourke, R.S., and Kimelberg, H.K. (1975) The effect of HCO_3^- on the swelling and ion uptake of monkey cerebral cortex under conditions of raised extracellular potassium. *Journal of Neurochemistry* **25**, 323–328.

Brown, D., and Kumpulainen, T. (1985) Immunocytochemical localization of carbonic anhydrase on ultrathin frozen sections with protein A-gold. *Histochemistry* **83**, 153–158.

Brown, D., Kumpulainen, T., Roth, J., and Orci, L. (1983) Immunohistochemical localization of carbonic anhydrase in postnatal and adult rat kidney. *American Journal of Physiology* **245**, F110–F118.

Brown, D., Zhu, X.L., and Sly, W.S. (1990) Localization of membrane-associated carbonic anhydrase type IV in kidney epithelial cells. *Proceedings of the National Academy of Sciences of the United States of America* **87**, 7457–7461.

Cabiscol, E., and Levine, R.L. (1995) Carbonic anhydrase III. Oxidative modification *in vivo* and loss of phosphatase activity during aging. *Journal of Biological Chemistry* **270**, 14742–14747.

Casini, A., Antel, J., Abbate, F., Scozzafava, A., David, S., Waldeck, H., Schäfer, S., and Supuran, C.T. (2003) Carbonic anhydrase inhibitors: SAR and x-ray crystallographic study for the interaction of sugar sulfamates/sulfamides with isozymes I, II and IV. *Bioorganic and Medicinal Chemistry Letters* **13**, 841–845.

Cammer, W. (1991) Carbonic anhydrase in myelin and glial cells in the mammalian central nervous system. In *The Carbonic Anhydrases — Cellular Physiology and Molecular Genetics*, Dodgson, S.J., Tashian, R.E., and Carter, N.D. (Eds.), Plenum Press, New York, pp. 325–332.

Cammer, W., Bieler, L., Freman, T., and Norton, W.T. (1977) Quantitation of myelin carbonic anhydrase: Development and subfractionation of rat brain myelin and comparison with myelin from other species. *Brain Research* **138**, 17–28.

Cammer, W., Fredman, T., Rose, A.L., and Norton, W.T. (1976) Brain carbonic anhydrase: activity in isolated myelin and the effect of hexachlorophene. *Journal of Neurochemistry* **27**, 165–171.

Cammer, W., and Tansey, F.A. (1988) The astrocyte as a locus of carbonic anhydrase in the brains of normal and dysmyelinating mutant mice. *Journal of Comparative Neurology* **275**, 65–75.

Cammer, W., and Zhang, H. (1992) Carbonic anhydrase in distinct precursors of astrocytes and oligodendrocytes in the forebrains of neonatal and young rats. *Developmental Brain Research* **67**, 257–263.

Clarke, C. (1988) High altitude cerebral oedema. *International Journal of Sports Medicine* **9**, 170–174.

Davenport, H.W. (1939) Gastric carbonic anhydrase. *Journal of Physiology (London)* **97**, 32–43.

Davenport, H.W. (1946) In memoriam: The carbonic anhydrase theory of gastric acid secretion. *Gastroenterology* **7**, 374–375.

Davenport, H.W., and Fisher, R.B. (1938) Carbonic anhydrase in the gastrointestinal mucosa. *Journal of Physiology (London)* **94**, 16P–17P.

Dodgson, S.J. (1991) Liver mitochondrial carbonic anhydrase (CA V), gluconeogenesis, and ureagenesis in the hepatocyte. In *The Carbonic Anhydrases — Cellular Physiology and Molecular Genetics*, Dodgson, S.J., Tashian, R.E., and Carter, N.D. (Eds.), Plenum Press, New York, pp. 297–306.

Dodgson, S.J., Forster, R.E., Schwed, D.A., and Storey, B.T. (1983) Contribution of matrix carbonic anhydrase to citrulline synthesis in isolated guinea pig liver mitochondria. *Journal of Biological Chemistry* **258**, 7696–7701.

Dodgson, S.J., Shank, R.P., and Maryanoff, B.E. (2000) Topiramate as an inhibitor of carbonic anhydrase isoenzymes. *Epilepsia* **41**, S35–S39.

DuBose, T.D. Jr., and Lucci, M.S. (1983) Effect of carbonic anhydrase inhibition on superficial and deep nephron bicarbonate reabsorption in the rat. *Journal of Clinical Investigation* **71**, 55–65.

Erdei, A., Gyori, I., Gedeon, A., and Szabo, I. (1990) Successful treatment of intractable gastric ulcers with acetazolamide. *Acta Medica Hungarica* **47**, 171–178.

Feldman, G.M., Koethe, J.D., and Stephenson, R.L. (1990) Base secretion in rat distal colon: ionic requirements. *American Journal of Physiology* **258**, G825–G832.

Feldstein, J.B., and Silverman, D.N. (1984) Purification and characterization of carbonic anhydrase from the saliva of the rat. *Journal of Biological Chemistry* **259**, 5447–5453.

Fernley, R.T. (1991) Secreted carbonic anhydrases. In *Carbonic Anhydrase: From Biochemistry and Genetics to Physiology and Clinical Medicine*, Botré, F., Gros, G., and Storey, B.T. (Eds.), VCH Verlagsgesellschaft, Weinheim, pp. 178–185.

Fernley, R.T., Wright, R.D., and Coghlan, J.P. (1979) A novel carbonic anhydrase from the ovine parotid gland. *FEBS Letters* **105**, 299–302.

Fernley, R.T., Wright, R.D., and Coghlan, J.P. (1988) Complete aminoacid sequence of ovine salivary carbonic anhydrase. *Biochemistry* **27**, 2815–2820.

Fleming, R.E., Parkkila, S., Parkkila, A.-K., Rajaniemi, H., Waheed, A., and Sly, W.S. (1995) Carbonic anhydrase IV in rat and human gastrointestinal tract: Regional, cellular, and subcellular localization. *Journal of Clinical Investigation* **96**, 2907–2913.

Flemström, G., and Isenberg, J.I. (2001) Gastroduodenal mucosal alkaline secretion and mucosal protection. *News in Physiological Sciences* **16**, 23–28.

Frommer, J.P., Laski, M.E., Wesson, D.E., and Kurtzman, N.A. (1984) Internephron heterogeneity for carbonic anhydrase-independent bicarbonate reabsorption in the rat. *Journal of Clinical Investigation* **73**, 1034–1045.

Fujikawa-Adachi, K., Nishimori, I., Sakamoto, S., Morita, M., Onishi, S., Yonezava, S., and Hollingsworth, M.A. (1999a) Identification of carbonic anhydrase IV and VI mRNA expression in human pancreas and salivary glands. *Pancreas* **18**, 329–335.

Fujikawa-Adachi, K., Nishimori, I., Taguchi, T., and Onishi, S. (1999b) Human carbonic anhydrase XIV (CA 14): cDNA cloning, mRNA expression, and mapping to chromosome 1. *Genomics* **61**, 74–81.

Giacobini, E. (1961) Localization of carbonic anhydrase in the nervous system. *Science* **134**, 1524–1525.

Giacobini, E. (1962) A cytochemical study of the localization of carbonic anhydrase in the nervous system. *Journal of Neurochemistry* **9**, 169–177.

Ghandour, M.S., Langley, O.K., Vincendon, G., and Gombos, G. (1979) Double labeling immunohistochemical technique provides evidence of the specificity of glial cell markers. *Journal of Histochemistry and Cytochemistry* **27**, 1634–1637.

Ghandour, M.S., Langley, O.K., Vincendon, G., Gombos, G., Filippi, D., Limozin, N., Dalmasso, C., and Laurent, G. (1980) Immunochemical and immunohistochemical study of carbonic anhydrase II in adult rat cerebellum: A marker for oligodendrocytes. *Neuroscience* **5**, 559–571.

Ghandour, M.S., Langley, O.K., Zhu, X.L., Waheed, A., and Sly, W.S. (1992) Carbonic anhydrase IV on brain capillary endothelial cells: a marker associated with the blood-brain barrier. *Proceedings of the National Academy of Sciences of the United States of America* **89**, 6823–6827.

Ghandour, M.S., Parkkila, A.-K., Parkkila, S., Waheed, A., and Sly, W.S. (2000) Mitochondrial carbonic anhydrase in the nervous system: Expression in neuronal and glial cells. *Journal of Neurochemistry* **75**, 2212–2220.

Ghandour, M.S., and Skoff, R.P. (1991) Double labeling *in situ* hybridization analysis of mRNAs for carbonic anhydrase II and myelin basic protein: mRNAs expression in developing cultured glial cells. *Glia* **4**, 1–10.

Gustin, M.C., and Goodman, D.B.P. (1981) Isolation of brush-border membrane from the rabbit descending colon epithelium. Partial characterization of a unique K$^+$-activated ATPase. *Journal of Biological Chemistry* **256**, 10651–10656.

Hackett, P.H., and Roach, R.C. (2001) High-altitude illness. *New England Journal of Medicine* **345**, 107–114.

Hassel, B. (2001) Pyruvate carboxylation in neurons. *Journal of Neuroscience Research* **66**, 755–762.

Hauge, A., Nicolaysen, G., and Thoresen, M. (1983) Acute effects of acetazolamide on cerebral blood flow in man. *Acta Physiologica Scandinavica* **117**, 233–239.

Henkin, R.I., Martin, B.M., and Agarwal, R.P. (1999) Efficacy of exogenous oral zinc in treatment of patients with carbonic anhydrase VI deficiency. *American Journal of the Medical Sciences* **318**, 392–405.

Holthöfer, H., Schulte, B.A., Pasternack, G., Siegel, G.J., and Spicer, S.S. (1987) Immuno-cytochemical characterization of carbonic anhydrase-rich cells in the rat kidney collecting duct. *Laboratory Investigation* **57**, 150–156.

Ivanov, S.V., Kuzmin, I., Wei, M.-H., Pack, S., Geil, L., Johnson, B.E., Stanbridge, E.J., and Lerman, M.I. (1998) Down-regulation of transmembrane carbonic anhydrases in renal cell carcinoma cell lines by wild-type von Hippel-Lindau transgenes. *Proceedings of the National Academy of Sciences of the United States of America* **95**, 12596–12601.

Janowitz, H.D., Colcher, H., and Hollander, F. (1952) Inhibition of gastric secretion of acid in dogs by carbonic anhydrase inhibitor, 2-acetylamino-1, 3, 4-thiadiazole-5-sulfona-mide. *American Journal of Physiology* **171**, 325–333.

Janowitz, H.D., Dreiling, D.A., Rolbin, H.L., and Hollander, F. (1957) Inhibition of the formation of hydrochloric acid in the human stomach by Diamox: The role of carbonic anhydrase in gastric secretion. *Gastroenterology* **33**, 378–384.

Jeffery, S., Edwards, Y., and Carter, N. (1980) Distribution of CA III in fetal and adult human tissue. *Biochemical Genetics* **18**, 843–849.

Jeffrey, M., Wells, G.A.H., and Bridges A.W. (1991) Carbonic anhydrase II expression in fibrous astrocytes of the sheep. *Journal of Comparative Pathology* **104**, 337–343.

Juvonen, T., Parkkila, S., Parkkila, A.-K., Niemelä, O., Lajunen, L.H.J., Kairaluoma, M.I., Perämäki, P., and Rajaniemi, H. (1994) High activity carbonic anhydrase isoenzyme (CA II) in human gallbladder epithelium. *Journal of Histochemistry and Cytochem-istry* **42**, 1393–1397.

Kadoya, Y., Kuwahara, H., Shimazaki, M., Ogawa, Y., and Yagi, T. (1987) Isolation of a novel carbonic anhydrase from human saliva and immunohistochemical demonstration of its related isoenzymes in salivary gland. *Osaka City Medical Journal* **33**, 99–109.

Karhumaa, P., Leinonen, J., Parkkila, S., Kaunisto, K., Tapanainen, J., and Rajaniemi, H. (2001) The identification of secreted carbonic anhydrase VI as a constitutive glyco-protein of human and rat milk. *Proceedings of the National Academy of Sciences of the United States of America* **98**, 11604–11608.

Karhumaa, P., Parkkila, S., Waheed, A., Parkkila, A.-K., Kaunisto, K., Tucker, P.W., Huang, C.-J., Sly, W.S., and Rajaniemi, H. (2000) Nuclear NonO/p54nrb protein is a non-classical carbonic anhydrase. *Journal of Biological Chemistry* **275**, 16044–16049.

Katayama, F., Miura, H., and Takanashi, S. (2002) Long-term effectiveness and side effects of acetazolamide as an adjunct to other anticonvulsants in the treatment of refractory epilepsies. *Brain and Development* **24**, 150–154.

Kaunisto, K., Parkkila, S., Rajaniemi, H., Waheed, A., Grubb, J., and Sly, W.S. (2002) Carbonic anhydrase XIV: Luminal expression suggests key role in renal acidification. *Kidney International* **61**, 2111–2118.

Kimelberg, H.K., Stieg, P.E., and Mazurkiewicz, J.E. (1982) Immunocytochemical and bio-chemical analysis of carbonic anhydrase in primary astrocyte cultures from rat brain. *Journal of Neurochemistry* **39**, 734–742.

Kivelä, J., Parkkila, S., Parkkila, A.-K., and Rajaniemi, H. (1999) A low concentration of carbonic anhydrase isoenzyme VI in whole saliva is associated with caries prevalence. *Caries Research* **33**, 178–184.

Kivelä, A.J., Parkkila, S., Saarnio, J., Karttunen, T.J., Kivelä, J., Parkkila, A.-K., Pastoreková, S., Pastorek, J., Waheed, A., Sly, W.S., and Rajaniemi, H. (2000a) Expression of transmembrane carbonic anhydrase isozymes IX and XII in normal human pancreas and pancreatic tumors. *Histochemistry and Cell Biology* **114**, 197–204.

Kivelä, A., Parkkila, S., Saarnio, J., Karttunen, T.J., Kivelä, J., Parkkila, A.-K., Waheed, A., Sly, W.S., Grubb, J.H., Shah, G., Türeci, Ö., and Rajaniemi, H. (2000b) Expression of a novel transmembrane carbonic anhydrase isozyme XII in normal human gut and colorectal tumours. *American Journal of Pathology* **156**, 577–584.

Korhonen, L.K., Näätänen, E., and Hyyppä, M. (1964) A histochemical study of carbonic anhydrase in some parts of the mouse brain. *Acta Histochemica* **18**, 336–347.

Kumpulainen, T., Dahl, D., Korhonen, L.K., and Nyström, S.H.M. (1983) Immunolabeling of carbonic anhydrase isoenzyme C and glial fibrillary acidic protein in paraffin-embedded tissue sections of human brain and retina. *Journal of Histochemistry and Cytochemistry* **31**, 876–886.

Kumpulainen, T., and Korhonen, L.K. (1982) Immunohistochemical localization of carbonic anhydrase isoenzyme C in the central and peripheral nervous system of the mouse. *Journal of Histochemistry and Cytochemistry* **30**, 283–292.

Langley, O.K., Ghandour, M.S., Vincendon, G., and Gombos, G. (1980) Carbonic anhydrase: an ultrastructural study in rat cerebellum. *Histochemical Journal* **12**, 473–483.

Lewis, S.E., Erickson, R.P., Barnett, L.B., Venta, P.J., and Tashian, R.E. (1988) N-Ethyl-N-nitrosourea-induced null mutation at the mouse Car-2 locus: An animal model for human carbonic anhydrase II deficiency syndrome. *Proceedings of the National Academy of Sciences of the United States of America* **85**, 1962–1966.

Lucci, M.S., Pucacco, L.R., DuBose, T.D. Jr, Kokko, J.P., and Carter, N.W. (1980) Direct evaluation of acidification by rat proximal tubule: role of carbonic anhydrase. *American Journal of Physiology* **238**, F372–F379.

Lucci, M.S., Tinker, J.P., Weiner, I.M., and DuBose, T.D. Jr. (1983) Function of proximal tubule carbonic anhydrase defined by selective inhibition. *American Journal of Physiology* **245**, F443–F449.

Lönnerholm, G. (1973) Histochemical demonstration of carbonic anhydrase in the human kidney. *Acta Physiologica Scandinavica* **88**, 455–469.

Lönnerholm, G. (1983) Carbonic anhydrase in the monkey kidney. *Histochemistry* **78**, 195–209.

Lönnerholm, G., and Ridderstråle, Y. (1974) Distribution of carbonic anhydrase in the frog nephron. *Acta Physiologica Scandinavica* **90**, 764–778.

Lönnerholm, G., and Ridderstråle, Y. (1980) Intracellular distribution of carbonic anhydrase in the rat kidney. *Kidney International* **17**, 162–174.

Lönnerholm, G., Selking, Ö., and Wistrand, P.J. (1985) Amount and distribution of carbonic anhydrases CA I and CA II in the gastrointestinal tract. *Gastroenterology* **88**, 1151–1161.

Lönnerholm, G., Wistrand, P.J., and Barany, E. (1986) Carbonic anhydrase isoenzymes in the rat kidney: Effects of chronic acetazolamide treatment. *Acta Physiologica Scandinavica* **126**, 51–60.

Lönnerholm, G., and Wistrand, P.J. (1984) Carbonic anhydrase in the human kidney: A histochemical and immunocytochemical study. *Kidney International* **25**, 886–898.

Maren, T.H. (1967) Carbonic anhydrase: Chemistry, physiology, and inhibition. *Physiological Reviews* **47**, 595–781.

Maren, T.H. (1972) Bicarbonate formation in cerebrospinal fluid: Role in sodium transport and pH regulation. *American Journal of Physiology* **222**, 885–899.

McCarthy, K.D., and Reed, D.J. (1974) The effect of acetazolamide and furosemide on cerebrospinal fluid production and choroid plexus carbonic anhydrase activity. *Journal of Pharmacology and Experimental Therapeutics* **189**, 194–201.

McKinley, D.N., and Whitney, P.L. (1976) Particulate carbonic anhydrase in homogenates of human kidney. *Biochimica et Biophysica Acta* **445**, 780–790.

Metcalfe, H.K., Monson, J.P., Drew, P.J., Iles, R.A., Carter, N.D., and Cohen, R.D. (1985) Inhibition of gluconeogenesis and urea synthesis in isolated rat hepatocytes by acetazolamide. *Biochemical Society Transactions* **13**, 255.

Miyake, H., Ohta, T., Kajimoto, Y., and Deguchi, J. (1999) Diamox® challenge test to decide indications for cerebrospinal fluid shunting in normal pressure hydrocephalus. *Acta Neurochirurgica* **141**, 1187–1193.

Mori, K., Ogawa, Y., Ebihara, K., Tamura, N., Tashiro, K., Kuwahara, T., Mukoyama, M., Sugawara, A., Ozaki, S., Tanaka, I., and Nakao, K. (1999) Isolation and characterization of CA XIV, a novel membrane-bound carbonic anhydrase from mouse kidney. *Journal of Biological Chemistry* **274**, 15701–15705.

Murakami, H., and Sly, W.S. (1987) Purification and characterization of human salivary carbonic anhydrase. *Journal of Biological Chemistry* **262**, 1382–1388.

Nagao, Y., Platero, J.S., Waheed, A., and Sly, W.S. (1993) Human mitochondrial carbonic anhydrase: cDNA cloning, expression, subcellular localization, and mapping to chromosome 16. *Proceedings of the National Academy of Sciences of the United States of America* **90**, 7623–7627.

Nagata, Y., Mikoshiba, K., and Tsukada, Y. (1974) Neuronal cell body enriched and glial cell enriched fractions from young and adult rat brains: Preparation and morphological and biochemical properties. *Journal of Neurochemistry* **22**, 493–503.

Nakagawa, Y., Perentes, E., and Rubinstein, L.J. (1986) Immunohistochemical characterization of oligodendrogliomas: an analysis of multiple markers. *Acta Neuropathologica* **72**, 15–22.

Nakagawa, Y., Perentes, E., and Rubinstein, L.J. (1987) Non-specificity of anti-carbonic anhydrase C antibody as a marker in human neurooncology. *Journal of Neuropathology and Experimental Neurology* **46**, 451–460.

Neubauer, J.A. (1991) Carbonic anhydrase and sensory function in the central nervous system. In *The Carbonic Anhydrases — Cellular Physiology and Molecular Genetics*, Dodgson, S.J., Tashian, R.E., and Carter, N.D. (Eds.), Plenum Press, New York, pp. 319–323.

Nogradi, A. (1993) Differential expression of carbonic anhydrase isoenzymes in microglial cell types. *Glia* **8**, 133–142.

Nogradi, A., Kelly, C., and Carter, N.D. (1993) Localization of acetazolamide-resistant carbonic anhydrase III in human and rat choroid plexus by immunocytochemistry and *in situ* hybridisation. *Neuroscience Letters* **151**, 162–165.

O'Brien, P., Rosen, S., Trencis-Buck, L., and Silen, W. (1977) Distribution of carbonic anhydrase within the gastric mucosa. *Gastroenterology* **72**, 870–874.

Ortova Gut, M., Parkkila, S., Vernerová, Z., Rohde, E., Závada, J., Höcker, M., Pastorek, J., Karttunen, T., Gibadulinová, A., Zavadová, Z., Knobeloch, K.-L., Wiedenmann, B., Svoboda, J., Horak, I., and Pastoreková, S. (2002) Gastric hyperplasia in mice with targeted disruption of the carbonic anhydrase gene *Car9*. *Gastroenterology* **123**, 1889–1903.

Parkkila, A.-K., Herva, R., Parkkila, S., and Rajaniemi, H. (1995) Immunohistochemical demonstration of human carbonic anhydrase isoenzyme II in brain tumours. *Histochemical Journal* **27**, 974–982.

Parkkila, A.-K., Scarim, A.L., Parkkila, S., Waheed, A., Corbett, J.A., and Sly, W.S. (1998) Expression of carbonic anhydrase V in pancreatic β-cells suggests role for mitochondrial carbonic anhydrase in insulin secretion. *Journal of Biological Chemistry* **273**, 24620–24623.

Parkkila, S., Kaunisto, K., Rajaniemi, L., Kumpulainen, T., Jokinen, K., and Rajaniemi, H. (1990) Immunohistochemical localization of carbonic anhydrase isoenzymes VI, II and I in human parotid and submandibular glands. *Journal of Histochemistry and Cytochemistry* **38**, 941–947.

Parkkila, S., Kivelä, A.J., Kaunisto, K., Parkkila, A.-K., Hakkola, J., Rajaniemi, H., Waheed, A., and Sly, W.S. (2002) The plasma membrane carbonic anhydrase in murine hepatocytes identified as isozyme XIV. *BMC Gastroenterology* **2**, 13.

Parkkila, S., and Parkkila, A.-K. (1996) Carbonic anhydrase in the alimentary tract. Roles of the different isozymes and salivary factors in the maintenance of optimal conditions in the gastrointestinal canal. *Scandinavian Journal of Gastroenterology* **31**, 305–317.

Parkkila, S., Parkkila, A.-K., Juvonen, T., and Rajaniemi, H. (1994) Distribution of the carbonic anhydrase isoenzymes I, II, and VI in the human alimentary tract. *Gut* **35**, 646–650.

Parkkila, S., Parkkila, A.-K., Juvonen, T., Waheed, A., Sly, W.S., Saarnio, J., Kaunisto, K., Kellokumpu, S., and Rajaniemi, H. (1996) Membrane-bound carbonic anhydrase IV is expressed in the luminal plasma membrane of the human gallbladder epithelium. *Hepatology* **24**, 1104–1108.

Parkkila, S., Parkkila, A.-K., Lehtola, J., Reinilä, A., Södervik, H.-J., Rannisto, M., and Rajaniemi, H. (1997) Salivary carbonic anhydrase protects gastroesophageal mucosa from acid injury. *Digestive Diseases and Sciences* **42**, 1013–1019.

Parkkila, S., Parkkila, A.-K., Rajaniemi, H., Shah, G.N., Grubb, J.H., Waheed, A., and Sly, W.S. (2001) Expression of membrane-associated carbonic anhydrase XIV on neurons and axons in mouse and human brain. *Proceedings of the National Academy of Sciences of the United States of America* **98**, 1918–1923.

Parkkila, S., Parkkila, A.-K., Saarnio, J., Kivelä, J., Karttunen, T.J., Kaunisto, K., Waheed, A., Sly, W.S., Türeci, Ö., Virtanen, I., and Rajaniemi, H. (2000) Expression of the membrane-associated carbonic anhydrase isozyme XII in the human kidney and renal tumors. *Journal of Histochemistry and Cytochemistry* **48**, 1601–1608.

Parkkila, S., Rajaniemi, H., Parkkila, A.-K., Kivelä, J., Waheed, A., Pastoreková, S., Pastorek, J., and Sly, W.S. (2000a) Carbonic anhydrase inhibitor suppresses invasion of renal cancer cells in vitro. *Proceedings of the National Academy of Sciences of the United States of America* **97**, 2220–2224.

Pastorek, J., Pastoreková, S., Callebaut, I., Mornon, J.P., Zelník, V., Opavský, R., Zat'ovicová, M., Liao, S., Portetelle, D., Stanbridge, E.J., Závada, J., Burny, A., and Kettmann, R. (1994) Cloning and characterization of MN, a human tumor-associated protein with a domain homologous to carbonic anhydrase and a putative helix-loop-helix DNA binding segment. *Oncogene* **9**, 2877–2888.

Pastoreková, S., Parkkila, S., Parkkila, A.-K., Opavský, R., Zelnik, V., Saarnio, J., and Pastorek, J. (1997) Carbonic anhydrase IX, MN/CA IX: Analysis of stomach complementary DNA sequence and expression in human and rat alimentary tracts. *Gastroenterology* **112**, 398–408.

Perucca, E. (1997) A pharmacological and clinical review on topiramate, a new antiepileptic drug. *Pharmacological Research* **35**, 241–256.

Puscas, I., Hajdu, A., Buzas, G., and Bernath, Z. (1989) Prevention of non-steroidal antiinflammatory agents induced acute gastric mucosal lesions by carbonic anhydrase inhibitors. An endoscopic study. *Acta Physiologica Hungarica* **73**, 279–283.

Reiss, W.G., and Oles, K.S. (1996) Acetazolamide in the treatment of seizures. *Annals of Pharmacotherapy* **30**, 514–519.

Ridderstråle, Y., and Hanson, M. (1985) Histochemical study of the distribution of carbonic anhydrase in the cat brain. *Acta Physiologica Scandinavica* **124**, 557–564.

Roussel, G., Delaunoy, J.-P., Nussbaum, J.-L., and Mandel, P. (1979) Demonstration of a specific localization of carbonic anhydrase C in the glial cells of rat CNS by an immunohistochemical method. *Brain Research* **160**, 47–55.

Räisänen, S.R., Lehenkari, P., Tasanen, M., Rahkila, P., Härkönen, P.L., and Väänänen, H.K. (1999) Carbonic anhydrase III protects cells from hydrogen peroxide-induced apoptosis. *FASEB Journal* **13**, 513–522.

Saarnio, J., Parkkila, S., Parkkila, A.-K., Waheed, A., Casey, M.C., Zhou, X.Y., Pastorekova, S., Pastorek, J., Karttunen, T., Haukipuro, K., Kairaluoma, M.I., and Sly, W.S. (1998) Immunohistochemistry of carbonic anhydrase isozyme IX (MN/CA IX) in human gut reveals polarized expression in the epithelial cells with the highest proliferative capacity. *Journal of Histochemistry and Cytochemistry* **46**, 497–504.

Sapirstein, V.S., and Lees, M.B. (1978) Purification of myelin carbonic anhydrase. *Journal of Neurochemistry* **31**, 505–511.

Sato, A., and Spicer, S.S. (1982) Cell specialization in collecting tubules of the guinea pig kidney: carbonic anhydrase activity and glycosaminoglycan production in different cells. *Anatomical Record* **202**, 431–443.

Sato, A., Spicer, S.S., and Tashian, R.E. (1980) Ultrastructural localization of carbonic anhydrase in gastric parietal cells with the immunoglobulin-enzyme bridge method. *Histochemical Journal* **12**, 651–659.

Schwartz, G.J. (2002) Physiology and molecular biology of renal carbonic anhydrase. *Journal of Nephrology* **15**, S61–S74.

Schwartz, G.J., Kittelberger, A.M., Watkins, R.H., and O´Reilly, M.A. (2003) Carbonic anhydrase XII mRNA encodes a hydratase that is differentially expressed along the rabbit nephron. *American Journal of Physiology* **284**, F399–F410.

Sellin, J.H., and DeSoignie, R. (1990) Short-chain fatty acid absorption in rabbit colon in vitro. *Gastroenterology* **99**, 676–683.

Shin, R.K., and Balcer, L.J. (2002) Idiopathic intracranial hypertension. *Current Treatment Options in Neurology* **4**, 297–305.

Sinha, A.K., and Rose, S.P.R. (1971) Bulk separation of neurones and glia: a comparison of techniques. *Brain Research* **33**, 205–217.

Sly, W.S., Hewett-Emmett, D., Whyte, M.P., Yu, Y.-S.L., and Tashian, R.E. (1983) Carbonic anhydrase II deficiency identified as the primary defect in the autosomal recessive syndrome of osteopetrosis with renal tubular acidosis and cerebral calcification. *Proceedings of the National Academy of Sciences of the United States of America* **80**, 2752–2726.

Sly, W.S., and Hu, P.Y. (1995) Human carbonic anhydrases and carbonic anhydrase deficiencies. *Annual Review of Biochemistry* **64**, 375–401.

Snyder, D.S., Zimmerman, T.R. Jr., Farooq, M., Norton, W.T., and Cammer, W. (1983) Carbonic anhydrase, 5´-nucleotidase, and 2´,3´-cyclic nucleotide-3´-phosphodiesterase activities in oligodendrocytes, astrocytes, and neurons isolated from the brains of developing rats. *Journal of Neurochemistry* **40**, 120–127.

Spicer, S.S., Ge, Z.-H., Tashian, R.E., Hazen-Martin, D.J., and Schulte, B.A. (1990) Comparative distribution of carbonic anhydrase isozymes III and II in rodent tissues. *American Journal of Anatomy* **187**, 55–64.

Spicer, S.S., Sens, M.A., and Tashian, R.E. (1982) Immunocytochemical demonstration of carbonic anhydrase in human epithelial cells. *Journal of Histochemistry and Cytochemistry* **30**, 864–873.

Spicer, S.S., Stoward, P.J., and Tashian, R.E. (1979) The immunolocalization of carbonic anhydrase in rodent tissues. *Journal of Histochemistry and Cytochemistry* **27**, 820–831.

Springer, J.L. (2000) A comparison of topiramate and acetatolamide on seizure duration and paired-pulse inhibition in the dentate gyrus of the rat. *Epilepsy Research* **40**, 147–153.

Supuran, C.T., and Scozzafava, A. (2000) Carbonic anhydrase inhibitors and their therapeutic potential. *Expert Opinion on Therapeutic Patents* **10**, 575–600.

Supuran, C.T., and Scozzafava, A. (2001) Carbonic anhydrase inhibitors. *Current Medicinal Chemistry Immunologic, Endocrine, Metabolic Agents* **1**, 61–97.

Supuran, C.T., and Scozzafava, A. (2002) Applications of carbonic anhydrase inhibitors and activators in therapy. *Expert Opinion on Therapeutic Patents* **12**, 217–242.

Suzuki, Y., and Kaneko, K. (1987) Acid secretion in isolated guinea pig colon. *American Journal of Physiology* **253**, G155–G164.

Tansey, F.A., Thampy, K.G., and Cammer, W. (1988) Acetyl-CoA carboxylase in rat brain. II. Immunocytochemical localization. *Developmental Brain Research* **43**, 131–138.

Tawil, R., Moxley, R.T., III, and Griggs, R.C. (1993) Acetazolamide-induced nephrolithiasis: implications for treatment of neuromuscular disorders. *Neurology* **43**, 1105–1106.

Thatcher, B.J., Doherty, A.E., Orvisky, E., Martin, B.M., and Henkin, R.I. (1998) Gustin from human parotid saliva is carbonic anhydrase VI. *Biochemical and Biophysical Research Communications* **250**, 635–641.

Turnberg, L.A., Bieberdorf, F.A., Morawski, S.G., and Fordtran, J.S. (1970a) Interrelationships of chloride, bicarbonate, sodium, and hydrogen transport in the human ileum. *Journal of Clinical Investigation* **49**, 557–567.

Turnberg, L.A., Fordtran, J.S., Carter, N.W., and Rector, F.C., Jr. (1970b) Mechanism of bicarbonate absorption and its relationship to sodium transport in the human jejunum. *Journal of Clinical Investigation* **49**, 548–556.

Turner, J.R., Odze, R.F., Crum, C.P., and Resnick, M.B. (1997) NM antigen expression in normal, preneoplastic esophagus. *Human Pathology* **28**, 740–744.

Türeci, Ö., Sahin, U., Vollmar, E., Siemer, S., Göttert, E., Seitz, G., Pfreundschuh, M., Parkkila, A.-K., Shah, G., Grubb, J.H., and Sly, W.S. (1998) Human carbonic anhydrase XII: cDNA cloning, expression and chromosomal localization of a novel carbonic anhydrase gene that is overexpressed in some renal cell cancers. *Proceedings of the National Academy of Sciences of the United States of America* **95**, 7608–7613.

Wistrand, P.J., and Kinne, R. (1977) Carbonic anhydrase activity of isolated brush border and basal-lateral membranes of renal tubular cells. *Pflugers Archiv* **370**, 121–126.

Yu, A.C.H., Drejer, J., Hertz, L., and Schousboe, A. (1983) Pyruvate carboxylase activity in primary cultures of astrocytes and neurons. *Journal of Neurochemistry* **41**, 1484–1487.

Zhu, X.L., and Sly, W.S. (1990) Carbonic anhydrase IV from human lung: Purification, characterization, and comparison with membrane carbonic anhydrase from human kidney. *Journal of Biological Chemistry* **265**, 8795–8801.

11 Carbonic Anhydrase Inhibitors in Dermatology

Antonio Mastrolorenzo, Giuliano Zuccati,
Andrea Scozzafava and Claudiu.T. Supuran

CONTENTS

11.1 Introduction 303
11.2 CAs in the Skin 305
11.3 Role of CAs in the Skin 306
11.4 Role of CAs in Skin Diseases 307
11.5 Role of CAs in Skin Tumors 309
11.6 CAIs and Skin Cancer 310
References 311

Different carbonic anhydrase (CA) isozymes are highly expressed in the human skin, where they play a multitude of physiological functions. Isozymes CA I and CA II, present in the cytoplasm and the basolateral membranes of the epithelial cells of the suprabasal and basal layers, are probably involved in physiological cell proliferation and adhesion. The same isozymes present in the inner layer of endothelial cells of capillaries and sweat glands are involved in the complex phenomena of macro-molecular secretion into luminal structures or bicarbonate and other ion-exchange transport processes. Strong correlations have been demonstrated between CA activity and fenestrations in juxtaepithelial capillaries of several tissues, including psoriatic lesions of the human skin as well as other skin diseases such as tumors. In the last condition, the overexpression of isozyme CA IX has frequently been observed. The recent advent of potent CA IX inhibitors and other sulfonamide CA inhibitors (CAIs) known to interfere with tumor cell growth opens new vistas for novel types of therapies for such skin diseases.

11.1 INTRODUCTION

To date, 14 different carbonic anhydrase (CA) isoforms have been described in higher vertebrates, including humans (Table 11.1; Supuran 1994; Hewett-Emmett

TABLE 11.1
Higher-Vertebrate α-CA Isozymes: A Summary of Expression in Normal Skin and Orogenital Tissues

Isozyme	Subcellular Localization	Skin and Orogenital Tissues	References
CA I	Cytosol	Sweat glands Salivary glands	Briggman et al. 1983 Noda et al. 1986; Parkkila et al. 1990
CA II	Cytosol	Sweat glands Salivary glands	Briggman et al.1983 Noda et al. 1986; Parkkila et al. 1990
CA III	Cytosol		
CA IV	Membrane bound	Salivary glands	Fujikawa-Adachi et al. 1999
CA V	Mitochondria		Nagao et al. 1993; Fujikawa-Adachi et al. 1999
CA VI	Excreted into saliva	Salivary glands	Murakami and Sly 1987; Parkkila et al. 1990; Fujikawa-Adachi et al. 1999
CA VII	Cytosol		Montgomery 1991
CA-RPVIII	Probably cytosolic		
CA IX	Membrane bound	Basal cells of epidermis and squamous mucosa of uterine cervix negative; high level in the basal cells in and near the infundibulum and medulla of hair follicle; sweats glands negative	Ivanov et al. 2001
CA-RP X	Unknown		
CA-RP XI	Unknown		
CA XII	Membrane bound	Focal expression in the basal cells of the epidermis; squamous mucosa of uterine cervix negative; basal cells of hair follicle negative; diffuse reactivity of sweat and salivary glands	Ivanov et al. 2001
CA XIII	Unknown		
CA XIV	Membrane-bound		

2000; Parkkila et al. 1998; Pastorekova et al. 1997; Kivela et al. 2000), all of which are involved in crucial physiological processes because of their ability to catalyze the hydration of CO_2 to bicarbonate at physiological pH (Supuran et al. 2003). CAs have been found in all tissues, and their presence has been also demonstrated in the skin of various mammals, including humans (Briggman et al. 1983; Noda et al. 1986, 1987; Mukarami and Sly 1987; Nagao et al. 1993; Fujikawa-Adachi et al. 1999; Ivanov et al. 2001). The first attempts to demonstrate CA activity in human skin were made by the histochemical studies of Hansson (1967). Tissue sections obtained from human facial skin, the leg (a case of panniculitis) and the scalp showed CA positivity in the epidermal cells and in the capillary loop of the dermal papillae. The enzyme was detected in the cell membranes and nuclei, and, to a lesser extent, in the cytoplasm of the epithelial cells, giving a chicken-wire appearance. Positive detection decreased toward the lumen. Further pharmacological and dermatological studies led Hansson (1968) to demonstrate CA activity in human eccrine sweat glands. Eichhorn et al. (1994) gave additional evidence by demonstrating the presence of CA in normal skin. Although the study used histochemical techniques and not immunohistochemistry to detect CA activity, they achieved accurate and consistent results.

11.2 CAs IN THE SKIN

Currently little is known about the cutaneous expression of different CA isozymes. Recent immunohistochemical studies (Mastrolorenzo et al. 2003) have provided additional evidence for the previously reported demonstration of the presence of CA in normal skin by Eichhorn et al. (1994). Data presented from an immunohistochemical investigation on the expression of three isozymes CA I, CA II and CA IX in some specimens of normal human skin clearly showed that in the epidermis both CA I and CA II were expressed in the spinous layers of the skin. In particular, moderate staining was predominantly observed in the cytoplasm, basolateral and apical plasma membrane of the prickle cells, and the basal cells displayed CA activity in their cytoplasm and in the apical and lateral membranes. No staining was found in the stratum granulosum and horny layer. A different pattern similar to the first description by Hansson (1967) of CAs I and II activity is less frequently shown with the staining at the lateral cell borders and intercellular spaces (Mastrolorenzo et al. 2003). Other stratified squamous epithelia, buccal mucosa and ectocervix gave a similar skin pattern (Christie et al. 1995). In contrast to previous studies (Eichhorn et al. 1994), a detectable staining of the endothelial cells of the capillaries in the stratum papillare has been described. As regards CA localization in the sweat glands, a similar pattern of immunoreactivity has been described (Briggman et al. 1983; Noda et al. 1987); in particular, the immunohistochemical distributions of CAs I and II were similar for each skin specimen studied, with CA II being more intense and homogeneous than CA I. The secretory portion, composed of coiled secretory glands, was homogeneously and strongly stained whereas the ductal epithelium was more weakly stained.

In striking contrast to the expression of CAs I and II and according to Ivanov et al. (2001), CA IX is not expressed in normal human skin. All epithelial cells of the normal skin and all endothelial cells of capillaries and sweat glands remain totally unstained for CA IX. The same authors through immunohistochemical studies recently demonstrated the presence of CA IX in basal cells, in and around the infudibulum and medulla of the hair follicle (Ivanov et al. 2001). In line with these observations, more recent investigations demonstrate a faint but always present staining for CAs I, II and IX in the cells surrounding the hair shaft. However, whereas positive staining demonstrates presence of the enzyme, a negative result could arise from the failure to detect a relatively low level of the enzyme cytochemically. Sites regarded as containing only one isozyme conceivably possess an undetected low concentration of the other isozymes (Turner et al. 1997). Immunocytochemical results showing one and not the other isozymes presumably detect the physiologically predominant form (Spicer et al. 1982).

11.3 ROLE OF CAs IN THE SKIN

At present, not all the cytoplasmic and membrane-associated CA isozymes have been comprehensively identified in the human skin, even though the availability of specific antibodies and CAIs has helped considerably (Briggman et al. 1983; Noda et al. 1986, 1987; Mukarami et al. 1987; Nagao et al. 1993; Fujikawa-Adachi et al. 1999; Ivanov et al. 2001). However, little is known regarding the physiological consequences of their activation or inhibition on the metabolic processes within human skin cells. Based on the assumption that knowledge of the various cell types containing an abundance of the enzyme will contribute to the understanding of its biological significance and on the specific distribution pattern described previously, it is possible speculatively to advance an interesting and questionable view on the biological significance of these findings. In this view, it can be easily suggested that both isozymes I and II expressed in the cytoplasm and the basolateral membranes of the epithelial cells of the suprabasal and basal layers (where CA IX is negative) might potentially be involved in physiological cell proliferation and adhesion. Studies by Nogradi (1998) seemed to retract this theory, revealing no clear relationship of CAs I and II with tumorigenesis. In contrast to CA I and II expression, CA IX is not expressed in normal human skin (Ivanov et al. 2001), suggesting that it is not apparently implicated in physiological epithelial cell growth regulation and proliferation. Studies by Liao et al. (1994) have shown a distinct expression of CA IX protein in cervical displasia and carcinoma. This suggests that CA IX expression is more restricted to cell growth, thus strengthening the attractive hypothesis of its role in neoplastic cell proliferation, and, possibly, in malignant transformation (Liao et al. 1994; Saarnio et al. 1998a, 1998b; Vermylen et al. 1999). However, its function in reactive, hyperplastic processes as well as intercellular interactions and cell adhesions in different organs other than the skin cannot be completely ruled out (Pastorekova et al. 1997; McKiernan et al. 1997; Turner et al. 1997).

CA-positive epidermal cells generally fall into the class of cells engaged in transport of fluid and ions, but the positive immunohistochemical observation of isozymes I and II in the inner layer of endothelial cells of the capillaries and sweat glands might be related to the complex phenomena of macromolecular secretion into luminal structures or bicarbonate and other ion exchanges as well documented by other investigators in eccrine sweat and large salivary glands (Briggman et al. 1983; Noda et al. 1986, 1987; Mukarami et al. 1987; Nagao et al. 1993; Fujikawa-Adachi et al. 1999; Ivanov et al. 2001; Spicer et al. 1982).

Models that receive particular attention in providing further evidence and helping explain the evolving concepts of CAs physiology are those that report the clinical effects and systemic adverse events from the use of CAIs. Such interesting pharmacological agents, sulfonamide CAIs, have a firm place in medicine and are mainly useful as diuretics or to treat and prevent a variety of diseases such as glaucoma, epilepsy, congestive heart failure, mountain sickness, gastric and duodenal ulcers, neurological disorders and osteoporosis (Supuran and Scozzafava 2000a; Supuran et al. 2003). Cases of olygohydrosis, a potentially serious adverse event characterized by deficient production and secretion of sweat, were reported in six children treated with zonisamide, an antiepileptic drug chemically classified as a sulfonamide and first marketed in Japan in 1989 (Knudsen et al. 2003). The apparent increased risk of oligohydrosis in the pediatric age group might be related to the dose and resulting blood levels of zonisamide and its metabolites. The present understanding of the pathophysiology of drug-induced oligohydrosis in pediatric patients is limited. Diminished sudomotor responsiveness, hypohidrosis or oligohydrosis and heat and fever after 2 to 3 months of treatment have also been reported in patients treated with topiramate, another recently marketed antiepileptic drug that inhibits CA (Arcas et al. 2001). The cumulative researches on the CA present in the literature might provide clues to possible mechanisms. Zonisamide inhibition of CA could modify the acid–base (acid equivalents) and ionic (mainly Ca^{2+} availability) cellular environment of the eccrine sweat gland epithelial cells. The result would be a conformational change (alteration in membrane form) and a decrease in the responsiveness of eccrine sweat gland muscarinic receptors to cholinergic stimulation (Knudsen et al. 2003). The drug might, in part, mediate its effects by influencing pH dynamics, H^+ concentration and available Ca^{2+} transient at the level of ion channels and receptors.

While investigating the effectiveness of the most commonly used drug (acetazolamide) at high altitude for prevention and therapy of acute mountain sickness, Burtscher and Likar (2002) suspected a relationship between the medication and the development of leukonykia of all fingernails following high-altitude exposure. The authors suggested that a severe hypoxia induced by a CAI was responsible for the nail changes.

11.4 ROLE OF CAs IN SKIN DISEASES

During the past two decades, a strong correlation was demonstrated between CA activity and fenestrations in juxtaepithelial capillaries of several tissues, including psoriatic lesions of the human skin (Eichhorn et al. 1994; Lutjen-Drecoll et al. 1983, 1985; Jungkunz et al. 1992). Eichhorn et al. (1994) reported the histochemical demonstration of CA activity in a large proportion of the capillaries of cherry hemangiomas and that the occurrence of CA in these capillaries corresponds to the fenestrations of venous capillaries, which are numerously revealed by electron microscopy. In psoriatic skin lesions, the interpapillary capillaries also react for the presence of CA, and fenestrations are also expressed in the papillary portions of the vertical capillary loops. In psoriatic lesions, fenestrated capillaries are found physiologically adjacent to epithelia involved in extensive transport of fluid and molecules. Eichhorn et al. (1994) observed no capillary staining for CA in normal undiseased skin. Although the function of the combined presence of fenestration and CA in capillary endothelium is not clear, these authors hypothesize that CA might be involved in transendothelial transport proper and is required for the formation and maintainance of fenestrations (Eichhorn et al. 1994). Therefore the fenestration of capillaries in psoriatic lesions might facilitate the passing of factors derived from the serum with a proliferative effect on keratinocytes or fibroblasts (Jungkunz et al. 1992). In addition, the consistent presence of CA in the endothelium of cherry hemangioma argues for a special degree of differentiation of these cells (Eichhorn et al. 1994). However, a positive reaction for CA might serve as a sensitive and simple marker for fenestrated capillaries in skin tissue.

CAs have been implicated in the pathogenesis of human autoimmune diseases (Nishimori and Onishi 2001). The description of a disease called autoimmune exocrinopathy (Strand and Talal 1980) and dry gland syndrome (Epstein et al. 1982) in the past and autoimmune epithelitis (Moutsopulos 1994) recently was supported by the concept of an autoimmune reaction against a common target antigen, such as CA II, expressed by the ductal epithelial cells of exocrine organs. Serum antibodies reactive to CA II have been reported in patients with several autoimmune diseases of multiple exocrine organs, including systemic lupus erythematous and Sjögren's syndrome (Inagaki et al. 1991; Itoh and Reichlin 1992). Autoantibodies to CA I, primarily expressed in red blood cells, have also been observed in the sera of patients with Sjögren's syndrome, and in these patients anti-CA II and anti-CA I antibodies were not cross-reactive (Kino-Ohsaki et al. 1996). One possible explanation is that serum antibodies primarily induced to another still-unknown CA isozyme might cross-react with CAs I and II (Nishimori and Onishi 2001). The induction of experimental autoimmune sialoadenitis by immunization of PL/J mice with CA II suggests that immunological recognition of CA II in a MHC-restricted manner is a proposed mechanism underlying the autoimmune diseases of multiple exocrine organs, including Sjögren's syndrome (Nishimori et al. 1995). Although congenital CA deficiency is a potential problem for pathological conditions, the discovery of such a deficiency could serve as *experimentum naturae* to better understand the physiological role of CAs in the human skin. Until now, only a CA II deficiency syndrome has been clinically recognized, but no clinical findings have been reported for human skin diseases (Sly and Hu 1995).

Calcifying epithelioma of Malherbe, or pilomatrixoma, is a benign tumor composed of cells resembling those of the hair matrix, which undergo "mummification" and might calcify. Thus, in generally 15 to 20% of the cases, it might exhibit areas of ossification in its stroma (Fernandes et al. 2003). Such a biological process probably involves metaplastic changes in the stromal connective tissue. Bone induction in epithelial tumors has been associated with epithelial structures, and bone-forming areas in the stroma are usually adjacent to the basement membrane of the tumor epithelial focus. Kumasa and co-workers (1987) reported an immunohistochemically detectable and strongly positive staining for CA II, more intense than for CA I though the histochemical localization was nearly the same, in osteoclasts, giant cells and peripheral tumor epithelial zones in the bone-inducing areas. Stromal connective tissue fibers showed slight to moderate CA reactions, but the bone tissue itself gave a negative reaction. The pathogenesis of this metaplastic process is still unknown, but the authors postulate that in the ossification process of calcifying epithelioma of Malherbe, CA II initiates the deposition of minerals onto a matrix that contains fibronectin and complex carbohydrates (Kumasa et al. 1987).

Dry and wrinkled skin, sparse gray areas and a multiplicity of pigmented lesions are the common characteristics of aging skin. This might be caused by the hypo-function of the skin appendages, such as a reduced output per sweat gland and a decrease in the amount of sebum (Inomata 1973; Fang 1989; Teruka et al. 1994). The differentiation of epidermal cells of the facial skin in the young and the aged was immunohistochemically examined by Teruka et al. (1994), continuing the pre-vious study of Takahashi and Teruka (1989), and the suggested a role of a carbonic anhydrase-like protein (CAP) as a differentiation marker, being a constituent of keratohyalin granules and transferred to the stratum corneum cell membrane region. The staining patterns obtained also by using monoclonal antibodies anti-CAP showed no difference in the amount of CA in the facial skin of the young and the aged, and even to the present day the role and biological functions remain unknown (Takahashi and Teruka 1989).

11.5 ROLE OF CAs IN SKIN TUMORS

The role of CAs in cancer can be explained by considering the metabolic conditions required by a growing cancer cell that develops with a higher rate of replication than a normal cell. Such a circumstance requires a high flux of bicarbonate into the cell itself to provide substrate for the synthesis of either nutritionally essential components (nucleotides) or cell structural components (membrane lipids; Cheg-widden et al. 2001). At least two CA isozymes (CA IX and CA XII) have close connections with tumors (Pastorekova and Pastorek 2003; Chegwidden et al. 2001). Among the CAs, the cell-surface transmembrane CA IX and CA XII are over-expressed in many tumors, suggesting that this common feature of cancer cells might be required for tumor progression. Furthermore, it has been hypothesized that these enzymes might be involved in maintaining the extracellular acidic pH in tumors, thereby providing a conducive environment for tumor growth and spread. An acidic extracellular pH is a fundamental property of the malignant phenotype. Analysis of RNA samples isolated from melanoma cell lines MALME-3M revealed moderate levels of expression of CA IX, whereas LOX IMVI, M14 and SK-MEL-2 melanoma cell lines lacked CA IX and XII. Expression of these isozymes was also analyzed in cultured cells under hypoxic conditions. A high level of CA protein was expressed by neoplastic cells located in and adjacent to the hypoxic or necrotic foci of the melanoma (Pastorekova and Pastorek 2003).

High to moderate levels of CAs IX and XII expression were detected through immunohistochemistry in various common epithelial tumor types and the immun-ostaining was seen predominantly on the cell-surface membrane. CAs IX and XII expression was consistently seen in 82% ($n = 12$) and 50% ($n = 4$), respectively, of squamous cell carcinoma of the head and neck, with a diffuse staining (>40% of cells in the section) in 71% of cases for CA IX compared with 100% of cases for CA XII. There was coexpression of the two isozymes in only 50% of the cases. CA IX and CA XII proteins were expressed in 100% and 71%, respectively, of squamous and basal cell carcinomas of the skin, and 50% of these tumors showed diffuse and strong staining for only CA IX whereas for CA XII the positive staining was focal and very weak. When the distribution of CA IX and XII was evaluated in melanoma ($n = 18$), the tumor cells showed no immunoreactivity (Pastorekova and Pastorek 2003).

11.6 CAIs AND SKIN CANCER

The connections between CAs and tumors, and the development of specific inhibitors for some of the isozymes presumably involved in cancerous processes, have been investigated in great detail in the laboratory (Scozzafava et al. 2003; Supuran et al. 2003; Supuran and Scozzafava 2000b, 2000c). A program of screening several hundred strong sulfonamide CAIs (both aromatic and heterocyclic derivatives) for their tumor cell growth inhibition against a panel of 60 cell lines of the National Institutes of Health (NIH) afforded the identification of sulfonamide derivatives **11.1** to **11.3** as interesting leads (Supuran and Scozzafava 2000b, 2000c; Supuran et al. 2001; Table 11.2).

These compounds showed tumor growth inhibitory properties. Typically, they showed GI_{50} values in the low micromolar range (10 to 42 nM) against a wide number of tumors and melanoma (Supuran and Scozzafava 2000b, 2000c) and are considered among the most potent tumor cell growth inhibitors belonging to the sulfonamide CAIs to date (Table 11.2). The mechanism of tumor growth inhibition with these sulfonamide CAIs is not known at present, but several hypotheses have been proposed. Thus, as suggested by Chegwidden and Spencer (1995), these compounds, similarly to the classical inhibitors acetazolamide, methazolamide or ethoxzolamide, might reduce the provision of bicarbonate for the synthesis of nucleotides (mediated by carbamoyl phosphate synthetase II) and other cell components such as membrane lipids (mediated by pyruvate carboxylase). Such a mechanism would likely involve CA II and CA V. An alternative, or additional, mechanism might involve acidification of the intracellular milieu as a consequence of CA inhibition by these potent CAIs, as shown by Teicher et al. (1993). It is also possible that the sulfonamides reported here interfere with the activity of the CA isozymes CAs IX, XII and XIV known to be present predominantly in tumor cells. It has recently been shown that CA IX is strongly inhibited by many sulfonamide/sulfamate CAIs (Vullo et al. 2003; Ilies et al. 2003; Winum et al. 2003). A combination of several of the mechanisms proposed previously is also possible (Supuran et al. 2001). Thus, in the future, potent CAIs might be designed and used as new treatment approaches for cancers, including the types that affect the skin.

TABLE 11.2
In Vitro **Tumor (Melanoma) Growth Inhibition Data with the Sulfonamide CA Inhibitors 11.1, 11.2 and 11.3**

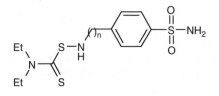

11.1: n = 0
11.2: n = 1
11.3: n = 2

Cell Line	GI_{50} (μM)[a]		
	11.1	**11.2**	**11.3**
LOX IMVI	39	—	7
MALME-3M	26	20	17
M14	20	10	11
SK-MEL-2	39	18	21
SK-MEL-28	34	28	7
SK-MEL-5	42	21	21

[a] Molarity of inhibitor producing a 50% inhibition of growth of the tumor cells after 48 h exposure to variable concentrations (10^{-4} to 10^{-8} M) of the test compound [Supuran, C.T. et al. (2001) *Bioorganic and Medicinal Chemistry* **9**, 703–714].

REFERENCES

Arcas, J., Ferrer, T., Roche, M., Bermejo-Martinez, A., and Lopez-Martin, V. (2001) Hypohidrosis related to the administration of topiramate to children. *Epilepsia* **42**, 1363–1365.

Briggman, J.V., Tashian, R.E., and Spicer, S.S. (1983) Immunohistochemical localization of carbonic anhydrase I and II in eccrine sweat glands from control subjects and patients with cystic fibrosis. *American Journal of Pathology* **112**, 250–257.

Burtscher, M., and Likar, R. (2002) Leukonichia following high altitude exposure (Letter to the editor). *High Altitude Medicine and Biology* **3**, 93–94.

Chegwidden, W.R., and Spencer, I.M. (1995) Sulphonamide inhibitors of carbonic anhydrase inhibit the growth of human lymphoma cells in culture. *Inflammopharmacology*, **3**, 231–239.

Chegwidden, W.R., Spencer, I.M., and Supuran, C.T. (2001) The roles of carbonic anhydrase isozymes in cancer. In *Gene Families: Studies of DNA, RNA, Enzymes and Proteins*, Xue, G., Xue, Y., Xu, Z., Holmes, R., Hammond, G.L., and Lim, H.A. (Eds.), World Scientific, Singapore, pp. 157–170.

Christie, K.N., Thompson, C., Morley, S., Anderson, J., and Hopwood, D. (1995) Carbonic anhydrase is present in human oesophageal epithelium and submucosal gland. *Histochemical Journal* **27**, 587–590.

Eichhorn, M., Jungkunz, W., Wörl, J., and Marsch, C.H. (1994) Carbonic anhydrase is abundant in fenestrated capillaries of cherry hemangioma. *Acta Dermato-Venereologica (Stockholm)* **74**, 51–54.

Epstein, O., Chapman, R.W., Lake-Bakaar, G., Foo, A.Y., Rosalki, S.B., Sherlock, S. et al. (1982) The pancreas in primary biliary cirrhosis and primary sclerosing cholangitis. *Gastroenterology* **83**, 1177–1182.

Fang, K.T. (1989) Comparison of changes of terminal differentiation in the facial skin between the young and the aged. *Acta Medica Kinki Universita* **14**, 161–169.

Fernandes, R., Holmes, J., and Mullenix, C. (2003) Giant pilomatricoma (Epithelioma of Melherbe): Report of a case and review of literature. *Journal of Oral and Maxillofacial Surgery* **61**, 634–636.

Fujikawa-Adachi, K., Nishimori, I., Taguchi, T., and Onishi, S. (1999) Human mitochondrial carbonic anhydrase VB: cDNA cloning, mRNA expression, subcellular localization, and mapping to chromosome X. *Journal of Biological Chemistry* **274**, 21228–21233.

Hansson, H.P.J. (1967) Histochemical demonstration of carbonic anhydrase activity. *Histochemie* **11**, 112–128.

Hansson, H.P.J. (1968) Histochemical demonstration of carbonic anhydrase activity in some epithelia noted for active transport. *Acta Physiologica Scandinavica* **73**, 427–434.

Hewett-Emmett, D. (2000) Evolution and distribution of the carbonic anhydrase gene families. In *Carbonic Anhydrase — New Horizons*, Chegwidden, W.R., Edwards, Y., and Carter, N. (Eds.), Birkhäuser, Basel, pp. 29–76.

Ilies, M.A., Vullo, D., Pastorek, J., Scozzafava, A., Ilies, M., Caproiu, M.T., Pastorekova, S., and Supuran, C.T. (2003) Carbonic anhydrase inhibitors. Inhibition of tumor-associated isozyme IX with halogenosulfanilamide and halogeno-aminophenylbenzolamide derivatives. *Journal of Medicinal Chemistry* **46**, 2187–2196.

Inagaki, Y., Jinno-Yoshida, Y., Hamasaki, Y., and Ueki, H. (1991) A novel autoantibody reactive with carbonic anhydrase in sera from patients with systemic lupus erythematosus and Sjögren's syndrome. *Journal of Dermatological Sciences* **2**, 147–154.

Inomata, N. (1973) The aging and sebaceous activity. *Japanese Journal of Dermatology* **83**, 517–519.

Itoh, S., and Reichlin, M. (1992) Autoantibodies to carbonic anhydrase in systemic lupus erythematosus and rheumatic diseases. *Arthritis and Rheumatology* **35**, 73–82.

Ivanov, S., Liao, S.Y., Ivanova, A., Danilkovitch-Miagkova, A., Tarasova, N., Weirich, G., Merrill, M.J., Proescholdt, M.A., Oldfield, E.H., Lee, J., Zavada, J., Waheed, A., Sly, W., Lerman, M.I., and Stanbridge, E.J. (2001) Expression of hypoxia-inducible cell-surface transmembrane carbonic anhydrases in human cancer. *American Journal of Pathology* **158**, 905–919.

Jungkunz, W., Eichhorn, M., Worl, J., Marsch, W.C., and Holzmann, H. (1992) Carbonic anhydrase: A marker for fenestrated capillaries in psoriasis. *Archives of Dermatological Research* **284**, 146–149.

Kino-Ohsaki, J., Nishimori, I., Morita, M., Okazaki, K., Yamamoto, Y., Onishi, S. et al. (1996) Serum autoantibodies to carbonic anhydrase I and II in patients with idiopathic chronic pancreatitis and Sjögren's syndrome. *Gastroenterology* **110**, 1579–1586.

Kivela, A., Parkkila, S., Saarnio, J., Karttunen, T.J., Kivela, J., Parkkila, A.K., Waheed, A., Sly, W.S., Grubb, J.H., Shah, G., Tureci, O., and Rajaniemi, H. (2000) Expression of a novel transmembrane carbonic anhydrase isozyme xii in normal human gut and colorectal tumors. *American Journal of Pathology* **156**, 577–584.

Knudsen, J.F., Thambi, L.R., Kapcala, L.P., and Racoosin, J.A. (2003) Oligohydrosis and fever in pediatric patients treated with zonisamide. *Pediatric Neurology* **28**, 184–189.

Kumasa, S., Mori, H., Tsujimura, T., and Mori, M. (1987) Calcifying epithelioma of Malherbe with ossification: Special reference to lectin binding and immunohistochemistry of ossified sites. *Journal of Cutaneous Pathology* **14**, 181–187.

Liao, Y., Brewer, C., Zavada, J., Pastorek, J., Pastorekova, S., Manetta, A., Barman, M.L., DiSaia, P.J., and Stanbridge, E.J. (1994) Identification of the MN antigen as a diagnostic biomarker of cervical intraepithelial squamous and glandular neoplasia and cervical carcinoma. *American Journal of Pathology* **145**, 598–609.

Lutjen-Drecoll, E., Lönnerholm, G., and Eichhorn, M. (1983) Carbonic anhydrase distribution in the human and monkey eye by light and electron microscopy. *Graefes Archives of Clinical and Experimental Ophthalmology* **220**, 285–291.

Lutjen-Drecoll, E., Eichhorn, M., and Barany, E.H. (1985) Carbonic anhydrase in epithelia and fenestrated juxtaepithelial capillaries of *Macaca fascicularis*. *Acta Physiologica Scandinavica* **124**, 295–307.

Mastrolorenzo, A., Zuccati, G., Massi, D., Gabrielli, M.G., Casini, A., Scozzafava, A., and Supuran, C.T. (2003) Immunohistochemical study of carbonic anhydrase isozymes in human skin. *European Journal of Dermatology*, **13**, 440–444.

McKiernan, J.M., Buttyan, R., Bander, N.H., Stifelman, M.D., Katz, A.E., Chen, M.W., Olsson, C.A., and Sawczuk, I.S. (1997) Expression of tumor-associated gene MN; a potential biomarker for human renal cell carcinoma. *Cancer Research* **57**, 2362–2365.

Moutsopulos, H.M. (1994) Sjögren's syndrome: Autoimmune epithelitis. *Clinical Immunology and Immunopathology* **72**, 162–165.

Mukarami, H.M., and Sly, W.S. (1987) Purification and characterization of human salivary carbonic anhydrase. *Journal of Biological Chemistry* **262**, 1382–1388.

Nagao, Y., Platero, J.S., Waheed, A., and Sly, W.S. (1993) Human mitochondrial carbonic anhydrase: cDNA cloning, expression, subcellular localization, and mapping to chromosome 16. *Proceedings of the National Academy of Sciences of the United States of America* **90**, 7623–7627.

Nishimori, I., Bratanova, T., Toshkov, I., Caffrey, T., Mogani, M., Shibata, Y. et al. (1995) Induction of experimental autoimmune sialoadenitis by immunization of PL/J mice with carbonic anhydrase II. *Journal of Immunology* **154**, 4865–4873.

Nishimori, I., and Onishi, S. (2001) Carbonic anhydrase isozymes in the human pancreas. *Digestive and Liver Disease* **33**, 58–74.

Noda, Y., Oosumi, H., Morishima, T., Tsujimura, T., and Mori, M. (1987) Immunohistochemical study of carbonic anhydrase in mixed tumours and adenomas of sweat and sebaceous glands. *Journal of Cutaneous Pathology* **14**, 285–290

Noda, Y., Takai, Y., Iwai, Y., Meenaghan, M.A., and Mori, M. (1986) Immunohistochemical study of carbonic anhydrase in mixed tumours from major salivary glands and skin. *Virchows Archives* **408**, 449–459.

Nogradi, A. (1998) The role of carbonic anhydrases in tumors. *American Journal of Pathology* **153**, 1–4.

Parkkila, S., Kaunisto, K., Rajoniemi, L., Kumpulainen, T., Jokinen, K., Rajaniemi, H. (1990) Immunohistochemical localization of carbonic anydrase VI, II, and I in human parotid and submandibular glands. *Journal of Histochemistry and Cytochemistry* **38**, 941–947.

Parkkila, A.K., Scarim, A.L., Parkkila, S., Waheed, A., Corbett, J.A., and Sly, W.S. (1998) Expression of carbonic anhydrase V in pancreatic beta cells suggests role for mitochondrial carbonic anhydrase in insulin secretion. *Journal of Biological Chemistry* **273**, 24620–24623.

Pastorekova, S., Parkkila, S., Parkkila, A.K., Opavsky, R., Zelnik, V., Saarnio, J., and Pastorek, J. (1997) Carbonic anhydrase IX, MN/CA IX: Analysis of stomach complementary DNA sequence and expression in human and rat alimentary tracts. *Gastroenterology* **112**, 398–408.

Pastorekova, S., and Pastorek, J. (2003) Cancer-related carbonic anhydrase isozymes and their inhibition. In *Carbonic Anhydrase, Its Inhibitors and Activators*, Supuran, C.T., Scozzafava, A., and Conway, J., (Eds.), Taylor & Francis, London, chap. 9.

Saarnio, J., Parkkila, S., Parkkila, A.K., Haukipuro, K., Pastorekova, S., Pastorek, J., Kairaluoma, M.I., and Karttunen, T.J. (1998a) Immunohistochemical study of colorectal tumors for expression of a novel transmembrane carbonic anhydrase, MN/CA IX, with potential value as a marker of cell proliferation. *American Journal of Pathology* **153**, 279–285.

Saarnio, J., Parkkila, S., Parkkila, A.K., Waheed, A., Casey, M.C., Zhou, X.Y., Pastorekova, S., Pastorek, J., Karttunen, T., Haukipuro, K., Kairaluoma, M.I., and Sly, W.S. (1998b) Immunohistochemistry of carbonic anhydrase isozyme IX (MN/CA IX) in human gut reveals polarized expression in the epithelial cells with the highest proliferative capacity. *Journal of Histochemistry and Cytochemistry* **46**, 497–504.

Scozzafava, A., Owa, T., Mastrolorenzo, A., and Supuran, C.T. (2003) Anticancer and antiviral sulfonamides. *Current Medicinal Chemistry* **10**, 925–953.

Sly, W.S., and Hu, P.Y. (1995) Human carbonic anhydrase and carbonic anhydrase deficiencies. *Annual Reviews in Biochemistry* **64**, 375–401.

Spicer, S.S., Sens, M.A., and Tashian, R.E. (1982) Immunohistochemical demonstration of carbonic anhydrase in human epithelial cells. *Journal of Histochemistry and Cytochemistry* **30**, 864–873.

Strand, V., and Talal, N. (1980) Advances in the diagnosis and concept of Sjögren's syndrome (autoimmune exocrinopathy). *Bulletin of Rheumatological Diseases* **30**, 1046–1052.

Supuran, C.T. (1994) Carbonic anhydrase inhibitors. In *Carbonic Anhydrase and Modulation of Physiologic and Pathologic Processes in the Organism*, Puscas, I. (Ed.), Helicon, Timisoara, Romania, pp. 29–111.

Supuran, C.T., Briganti, F., Tilli, S., Chegwidden, W.R., and Scozzafava, A. (2001) Carbonic anhydrase inhibitors: Sulfonamides as antitumor agents? *Bioorganic and Medicinal Chemistry* **9**, 703–714.

Supuran, C.T., and Scozzafava A. (2000a) Carbonic anhydrase inhibitors and their therapeutic potential. *Expert Opinion on Therapeutic Patents* **10**, 575–600.

Supuran, C.T., and Scozzafava, A. (2000b) Carbonic anhydrase inhibitor. Part 94. 1,3,4–thiadiazole-2-sulfonamide derivatives as antitumor agents? *European Journal of Medicinal Chemistry* **35**, 867–874.

Supuran, C.T., and Scozzafava, A. (2000c) Carbonic anhydrase inhibitors: Aromatic sulfonamides and disulfonamides act as efficient tumor growth inhibitors. *Journal of Enzyme Inhibition* **15**, 597–610.

Supuran, C.T., Scozzafava, A., and Casini, A. (2003) Carbonic anhydrase inhibitors. *Medicinal Research Reviews* **23**, 146–189.

Takahashi, M., and Teruka, T. (1989) Carbonic anhydrase-like activity in Ted-H-1 fraction obtained from human keratohyalin granules. *Japanese Journal of Dermatology* **99**, 315.

Teicher, B.A., Liu, S.D., Liu, J.T., Holden, S.A., and Herman, T.S. (1993) A carbonic anhydrase inhibitor is a potential modulator of cancer therapies. *Anticancer Research* **13**, 1549–1556.

Teruka, T., Qing, J., Saheki, M., Kusuda, S., and Takahashi, M. (1994) Terminal differentiation of facial epidermis of the aged: immunohistochemical studies. *Dermatology* **188**, 21–24.

Turner, J.R., Odze, R.D., Crum, C.P., and Resnick, M.B. (1997) MN antigen expression in normal preneoplastic and neoplastic esophagus: a clinicopathological study of a new cancer-associated biomarker. *Human Pathology* **28**, 740–744.

Vermylen, P., Roufosse, C., Burny, A., Verhest, A., Bosschaerts, T., Pastorekova, S., Ninane, V., and Sculier, J.P. (1999) Carbonic anhydrase IX antigen differentiates between preneoplastic malignant lesion in non-small cell lung carcinoma. *European Respiration Journal* **14**, 806–811.

Vullo, D., Franchi, M., Gallori, E., Pastorek, J., Scozzafava, A., Pastorekova, S., and Supuran, C.T. (2003) Carbonic anhydrase inhibitors: Inhibition of the tumor-associated isozyme IX with aromatic and heterocyclic sulfonamides. *Bioorganic and Medicinal Chemistry Letters* **13**, 1005–1009.

Winum, J.-Y., Vullo, D., Casini, A., Montero, J.-L., Scozzafava, A., and Supuran, C.T. (2003) Carbonic anhydrase inhibitors: Inhibition of cytosolic isozymes I and II and the transmembrane, tumor-associated isozyme IX with sulfamates including EMATE also acting as steroid sulfatase inhibitors. *Journal of Medicinal Chemistry* **46**, 2197–2204.

12 Carbonic Anhydrase Activators

Monica Ilies, Andrea Scozzafava
and Claudiu T. Supuran

CONTENTS

12.1 Introduction 318
12.2 Mechanism of Action of CAAs 319
12.3 CAA Types: Design Strategies, Structure–Activity Relationships
and Isozyme Specificity 325
    12.3.1 Background 326
    12.3.2 CA Activation with Amine Derivatives 328
    12.3.3 CA Activation with Azole Derivatives 329
    12.3.4 CA Activation with Histamine Derivatives 333
    12.3.5 CA Activation with Amino Acid Derivatives 338
    12.3.6 CA Activation with Anions 342
12.4 Connections of CA Activators with Physiopathology: Possible Clinical
Developments 343
12.5 Concluding Remarks 346
References 347

Activation of carbonic anhydrases (CAs) has been a controversial phenomenon for a long time. Recently, kinetic, spectroscopic and x-ray crystallographic data have offered a clear-cut explanation for this phenomenon, based on the catalytic mechanism of these enzymes. It has been demonstrated that molecules acting as carbonic anhydrase activators (CAAs) bind at the entrance of the enzyme active-site cavity and participate in assisted proton transfer processes between the active site and the reaction medium, thereby facilitating the rate-determining step of the CA catalytic cycle. In addition to CA II–activator adducts, x-ray crystallographic studies have been reported for ternary complexes of this isozyme with activators and anion inhibitors. Drug design studies have been successfully performed to obtain strong CAAs belonging to several chemical classes: amines and their derivatives, azoles, amino acids and oligopeptides, etc. Structure–activity correlations for diverse classes of activators are discussed for the isozymes for which the phenomenon has been studied. The physiological relevance of CA activation and the possible application of CAAs in Alzheimer's disease and for other memory therapies are also reviewed.

12.1 INTRODUCTION

About half of the 14 different isoforms of carbonic anhydrase (CA) isolated until the present time in higher vertebrates (Hewett-Emmett 2000) act in reversible CO_2 hydration under physiological conditions as some of the most efficient enzymatic catalysts. Detailed studies based on kinetics, spectroscopy, x-ray crystallography and site-specific mutagenesis have clearly revealed that CA II, CA VII and CA IX are among the fastest α-CAs, possessing turnover numbers $k_{cat} = 10^6$ s^{-1} and second-order steady-state rate constants close to the diffusion-controlled processes (e.g., for hCA II, $k_{cat}/K_M \cong 10^8$ M^{-1}s^{-1}; Lindskog and Silverman 2000; Stams and Christianson 2000; Mauksch et al. 2001).

A simple question arises naturally: Does such a rapid enzyme really need to be activated? It has taken a couple of decades to formulate a clear-cut positive answer. Chronological insights into the historical backgrounds of CA activation (Supuran and Puscas 1994; Supuran and Scozzafava 2000a) pointed out that although CA activation had been reported (Leiner 1940) simultaneously with its inhibition (Mann and Keilin 1940), the development of research in these two related fields did not evolve at the same pace. Whereas CA inhibitors (CAIs) have been constantly and extensively studied as well as fruitfully exploited clinically (reviewed in Supuran and Scozzafava 2000b, 2001), CA activation became after its initial report an extremely controversial issue. Consequently, little progress has been made in CA activator (CAA) research, the topic having been kept for a long period in an undeserved scientific shadow.

In the early 1990s, research in the area was practically relaunched by looking at old theories in the new light of some relevant discoveries: (1) the report of CA III anionic activators (Shelton and Chegwidden 1988) shortly followed by the report of a large variety of diverse structural classes of organic compounds with an efficient activating effect against isozymes I, II, and IV (reviewed in Supuran and Puscas 1994; Supuran and Scozzafava 2000a; Supuran et al. 2003); (2) the complete elucidation of the activation mechanism at the molecular level by kinetic, spectroscopic and crystallographic data (reviewed in detail in Supuran and Scozzafava 2000a); (3) the provision of compelling evidence that CA genetic deficiency, although much less encountered than enzyme excessive activity, causes severe physiological and pathological disorders, which might be corrected or treated through an appropriate activation of the isozymes involved (Roth et al. 1992; De Coursey 2000; Rousselle and Heymann 2002; Sun and Alkon 2002; Alper 2002); (4) the establishment of some linear chemical structure–biological activity correlations for CAAs, which could be usefully exploited in rational drug design with pharmacological applications (Clare and Supuran 1994; Supuran et al. 2003).

At present, the previously mentioned data are considered by the scientific community as a strong argument for (1) existence of CAAs, (2) molecular mechanism of action of CAAs, (3) involvement of CAAs in physiopathology, (4) possible development of CAAs as target-synthons in the projecting of drug diagnostic tools to manage some specific disorders.

12.2 MECHANISM OF ACTION OF CAAs

The model proposed for the mechanism of CA activation (Rowlett et al. 1991; Supuran 1992; Briganti et al. 1997a, 1997b, 1998) is closely related with that for the catalytic cycle. As discussed in Chapter 1, catalysis of α-CAs occurs in two well-defined stages that follow a ping-pong type of kinetics. The rate-limiting second step (Equation 12.1) is the intramolecular proton transfer between the zinc-bound water molecule and the reaction medium through the so-called proton shuttle residue, widely assumed to be the imidazole ring of His 64 in isozyme II (and several other efficient isozymes), as shown in Figure 12.1.

Crystallographic data (Nair and Christianson 1991) show that the His 64 imidazole side chain exhibits a pH-dependent conformational mobility, changing gradually its orientation related to the metal site through a 64° ring-flipping. Thus, at pH 8.5, conformer "in" predominates, the imidazole moiety being directed toward the zinc ion and linked to it through hydrogen bond bridges, whereas after proton catching the "out" conformation, in which the heterocyclic ring points toward the enzyme surface and engages no hydrogen bond contacts, becomes increasingly predominant as the pH decreases, reaching the maximum conformational percentage at a pH of 5.7. The high flexibility between the two conformations is crucial for the catalytic proton shuttling (Christianson and Fierke 1996; Duda et al. 2001; An et al. 2002).

It has been proposed that a CAA interferes directly in the proton transfer step of the catalytic cycle, facilitating such an intramolecular transfer by a transient enzyme–activator complex (Equation 12.1). Because intramolecular reactions are much faster than the intermolecular ones (Page and Williams 1989), the result is a significantly enhanced catalytic rate:

$$EZn^{2+}\text{-}OH_2 + A \rightleftharpoons [EZn^{2+}\text{-}OH_2 \text{-----} A \rightleftharpoons EZn^{2+}\text{-}OH^- \text{-----} AH^+] \rightleftharpoons EZn^{2+}\text{-}OH^- + AH^+ \quad (12.1)$$

$$\text{enzyme–activator complexes}$$

The first x-ray structure obtained for a CAA complex was that of the adduct between human isozyme II (hCA II) and histamine (Briganti et al. 1997b). This structure has confirmed the previous hypothesis on the activation mechanism (Supuran 1991, 1992), revealing that histamine binds in the hydrophilic region located at the entrance of the active site, establishing through the nitrogen atoms of the imidazole ring new hydrogen bonds with water molecules and polar amino acids residues, whereas the aliphatic amino group remains free in the solvent (Figure 12.2 and Figure 12.3).

The supplementary hydrogen bond pathways generated by the activator binding have two key consequences for the rate-determining step of catalysis: (1) stabilize the His 64 "in" conformation, which is a steric requirement for the proton shuttling, and (2) offer adjacent routes for the proton transport from the zinc-bound water molecule to the external medium. Furthermore, histamine shows few contacts with

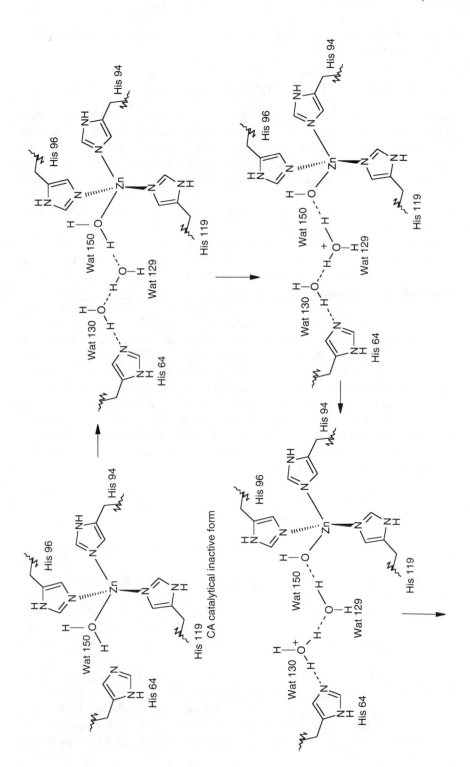

CA catalytical inactive form

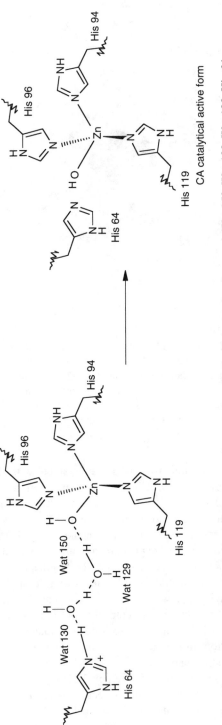

FIGURE 12.1 Schematic representation of proton transfer steps through the hydrogen bond network Wat 150–Wat 129–Wat 130–His 64.

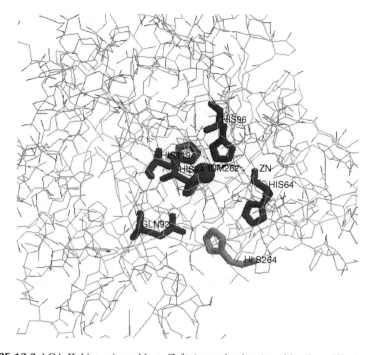

FIGURE 12.2 hCA II–histamine adduct. Zn^{2+} (central sphere) and its three His ligands are placed in the center of the active site. Histamine (numbered as His 264) is anchored at the active site entrance, between residues His 64 and Gln 92. The figure was generated with the program RasMol for Windows 2.6, by using the x-ray crystallographic coordinates available in the Brookhaven Protein Database (PDB entry 4TST). (Reproduced from Briganti, F. et al. (1997b) *Bioorganic and Medicinal Chemistry Letters* **9**, 2043–2048. With permission from Elsevier.)

the enzyme, limited to the imidazole ring that interacts with three protein amino acids residues (Asn 62, Asn 67, Gln 92). This interaction is favorable for the CA–histamine complex dissociation in the last step of the activation mechanism, in the same way by which the conformational flexibility of His 64 confers to this residue the ability to easily pendulate between the cavity and the opening of the active site during the catalytic mechanism. It can thus be concluded that the activator actually acts as an efficient second proton shuttle, besides the native one, His 64.

hCA II binds histamine with the displacement of at least three water molecules from the active-site cavity, followed by a substantial rearrangement of the hydrogen bond network within the cavity. The entropic contribution to the histamine free energy of binding provided by the release of water molecules favors energetically CAA complex formation, in addition to its kinetic and steric stabilization through the new hydrogen bond bridges.

Structural details of active sites of different CA isozymes have provided a further interesting explanation for the activation effect (Briganti et al. 1997b). The CA II active site possesses a well-defined histidine cluster, which starts from His 64 and

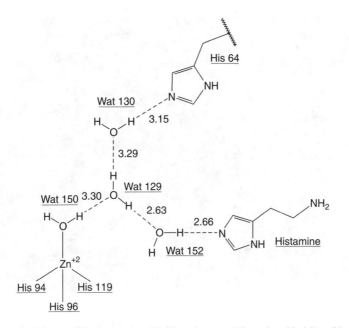

FIGURE 12.3 Scheme of hydrogen bond bridges between histamine, histidine 64 and zinc-bound water molecule (Wat 150); the hydrogen bonds lengths are in Å. (Reproduced from Briganti, F. et al. (1997a) *Biochemistry* **36**, 10384–10392. With permission.)

extends to the vicinity of the active-site entrance through the His residues in positions 4, 3, 10, 15 and 17, such that it surrounds the entrance and the edge of the active site (Figure 12.4). This cluster acts as a kind of rapid proton release channel, and might explain the high catalytic efficiency of the isozyme. Histamine binds within the hCA II active site at the beginning of this channel, adding a particular extension to it and thereby accentuating the tunnelling effect.

The CA I active site lacks the histidine cluster. Moreover, the histidine residues able to participate in the proton transfer are placed at bifurcating positions, creating divergent pathways for the proton release and thereby significantly slowing down the overall catalytic process (Briganti et al. 1997a). It is supposed that these totally different structural features explain the experimental results according to which, at low concentrations, histamine modulates more strongly the hCA I activity than the hCA II one; isozyme II seems to be already activated by its unique histidine cluster.

The ternary adduct of hCA II with the activator phenylalanine and the inhibitor azide is the only other structure of the CA activation complex reported up to the present (Briganti et al. 1998). From the analysis of the refined atomic model of this adduct by electronic spectroscopy and x-ray crystallography, it is clear that azide replaces the hydroxide anion from the Zn(II) native enzyme coordination sphere, binding directly to the metal site (with the maintenance of the overall tetrahedral geometry) and extending into the hydrophobic half of the active-site cavity. Unlike the inhibitor molecule and similarly to histamine, phenylalanine experiences no contacts with the zinc ion, being anchored through hydrogen bond bridges in the

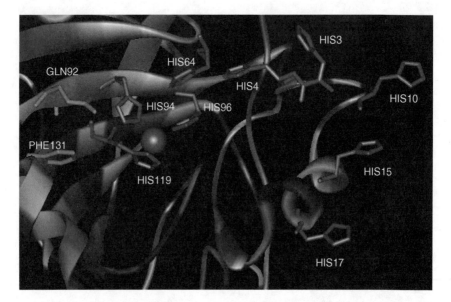

FIGURE 12.4 (See color insert following page 148.) hCA II active site. Zinc ion (pink); its three histidine ligands — His 94, His 96, His 119 (green); and the histidine cluster — His 64, His 4, His 3, His 17, His 15, His 10 (orange) are shown. The figure was generated from the diffraction coordinates reported by Briganti et al. (1997a) (PDB entry 4TST). (Reproduced from Scozzafava and Supuran (2002a). *Bioorganic and Medicinal Chemistry Letters* **12**, 1177–1180. With permission from Elsevier.)

proximity of the active-site entrance, with the phenyl ring oriented toward this entrance. A water molecule (Wat 73) links simultaneously the inhibitor and activator molecules by two strong hydrogen bonds involving the zinc-bound nitrogen atom and the amino group, respectively (Figure 12.5).

Not surprisingly, as in the hCA II–histamine complex, the His 64 imidazole side chain appears with the "in" conformation. The strong, charge-assisted hydrogen bond (2.4 Å) between the phenylalanine carboxylate group and the $N\varepsilon1$ hydrogen of His 64 is undoubtedly one of the most important factors stabilizing this conformation.

Although belonging to diverse classes of compounds, both histamine and phenyl-alanine have a similar location when bound to the enzyme, being fixed at the active-site entrance. It is assumed that this particular active-site opening stop experienced by the activator molecules is also a consequence of their weak Lewis base character, which makes them unable to replace the stronger HO^- zinc ligand. All these findings led to the conclusion that, in addition to the CO_2 substrate binding site (the hydrophobic pocket consisting of the amino acid residues Val 121, Val 143, Leu 198, Thr 199, Val 207 and Trp 209) and the inhibitor binding site (the zinc ion), CAs possess a third site, the activators binding site, located at the entrance of the active-site cavity, between the His 64, Gln 92, Asn 62 and Asn 67 side chains. This site, although an enzymatic one, does not match strictly with the lock and key principle that generally governs the physicochemical interactions of proteins; that is, it is able to accommodate

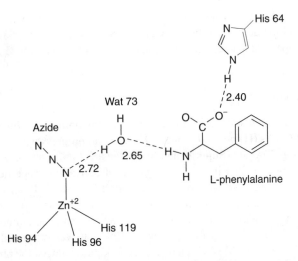

FIGURE 12.5 Scheme of hydrogen bond network that links azide, water molecule at position 73, phenylalanine, and the His 64 residue (hydrogen bonds lengths in Å also indicated). (Reproduced from Briganti, F. et al. (1998) *Inorganica Chimica Acta* 275/276, 295–300. With permission from Elsevier.)

molecules from diverse classes of compounds, with no *a priori* substrate specificity. The only thing these molecules have in common is the presence in their structure of one or more functional groups by which they can be efficiently involved in hydrogen bonds with well-defined strength and orientation so that they can participate in proton transfer processes between the active site and the environment. As previously shown, the activator binding mode or, in other words, stability of the enzyme–activator complex, is precisely tuned kinetically (through the binding affinity between the protein and its activator), sterically (through an advanced stabilization of His 64 "in" conformation) and energetically (through the rearrangement of the active-site water structure) to achieve the most appropriate alternative hydrogen bond network for proton transfer from the zinc-bound water molecule to the bulk solvent.

Although this activation mechanism was initially proposed and then investigated in great detail for isozyme II only, subsequent activation studies for isozymes III, IV and V (see later) have suggested its general validity for all CA isoforms investigated until the present time.

12.3 CAAs TYPES: DESIGN STRATEGIES, STRUCTURE–ACTIVITY RELATIONSHIPS AND ISOZYME SPECIFICITY

The drug design of CAAs has three major aims: (1) clarifying aspects of the catalytic mechanism; (2) elucidating the *in vivo* role of CAAs both in physiological and pathological conditions; and (3) obtaining biologically active compounds possessing isozyme specificity, or, at least, organ and tissue selectivity, which might be used in the clinical development of pharmacological agents devoid of severe side effects.

The first goal has been achieved almost completely through detailed studies that have explained the activation mechanism (see Section 12.2), but the other two still represent an open challenge, although much progress has been made in the past few years.

12.3.1 BACKGROUND

The biogenic amine histamine and the corresponding amino acid histidine were the first reported CAAs (Leiner and Leiner 1941), but no progress was made in this area in the following 30 years (Supuran and Scozzafava 2000a). The early 1970s marked the beginning of a revival in the investigation of activating properties of this type of compound. Narumi and Kanno (1973) reported *in vivo* CA II activation by some gastric acid stimulants (such as histamine, tetra- and pentagastrin and carbachol). These results were supported later by the finding that catecholamines enhance the activity of this red cell isozyme (Igbo et al. 1994). It was presumed (Narumi and Miyamoto 1974) that a cAMP-dependent protein kinase induced the phosphorylation of an active-site threonine residue, being hypothesized that the phosphorylated enzyme is more active than the unphosphorylated one. It is now clear that the explanation for the CA activating effect with such compounds is different and that phosporylation of CA active-site residues does not occur (Supuran and Scozzafava 2000a).

The CAA behavior of histamine continued to be further investigated. Puscas (1978) showed that histamine is strongly involved in gastric ulcers and proved its *in vitro* and *in vivo* activating properties against CA II.

Silverman's laboratory reported hCA II activation by histidine (Silverman et al. 1978), but subsequently considered their discovery an artefact owing to the ethyl-enediaminotetraacetic acid (EDTA) added in the buffer, which might form complexes with the adventitious Cu^{2+} possibly present in the protein preparation, thus restoring the activity of the enzyme (heavy metal ions act as micromolar inhibitors of most CAs; Tu et al. 1981).

Seeking for structural requirements that could determine CA activation, Supuran and coworkers (Supuran et al. 1991; Puscas et al. 1990; Supuran 1991, 1992) investigated CA activity in the presence of various natural and synthetic compounds of biological relevance, such as biogenic amines, amino acids and their derivatives, oligopeptides and several pharmacological agents (*vide infra*). It was shown (Supuran 1991, 1992) that all these compounds fit well in a general structure of type **12.1**, incorporating as main structural elements a proton-accepting moiety (a primary or secondary amino group) attached to a bulky aromatic/heterocyclic ring through an aliphatic carbon chain linker. Table 12.1 presents the activation data with some of these compounds and their detailed structural identity.

In the light of this study (Supuran et al. 1991), the initial report by Silverman's group regarding hCA II activation with histidine (Silverman et al. 1978) was proposed to be authentic. It was explained that the controversial results of Tu et al. (1981) were because of the experimental protocols used, EDTA probably possessing a competitive suppressing effect for the histidine activating effect (Supuran and Puscas 1994).

TABLE 12.1

hCAs I and II Activation by Different Biologically Active Compounds for CO_2 Hydration at 10^{-5} M of Activator

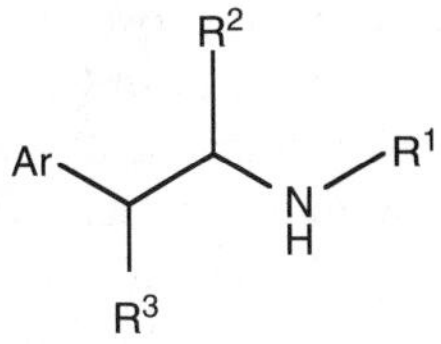

(12.1)

Compound[a]	Ar	R^1	R^2	R^3	% CA Activity[b]	
					hCA I	hCA II
Phenethylamine	Ph	H	H	H	114	110
Dopamine	3,4-Di(OH)Ph	H	H	H	138	141
Noradrenaline	3,4-Di(OH)Ph	H	H	OH	140	143
Adrenaline	3,4-Di(OH)Ph	Me	H	OH	145	153
Isoprotenerol	3,4-Di(OH)Ph	i-Pr	H	OH	143	146
Histamine	4-Imidazolyl	H	H	H	180	173
2-Pyridyl-ethylamine	2-Pyridyl	H	H	H	134	120
Serotonin	5-OH-Indol-3-yl	H	H	H	128	115
Phenylalanine	Ph	H	COOH	H	170	196
4-Hydroxyphenylalanine	4-OH-Ph	H	COOH	H	174	202
3,4-Dihydroxyphenylalanine	3,4-Di(OH)Ph	H	COOH	H	164	142
4-Fluorophenylalanine	4-F-Ph	H	COOH	H	169	175
3-Amino-4-hydroxyphenylalanine	3-NH_2-4-OH-Ph	H	COOH	H	171	177
4-Aminophenylalanine	4-NH_2-Ph	H	COOH	H	152	163
Histidine	4-Imidazolyl	H	COOH	H	153	149
Tryptophane	3-Indolyl	H	COOH	H	129	124

[a] Amino acids were L-enantiomers.

[b] CA activity without activator added is taken as 100%.

Source: Adapted from Supuran, C.T., and Puscas, I. (1994) In *Carbonic Anhydrase and Modulation of Physiologic and Pathologic Processes in the Organism*, Puscas, I. (Ed.), Helicon, Timisoara, Romania, pp. 113–145.

The data given in Table 12.1 afforded the first structure–activity relationships (SARs) for CAAs. Thus, consistent with the activation mechanism previously described, substitution of the parent structure (**12.1**) with polar moieties able to engage in hydrogen bonds (such as hydroxyl and carboxyl groups or fluorine atoms) increased activator potency. Most of the arylalkylamines and all the amino acids from Table 12.1 were also included in a quantitative structure–activity relationship (QSAR) study (Clare and Supuran 1994). This is the only QSAR investigation on hCA II activators available until the present time. By regression analysis and partial least squares method, equations for quantitative correlation were derived, which translated into mathematical terms the interdependence between biological activity and two main physicochemical parameters — the molecular charge distribution and the molecular volume. The activating properties were expressed as log A, where A is the hCA II activation percentage. The electronic density and the size of the activator molecule were expressed as a large number of different descriptors calculated by the complete neglect of differential overlap (CNDO) approximation. Such optimized equations show that the efficiency of CAAs is determined both by steric and electronic factors. Thus, consideration of a molecular model projection on a three-dimensional axis clearly showed that a decrease in the two smaller orthogonal dimensions of the CAA molecule yields an intensification of its biological activity. This fact has forwarded for the first time the concept that CAAs bind to the enzyme in a site of limited size, as has been subsequently confirmed through x-ray crystallographic studies (Briganti et al. 1997b). Alternatively, the higher the charge of the most electronegative atom in the molecule, the more active the compound is as a CAA. The second theoretical result explained experimental data that showed that molecules bearing highly charged oxygen or nitrogen atoms (such as amino acids and amino-azole derivatives) were more powerful CAAs than were analogous compounds in which the amino group is attached to a molecular skeleton containing only carbon atoms (Supuran et al. 1991, 1993a). This first QSAR of CAAs afforded two main conclusions: (1) the activating properties are conditioned by a relatively compact structure of the molecule, strictly depending on the dimensions of the binding site of activators, and (2) the activating properties are significantly increased by the presence in the molecule of some oxygen-/nitrogen-containing moieties that are strongly polarized. These conclusions, together with the chemical structure of compounds given in Table 12.1, constituted the starting point for different directions in the design and synthesis of CAAs.

12.3.2 CA ACTIVATION WITH AMINE DERIVATIVES

Three premises have been considered in the development of CAAs of this type: (1) aliphatic amines with the general formula **12.2** have no activating effect on CA II (Puscas et al. 1990); (2) 2-amino-5-(2-aminoethyl)-1,3,4-thiadiazole (**12.3**), reported as a histamine agonist and proved to possess gastric-acid-stimulating effects, was later recognized as a powerful CAA (Supuran and Puscas 1994); (3) the pyridinium ring presents several features (i.e., aromatic character, permanent positive charge, strong chemical stability, especially when it is alkyl/aryl substituted) that recommend it as a candidate moiety for obtaining biologically active compounds with pharmacological applications. Efficient positively charged CAAs of type **12.4** to **12.6** were thus designed and synthesized (Supuran et al. 1993b, 1996a), by condensing different 2,4,6-trisubstituted pyrylium salts with the corresponding amine derivatives under the classical Baeyer–Piccard reaction conditions.

The following SAR conclusions were derived: (1) The polymethylene bridge between the amino group and the heterocyclic moiety might incorporate either two or three carbon atoms. A higher number than this leads to a decrease in activating properties. (2) The derivatives bearing methyl or *t*-butyl groups in the 2- and 6-positions of the pyridinium ring are much stronger CAAs than the 2,6-diphenylsubstituted

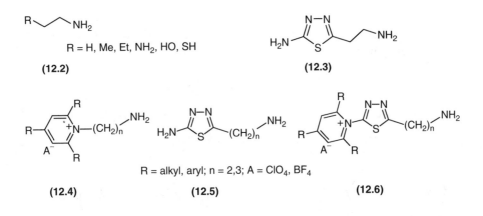

R = H, Me, Et, NH₂, HO, SH (12.2)

(12.3)

R = alkyl, aryl; n = 2,3; A = ClO₄, BF₄

(12.4)   (12.5)   (12.6)

analogues are, probably because of steric impairment of activator access into the enzymatic site. On the other hand, 2,6-dialkyl-4-phenyl derivatives were highly active, supporting the previous QSAR finding that the molecular elongation along the axis passing through the enzyme Zn(II) ion, the CAA molecule itself and the active-site exit is the most beneficial substitution pattern for activating properties. (3) The complete derivatization of the ω-amino alkyl group eliminates the activating effect. Furthermore, these studies (Supuran et al. 1993b, 1996a) first provided the suggestion that selective CAAs could be obtained. Because of their cationic nature, pyridinium derivatives are membrane impermeant, so that theoretically they should discriminate between the membrane-bound CA IV and the very similar cytosolic CA II (Ilies et al. 2002).

12.3.3 CA ACTIVATION WITH AZOLE DERIVATIVES

Imidazole is widely used as a buffer, and the imidazole moiety is present in both histamine (the prototypical CAA) and histidine (an amino acid of crucial importance for the CA catalytic cycle, and also a CAA). Consequently, research in the field began by investigating CA activation by imidazoles. A major difference between isozymes I and II was thus unveiled: imidazole is a unique competitive inhibitor with CO_2 as substrate for hCA I (Khalifah 1971), whereas it behaves as a very efficient activator for isozyme II (Parkes and Coleman 1989; Supuran 1992). Subsequent studies (Supuran et al. 1993a, 1996b) afforded CA II activation data for a large series of azole derivatives (**12.7** to **12.11**) as well as some interesting SARs.

A strong relationship between activatory efficiency of these compounds and the pK_a value of their proton shuttle moiety was observed (Table 12.2). pK_a values of 5.5 to 8.0 were optimal for the activator potency. The same empirical conclusion was found for compounds of types **12.1** and **12.4** (Supuran and Balaban 1994). The graphical representation of pK_a of the activator vs. % CA activation showed a Gaussian-type variation, with a maximum in the previously mentioned pK_a range (Figure 12.6). These facts can be correlated either with a pK_a value of 7.1, which characterizes His 64 — the native CA proton shuttle, or with a pK_a value of 6.8, which is that of the zinc-bound water molecule in the CA catalytically inactive form.

(12.7a–d)

(12.8a–c)

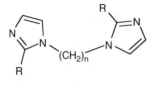

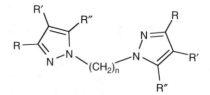

(12.9a–d)

(12.10a–e)

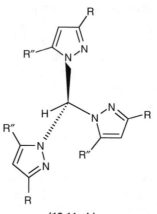

(12.11a,b)

Taking into account that (1) amino-pyridinium derivatives (Supuran et al. 1993b, 1996a) as well as azole compounds (Supuran et al. 1993a, 1996b) are efficient activators and (2) the salt-like character induced by the pyridinium group has been exploited successfully in generating CA IV-specific inhibitors (Supuran et al. 1998; Scozzafava et al. 2000a), Ilies and coworkers (1997, 2002) extended the investigated CAAs series to pyridinium azoles of types **12.12** to **12.15**. This research has led to (1) the first investigation of the activation of the bovine isozyme IV (bCA IV; Ilies et al. 1997); (2) a study of comparative activation for several CA isozymes (hCA I, hCA II, and bCA IV; Ilies et al. 1997); and (3) the first membrane-impermeant CAAs (Ilies et al. 2002; Ilies 2002).

Two structural variations in the choice of the reactants were followed to obtain insights into the SARs for this class of activators: (1) the number of azole nitrogen atoms (two, three or four) and their position (pyrazole/imidazole), and (2) the

TABLE 12.2
CA II Activation Data with Compounds 12.7 to 12.11
at 10^{-5} M for the Hydrolysis of 4-Nitrophenyl Acetate
and the Corresponding pK_a Values of the Activator

Compound	n	R	R′	R″	pK_a[a]	% CA Activation[b]
12.7a	—	H	H	—	7.08	190
12.7b	—	Me	H	—	7.12	194
12.7c	—	Et	H	—	7.19	203
12.7d	—	Me	Me	—	8.00	247
12.8a	—	H	H	H	2.5	100
12.8b	—	Me	H	H	2.06	100
12.8c	—	Me	Me	Me	3.74	118
12.9a	1	H	—	—	5.56	140
12.9b	1	Me	—	—	6.64	169
12.9c	2	H	—	—	6.41	154
12.9d	2	Me	—	—	7.28	231
12.10a	1	H	H	H	0.12	100
12.10b	2	H	H	H	1.67	100
12.10c	1	Me	H	Me	2.14	100
12.10d	2	Me	H	Me	3.30	112
12.10e	2	H	Br	H	0.00	100
12.11a	—	H	—	H	—	142
12.11b	—	Me	—	Me	—	153

[a] *Bis*-azoles pK_a values are for the second deprotonation step. (From Catalan et al. (1987) *Advances in Heterocyclic Chemistry* **41**, 187.)
[b] CA II activity in the absence of activator is taken as 100%.

pyridinium ring substitution pattern (alkyl/aryl moieties as well as different combinations of them), because this was one of the main parameters that strongly influenced the efficacy of some previously reported pyridinium derivatives, both as CAAs (Supuran et al. 1993a, 1996b) and CAIs (Supuran et al. 1998; Scozzafava et al. 2000a). The following findings emerged (Ilies et al. 1997, 2002; Ilies 2002):

1. Presuming an identical pyridinium ring substitution pattern and the same investigated isozyme, the nature of the azole ring determined the following order of the CA activation effect: imidazole > triazole > pyrazole » tetrazole. In keeping with earlier assumptions (Supuran et al. 1996b; Supuran and Balaban 1994), the pK_a values of the azolyl moieties in aqueous system might provide an explanation. Thus, it seems that Mother Nature always makes the best choice by giving to the His 64 imidazolyl residue the proton shuttle role in the native CA, and the imidazole ring system has been proved to be the best option for designing CAAs of the azole type. Proceeding to triazole and pyrazole, the pK_a becomes lower, reaching a quite acidic range for tetrazole, which has an inappropriate

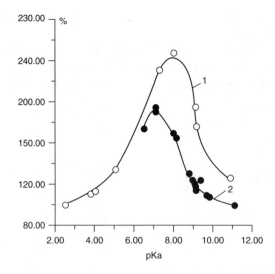

FIGURE 12.6 Percent CA activation vs. pK_a value for amino acid and amine derivatives of type **12.1** and **12.4** (Curve 1) and for azole derivatives of type **12.7 to 12.11** (Curve 2; enzyme concentration of 10 μM). [Adapted from Supuran, C.T., and Balaban, A.T. (1994) *Revue Roumaine de Chimie* **39**, 107–113. With permission.]

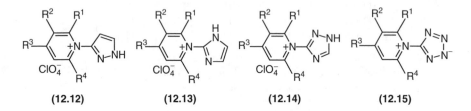

value for proton acceptance. Triazole and tetrazole were found to behave as CAIs (Alberti et al. 1981). It can be concluded that polarization of the molecule induced by the electron-attracting pyridinium group is practically responsible for the activating properties in tri- and tetrazole derivatives. A strong argument for this claim is the zwitterionic form in which pyridinium tetrazoles of type **12.15** have been isolated. By acquiring for the NH proton a pK_a value similar to that for compounds **12.13** (pK_a 6 to 8), most of the triazole derivatives (**12.14**) were efficient CAAs, although they were only micromolar activators, whereas the major part of their analogues (**12.13**) showed activatory effects in the nanomolar range.

2. The influence of the pyridinium moiety substituents on the biological activity has confirmed the corresponding SAR established previously for amino-pyridinium CAAs of types **12.4** and **12.6** (Supuran et al. 1993b, 1996a). Interesting extensions of such correlations have been made:

   a. Among the most active 2,6-dialkyl-4-phenyl-pyridinium-azoles, those substituted with the isopropyl radical have arguably been the best *in*

TABLE 12.3
In Vitro **Activation Pattern of hCA I, hCA II and bCA IV with Histamine and Compounds 12.12 to 12.15**

Activator	Activation Pattern	Affinity
Histamine	hCA I » bCA IV > hCA II	Micromolar
12.14	bCA IV » hCA I > hCA II	Micromolar
12.12, 12.13, 12.15	hCA I $\cong$ bCA IV > hCA II	Nanomolar for hCA I and bCA IV; micromolar for hCA II

     vitro activators in all the four new series (**12.12** to **12.15**) and against all three assayed isozymes.

b. Newly introduced in this study (Ilies et al. 2002), the substitution of the 2,6-dialkyl-pyridinium ring with a high lipophilic and flexible looped side chain the of 3,5-nonamethylen type has led to a remarkable enhancement in CA activity, such compounds being only slightly less effective than the 2,6-dialkyl-4-phenyl substituted azoles. These results show that the activator efficacy can be successfully fine-tuned by modulating its structural hydrophilic–hydrophobic balance through an appropriate substitution pattern.

3. Significant differences in isozyme specificity with both classical CAAs, such as histamine, and the newly synthesized compounds have been revealed, the *in vitro* activation pattern Shown in Table 12.3. The particular histidine cluster in the hCA II active site (Briganti et al. 1997b) explains well its constant lowest sensitivity to activation. The triazoles derivatives **12.14** maintain the general histamine activation pattern (probably a consequence of their similar pK_as for the NH proton), except for the membrane-bound bCA IV, which is much more susceptible to the new type of activators (presumably because of their cationic nature).

The best *in vitro* activators (i.e., 2,6-di-*i*-propyl-4-phenyl-pyridinium azoles) strongly enhance *ex vivo* red cell lysate CA activity, being about twice more efficient than the standard drug histamine. The same type of experiments performed with unlyzed erythrocytes clearly demonstrate that pyridinium pyrazole/imidazole derivatives have no influence on the cytosolic hCA I and hCA II, whereas the zwitterionic nature of the corresponding pyridinium tetrazoles permits them to cross the plasma membrane and thus activate the CA isozymes with an internal cellular localization, similar to that by histamine.

12.3.4 CA Activation with Histamine Derivatives

Histamine belongs to the amino-azole class of CAAs. Therefore, the CAAs derived from this lead molecule can be considered an independent and particular category of such compounds.

The rationale for designing novel activators of this type was based on some recent structural aspects of the hCA II–histamine interaction revealed by x-ray crystallography: (1) the unique histidine cluster of the hCA II active site surrounds its opening and consists of six residues, some of them possessing flexible conformations, which can easily participate in hydrogen bonds with activators or inhibitors bound within the active site (Briganti et al. 1997b, 1998); (2) the histamine molecule binds at the hCA II active-site entrance, being anchored to amino acid side chains and to water molecules through hydrogen bonds involving only the nitrogen atoms of the imidazole moiety, whereas its aliphatic amino group extends into the solvent, making no contact with the enzyme (Briganti et al. 1997b). Thus, it was of interest to derivatize the amino aliphatic end of the histamine molecule with highly polarized moieties. Such groups might interfere in an energetically favorable way with the polar amino acid residues at the rim of the active site. By decreasing the activators' whole energy of binding, these additional contacts will lead to more stable enzyme–activator adducts and hence to a more efficient activation process. This approach, based on improvement in the energy of binding of certain compounds to the enzyme through chemically modified units of a given parent structure, was also reported both by the groups of Whitesides (Gao et al. 1995) and Supuran (Scozzafava et al. 1999) for the design of tight-binding sulfonamide CAIs.

To obtain a much better affinity than histamine for the protein, the following concepts have been considered while choosing the polar derivatizing groups: (1) as shown by earlier OSAR calculations (Clare and Supuran 1994), the oxygen and nitrogen atoms have a stimulating effect on the activating properties; (2) a peptide-type side chain, fulfilling the previous condition, will additionally confer to the compound a specific embedded structure through which it can better interact with amino acid residues at the active site (Supuran and Scozzafava 1999); (3) according to the activation mechanism described previously, a free amino or imidazole moiety acting as a second proton transferring group (besides the imidazole ring from the lead histamine) might positively influence efficacy of the CAAs (Supuran and Scozzafava 1999, 2000c).

The synthetic strategy involved treating histamine with tetrabromophthalic anhydride and protecting its imidazole residue with tritylsulfenyl chloride, followed by hydrazinolysis of the phthalimido moiety under mild conditions. The obtained N-1-tritylsulfenyl histamine, the key intermediate, was further derivatized at its terminal amino ethyl function by a variety of procedures to introduce in the molecule the desired polar group, as summarized in Table 12.4. The final removal of the tritylsulfenyl moiety in an acidic medium afforded the target compounds 12.16 to 12.22 (Supuran and Scozzafava 1999, 2000c; Briganti et al. 1999; Scozzafava and Supuran 2000a, 2000b; Scozzafava et al. 2000b).

Thus, a large series of alkyl/arylsulfonamido derivatives of type 12.16 was obtained by reacting the key intermediate with the corresponding sulfonyl halides, whereas an alternative route involving its treatment with arylsulfonylisocyanates gave some arylsulfonylureido compounds with the general formula 12.17 (Briganti et al. 1999). Alkyl/arylcarboxamido structures of type 12.18 were generated by reacting the same key intermediate with acyl chlorides or carboxylic acid anhydrides

TABLE 12.4
General Derivatizing Procedures for the Key Intermediate
N-1-Tritylsulfenyl Histamine

Reagents and Reaction Conditions	Target Compound
RSO$_2$Cl/NEt$_3$/MeCN	**12.16**
ArSO$_2$NCO/MeCN/0°C	**12.17**
RCOCl/NEt$_3$/MeCN	
(R$_2$CO)O/NEt$_3$/MeCN	
RCOOH/iPr-N=C=N-iPr/MeCN	**12.18**
RNCO/MeCN	
RNCS/NEt$_3$/MeCN	
H$_2$N-CN/EtOH-H$_2$O/HCl	
H$_2$N-C(=NH)NH-CN/EtOH-H$_2$O/HCl	**12.19**
N-Boc-aa/iPr-N=C=N-iPr/MeCN or Me$_2$CO	**12.20**
X-C$_6$H$_4$SO$_2$NHCO-aa-COOH/iPr-N=C=N-iPr/MeCN	**12.21**
X-C$_6$H$_4$SO$_2$NHCO-aa$_1$-aa$_2$-COOH/iPr-N=C=N-iPr/MeCN	**12.22**

Note: R = alkyl; substituted-alkyl; aryl; substituted-aryl; hetaryl; Ar = substituted-phenyl; Boc = t-butoxy-carbonyl; X = 2-Me, 4-Me, 4-F, 4-Cl; aa, aa$_1$, aa$_2$ = amino acid/dipeptide residues.

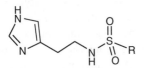

(12.16)

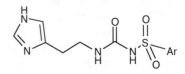

(12.17)

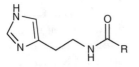

(12.18)

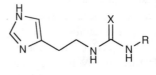

(12.19)

R = alkyl; substituted-alkyl; aryl; substituted-aryl; hetaryl;
Ar = substituted-phenyl;
X = O, NH, S.

in the presence of triethylamine, or carboxylic acids in the presence of diisopropyl-carbodiimide (DIPC; Scozzafava and Supuran 2000a). The use of aryl/alkyl isocyanates/isothiocyanates, cyanamide or dicyandiamide as derivatizing reagents led to

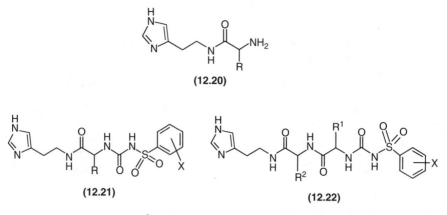

(12.20)

(12.21)   **(12.22)**

R, R¹, R² = amino acid/dipeptide residues;
X = 2-Me; 4-Me; 4-F; 4-Cl.

aryl/alkylureido/thioureido or guanidino compounds of type **12.19** (Scozzafava and Supuran 2000a). Novel CAAs with a **12.20** scaffold were prepared by treating the previously mentioned *N*-1-tritylsulfenyl histamine with Boc-protected amino acids/dipeptides in the presence of DIPC (Supuran and Scozzafava 1999). The previous strategies were combined and extended by coupling the key intermediate with arylsulfonylureido amino acids/dipeptides resulting from the reaction of the corresponding amino acids/dipeptides with arylsulfonyl isocyanates. A large series of arylsulfonylureido amino acyl/dipeptidyl histamine derivatives of types **12.21** and **12.22** were thereby obtained (Scozzafava and Supuran 2000b; Supuran and Scozzafava 2000c; Scozzafava et al. 2000b). This last synthetic approach yielded some very efficient activators, such as compounds **12.23**, **12.25** (Supuran and Scozzafava 1999) or **12.24** and **12.26** (Scozzafava and Supuran 2000b; Supuran and Scozzafava 2000c; Scozzafava et al. 2000b). This was explained by the authors as being because such molecules, in addition to the histamine imidazole, possess multiple moieties able to shuttle protons (i.e., NH from imidazole ring or free amino groups).

Different histamine-derived CAAs, i.e., compounds **12.27** and **12.28**, were obtained through another synthetic method involving direct acylation of the lead molecule with thylenediaminotetraacetic and diethylenetriaminopentaacetic acid dianhydride, respectively, in a molar ratio of 2:1 (Scozzafava and Supuran 2000a).

Enzymatic assays of the novel activators **12.21** to **12.28** against hCA I, hCA II and bCA IV led to some new SARs (Supuran and Scozzafava 1999, 2000c; Briganti et al. 1999; Scozzafava and Supuran 2000a, 2000b; Scozzafava et al. 2000b):

1. The most interesting finding in these series was the constant high affinity of hCA II (in the low nanomolar range) for many of the new derivatives, i.e., ca. 1000 times more affinity than the lead histamine. This might be explained by the stronger interactions with the hCA II histidine cluster assured by the specific structure of these compounds. Isozymes I and IV were also very susceptible to the new CAAs, in the nanomolar range for

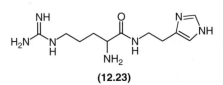

(12.23)

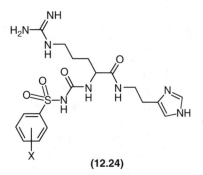

(12.24)

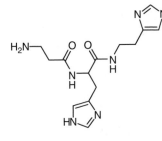

(12.25)

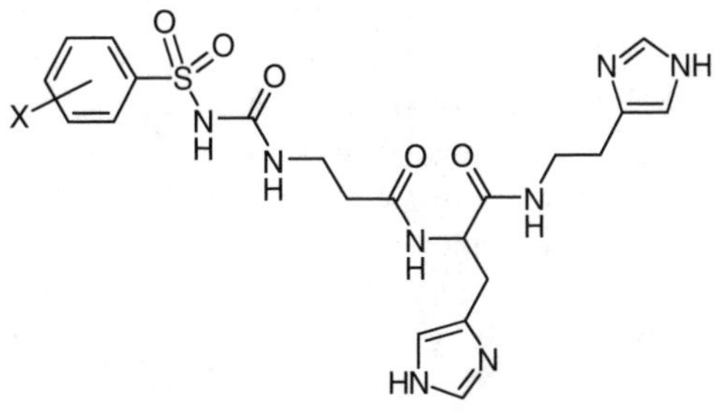

(12.26)

X = 4-F; 4-Cl; 4-Me; 2-Me.

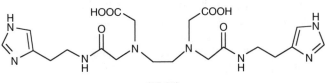

(12.27)

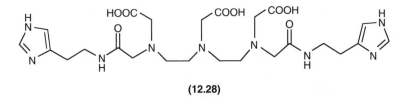

(12.28)

the best activators, the activation pattern being hCA I > bCA IV > hCA II, similar to that of histamine (see Section 12.3.3). The differences between the polar amino acid residues at the active-site edge were considered to determine the diverse sensitivity of isozymes for the same modulators of activity (Lindskog 1997; Briganti et al. 1997b, 1998).

2. As presumed in the design strategy, sticky tails like those from derivatives **12.24** and **12.26** or the polyamino-polycarboxamido chains from compounds **12.27** and **12.28** improved significantly the activating properties, probably because of particular contacts with the enzyme.

3. The presence in the molecule of a second proton shuttle moiety besides the parent histamine one powerfully enhanced the CA activation effect. This was another design strategy assumption confirmed by experimental results.

4. The following substitution patterns led to the most efficient CAAs in the series **12.16** to **12.19**: (1) Perfluoroalkyl (longer the perfluoroaliphatic chain, stronger the activatory effect) and perfluoroaryl; (2) 4-halogeno-phenyl; and (3) 2-, 3-, or 4-pyridyl. For the series **12.20** to **12.22**, the best activators were those derived from basic amino acids such as arginine, lysine or histidine, or from dipeptides such as glycyl-histidine, β-alanyl-histidine (carnosine), phenyl-proline or prolyl-glycine. At first sight, this experimental observation can be linked either to the molecule's ability for engaging some electrostatic interactions with the enzyme or with the pK_a of its proton shuttle moieties.

5. Some compounds of types **12.21** and **12.22** that produced strong *in vitro* CA activity enhancement acted also as effective *ex vivo* activators for the human red cell isozyme II, surpassing by ca. 1.5 to 2 times the effect of the standard drug histamine.

12.3.5 CA ACTIVATION WITH AMINO ACID DERIVATIVES

Supuran et al. (1991) reported the first systematic study of CA activation with amino acids and related compounds. A variety of such compounds, including natural or synthetic amino acids, their esters, *N*-alkyl or *N*-acyl derivatives, as well as some pyridinium derivatives of type **12.29**, were assayed against isozymes I and II for the reversible CO_2 hydration reaction.

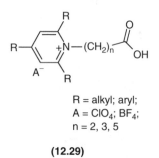

R = alkyl; aryl;
A = ClO_4; BF_4;
n = 2, 3, 5

(12.29)

The following SAR emerged: (1) The most powerful CAAs in these series were proline, homoproline, histidine and many of the aromatic amino acids, especially those possessing the general formula **12.1**, such as phenylalanine. X-ray crystallographic data (Briganti et al. 1997b, 1998) clearly demonstrated that the aryl/hetaryl

TABLE 12.5
bCA II Activation Data with Some Amino Acids
for CO_2 Hydration and the Corresponding
pK_a Values

Activator[a]	pK_a		% CA Activity[b]
	α-NH$_2$	Side chain	
Asn	8.80	—	130
Lys	8.95	10.53	124
Arg	9.04	12.48	121
Gln	9.13	—	118
Ser	9.15	—	113
His	9.17	6.20	149
Trp	9.39	—	124
Glu	9.67	4.25	109
Asp	9.82	3.86	108

[a] Amino acids had the L-configuration. Standard abbreviations from the amino acids IUPAC nomenclature were used.

[b] bCA II activity in the absence of activator was considered as 100%. Activation data were from Supuran, C.T., and Puscas, I. (1994) In *Carbonic Anhydrase and Modulation of Physiologic and Pathologic Processes in the Organism*, Puscas, I. (Ed.), Helicon, Timisoara, Romania, pp. 113–145.

Source: Adapted from Supuran, C.T., and Puscas, I. (1994) In *Carbonic Anhydrase and Modulation of Physiologic and Pathologic Processes in the Organism*, Puscas, I. (Ed.), Helicon, Timisoara, Romania, pp. 113–145.

moieties of these compounds increase the stability of the enzyme–activator complexes. (2) Derivatization of the amino or carboxyl groups strongly diminished the activating profile, mainly because of the decrease in charge on the most electronegative atom of the molecule (through the induced electronic effects), as was proved later by QSAR calculations (Clare and Supuran 1994). (3) As with the azole derivatives (Supuran et al. 1996b) and pyridinium amine derivatives of type **12.4** (Supuran and Balaban 1994), a strong correlation between pK_a value and activator strength was found for the investigated amino acids (Supuran and Balaban 1994). Table 12.5 gives some such examples. As seen from these data, again a pK_a of 5.5 to 8 for at least one deprotonatable moiety from the amino acid molecule seems to be essential for good activating properties.

No isozyme specificity was detected for this CAA class; some of the investigated amino acids possessed an increased affinity for CA I whereas others preferred CA II. In summary, the previously mentioned studies (Supuran et al. 1991, Supuran and Balaban 1994) delineated two major factors that modulate the activation behavior of the amino acid derivatives: (1) the presence of ring structural units in the molecule favors the stabilizing enzyme–activator adducts and (2) the pK_a value of the proton shuttle moiety strongly influences the activator's efficiency. These parameters were selected for further investigation by using an original series of supramolecular complexes obtained by coupling diverse amino acids with 2,3,11,12-bis[4-(10-aminodecylcarbonyl)]benzo-18-crown-6 ether (**12.30**; Barboiu et al. 1997).

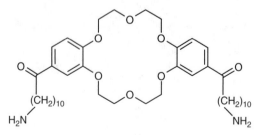

(**12.30**)

The new activators proved to be more efficient than the corresponding lead amino acids against hCA I and hCA II for the hydrolysis of 4-nitrophenylacetate and strictly followed the same order of potency (see Figure 12.7).

The interesting results of CA activation with amino acids (Supuran et al. 1991), as well as the report on pentagastrin (Puscas et al. 1996) and hemoglobin (Parkes and Coleman 1989) as CAAs, motivated a further search for similar properties in polypeptide derivatives, such as angiotensins (Puscas et al. 1994). The activatory efficacy of these compounds against bCA II for the CO_2 hydration varied over a large range. CA activation with tri- and tetrapeptides incorporating alanine and aspartic acid residues has recently been investigated in detail (Scozzafava and Supuran 2002a). These novel activators exhibited a significantly increased affinity for hCA I, hCA II and bCA IV as compared with the parent amino acids. The following were SAR conclusions from the study (Scozzafava and Supuran 2002a): (1) The first CAAs for which hCA II possesses the highest sensitivity among the three investigated isozymes were found. The activation pattern was hCA II > hCA I > bCA IV, whereas the major part of the activators tested till date (including the classical histamine) has shown a general range of isozyme affinity in the order hCA

L-Ala < 11-aminoundecanoic acid < L-Asp < L-Leu < L-Cis < L-Trp < L-His < L-Phe

(L-Ala)$_c$ < (11-aminoundecanoic acid)$_c$ < (L-Asp)$_c$ < (L-Leu)$_c$ < (L-Cis)$_c$ < (L-Trp)$_c$ < (L-His)$_c$ < (L-Phe)$_c$

weak CAAs   medium CAAs   strong CAAs

FIGURE 12.7 The activating properties against hCA I and hCA II for some amino acids and the corresponding supramolecular complexes with macrocyclic ligands of type **12.30**. Activation data for the hydrolysis of 4-nitrophenylacetate were used (from Barboiu, M. et al. (1997) *Liebigs Annalen der Chemie/Recueil* 1853–1859). Standard abbreviations from the amino acids IUPAC nomenclature are used and ()$_c$ symbol denotes their supramolecular chelates.

I > bCA IV » hCA II (Ilies et al. 1997, 2002; Briganti et al. 1999; Supuran and Scozzafava 1999, 2000c; Scozzafava et al. 2000b; Scozzafava and Supuran 2000a, 2000b, 2002b). Thus, the positively charged imidazolium moieties present in the hCA II histidine cluster presumably engage in appropriate contacts with the carboxylate groups of the previously mentioned oligopeptides, this favored interaction not being available for isozymes I and IV, which do not have such a cluster. (2) Aspartyl-alanyl-aspartyl-aspartic acid (DADD), the best CAA in the new series, possesses the sequence recently shown (Vince and Reithmeier 2000) to be essential for the binding of hCA II to the C-terminal region of the bicarbonate/chloride anion exchanger AE1. Consequently, this study (Scozzafava and Supuran 2002a) has advanced the idea that in the metabolon formed by these two proteins (Reithmeier 2001) the AE1 actually acts as a natural activator of isozyme II. In other words, it seems that AE1 facilitates the formation/transport of bicarbonate in erythrocytes by efficiently promoting the hCA II catalysis for CO_2 hydration.

A combined strategy for the synthesis of CAAs has been adopted, leading to another new class of tight-binding compounds derived from histamine and histidine as lead molecules, i.e., arylsulfonylureido-amino acyl/dipeptidyl carnosine derivatives of types **12.31** and **12.32**.

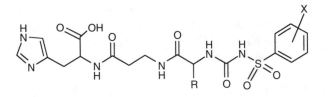

(12.31)

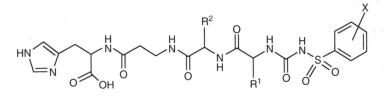

(12.32)

R, R^1, R^2 = amino acid/dipeptide residues;
X = 4-F; 2-Me.

This strategy has been based on the following considerations: (1) histidine is a proven hCA I and hCA II activator (Supuran and Puscas 1994), with a profile quite similar to that of histamine and phenylalanine — the two CAAs for which the first x-ray crystallographic data of their adducts with the enzyme have been provided (Briganti et al. 1997b, 1998); (2) carnosine, a natural dipeptide occurring especially in the brain and muscle (Quinn et al. 1992), is a much stronger CAA than histidine (Scozzafava and Supuran 2002b), confirming earlier QSAR conclusions that have stated that the elongation of the amino acid molecule improves its activating properties (Clare and Supuran 1994). Furthermore, in contrast to histidine or phenylalanine, carnosine is not readily metabolized *in vivo* (Quinn et al. 1992), an essential feature that recommends it strongly as a suited precursor for putative pharmacological agents; and (3) the best histamine-based tighter-binding activators have been obtained by using N-1-tritylsulfenyl histamine as key a intermediate and alkyl/aryl-sulfonyl/carboxamido/ureido moieties as derivatizing polar groups (Supuran and Scozzafava 1999, 2000c; Scozzafava and Supuran 2000a, 2000b; Scozzafava et al. 2000b).

Compounds **12.31** and **12.32** have thus been prepared in a three-step reaction involving (1) the coupling of β-alanine with 4-fluorophenyl-/2-tolyl-sulfonylureido amino acids; (2) the reaction of these derivatized β-alanines with N-1-tritylsulfenyl-*t*-butyl-histidine, the key intermediate, in the presence of carbodiimides; and (3) the removal of the various protecting groups under standard conditions.

The activatory effect of this last type of CAA against hCA I, hCA II and bCA IV has sustained the previous SAR conclusions (Scozzafava and Supuran 2000b; Supuran and Scozzafava 2000c; Scozzafava et al. 2000b) for the related compounds **12.21** and **12.22**: (1) the same isozyme specificity pattern (i.e., hCA I > bCA IV > hCA II) and the same range for the values of the activation constants are maintained; (2) the more the number of the moieties able to shuttle protons (except for the imidazole ring of the carnosine residue), the more efficient the molecule as a CAA; (3) as regards the influence of the phenyl ring substituents, the 4-fluorine atoms slightly induce increased activating properties when compared with the analogues with a methyl group in position-2, probably due either to the more elongated molecular profile or to an improved balance between the lipophilic and the hydrophilic characters of such derivatives; and (4) *ex vivo* CA activation with the best *in vitro* activators of types **12.31** and **12.32** leads to results similar to those obtained in the same experiments performed with the **12.21** and **12.22** derivatives.

A recent approach in the design of CAAs should be finally discussed for a complete view of the achievements in the area. Silverman's group (Earnhardt et al. 1999) has reported CA V activators of the histidine analogs type. Such compounds were linked covalently to cysteine residues at the edge of the active site; some of these derivatized enzymes possessed a threefold enhanced catalytic activity over the native one.

12.3.6 CA ACTIVATION WITH ANIONS

CA III possesses a very low catalytic efficiency for CO_2 hydration at a physiological pH of 7.4 (300-fold less than that observed for CA II; Tu et al. 1983; Engberg et al. 1985), although these two isozymes have a similar backbone conformation, especially in the active-site region (Eriksson and Liljas 1993; Mallis et al. 2000). From this point of view, the isozyme III activation phenomena might be enlightening for better understanding of both its catalytic mechanism and *in vivo* role, which might explain the maintaining in the phylogenetic tree of this low-activity cytosolic protein despite the presence of the perfectly evolved catalyst CA II. Accordingly, the first observation of chicken and hCA III activity enhancement by HPO_4^{2-} at millimolar concentration (Shelton and Chegwidden 1988) was a major clue for more advanced investigations. Other dianionic species, such as sulfite, pyrophosphate, 3-phospho-glycerate, 1,3-biphosphoglycerate and ATP, in millimolar concentrations, were found to increase hCA III catalysis by facilitating as much as 20-fold the proton transfer between the enzyme and the surrounding medium (Paranawithana et al. 1990; Rowlett et al. 1991; Shelton and Chegwidden 1996). An interesting observation was that some anions such as SO_3^{2-} or HPO_4^{2-} activate hCA III and inhibit hCA I and hCA II, whereas others such as SO_4^{2-} and oxalate are inhibitors of all three isozymes (Rowlett et al. 1991). These opposite influences were considered to be a consequence of the different base strengths of the dianions directly linked to (1) their own proton acceptance ability, and (2) their capacity for developing electrostatic interactions with the protein (Rowlett et al. 1991), because the CA III active site is more positively charged (because of the presence of Lys 64, Arg 67 and Arg 91 residues) as compared with the other two isozymes (Eriksson and Liljas 1993).

Detailed experiments carried out by Silverman's group (Tu et al. 1990) pointed out that buffers of small size, especially imidazole, also increase hCA III and hCA II activity in bicarbonate dehydration, similar to that by phosphate. Bulkier buffers such as 2,3-lutidine and *N*-methylmorpholine enhance hCA II catalysis only, having no effect on hCA III activity. This difference in the affinities of isozymes II and III for this particular type of activator was assumed to be a consequence of the steric impairments strongly induced in the hCA III active site by phenylalanine and isoleucine residues at positions 198 and 207, respectively.

On the other hand, the same study (Tu et al. 1990) revealed that pyrazole, which has a molecular volume and a chemical structure very similar to that of imidazole but a totally different pK_a value ($pK_{a\,imidazole} = 7.1$, $pK_{a\,pyrazole} = 2.5$), has no activating properties under identical experimental conditions. These results thus delineated the two main parameters that determine or modulate the activatory efficiency against CA isozymes: (1) an appropriate molecular size imposed by the limited amount of space available in the binding site; (2) a pK_a value well matched with that which dictates the proton transfer in the active-site environment. As shown earlier, subsequent works confirmed, detailed and extended such a hypothesis, supporting well-argued SAR/QSAR remarks (Supuran et al. 1991, 1993b, 1996b; Supuran and Balaban 1994; Clare and Supuran 1994; Ilies et al. 1997, 2002; Briganti et al. 1997b).

12.4 CONNECTIONS OF CAAs WITH PHYSIOPATHOLOGY: POSSIBLE CLINICAL DEVELOPMENTS

Many naturally occurring compounds found in the human body activate CAs at physiological concentrations (Supuran et al. 1991; Supuran and Balaban 1994; Scoz-zafava and Supuran 2002b). Examples are histamine, serotonin, dopamine, adrenaline, noradrenaline and other catecholamines, which are important autacoids; some amino acids and oligopeptides, some of which are hormones (Supuran and Puscas 1994); and others (such as β-alanyl-histidine) have been proved to possess different physiological roles (Quinn et al. 1992; Hipkiss and Chana 1998). Furthermore, their mechanism of action at the molecular level has been clearly explained (Briganti et al. 1997b, 1998; Supuran and Scozzafava 2000a). It has been also shown very recently that the AE1 Cl⁻/HCO₃⁻ exchanger is a natural hCA II activator (Scozzafava and Supuran 2002a). These results support the assumption (Supuran and Puscas 1994) that CAAs possess important physiological functions as metabolic modulators and intra- or intercellular signal-transducing systems, but more detailed studies are needed for clarifying in detail the *in vivo* role of such compounds.

The CA deficiency syndrome was recognized as a distinct genetic disorder translated into severe pathological modifications of the bone, brain, kidney and lung metabolism (Sly and Hu 1995a; Sato et al. 1990; Venta 2000). Clinical data revealed that absence of hCA II in the blood or in the previously mentioned tissues and organs is the primary defect in inherited osteopetrosis, associated with renal tubular acidosis, cerebral calcification (Sly et al. 1983; Sly 1991; Sly and Hu 1995b) and mental retardation (Roth et al. 1992; Alper 2002) or respiratory acidosis (De Coursey 2000). hCA IV deficiency is considered responsible for the particular pure proximal renal tubular acidosis that appears in infancy or childhood and tends to improve with age (Sato et al. 1990). The inherited lack of hCA I, although undoubtedly shown, seems to cause no phenotypic effect (Sly 1991; Venta 2000). It has been noted (Sato et al. 1990; Sly 1991) that in the pathological conditions previously described, only a certain isozyme is lacking, most often CA II, but also CA IV or CA I, or even CA VI (Murakami and Sly 1987), the others being found in different tissues at their normal concentrations. In such cases, the missing isozyme gene is either not expressed (Sato et al. 1990; Bergenhem et al. 1992; Murakami and Sly 1987) or its protein product is not stable because of deleterious mutations in the amino acid sequence (Tanis et al. 1973; Venta et al. 1991; Roth et al. 1992; Tu et al. 1993; Soda et al. 1996). All clinical abnormalities described as a CA deficiency syndrome are actually a direct consequence of the pH homeostasis disequilibrium created by the dysfunctions of this enzyme. Indeed, catalyzing the reversible CO_2 hydration, CA acts as a proton/bicarbonate source for osteoclastic acidification during bone resorption (Sly 1991; Rousselle and Heymann 2002), for H⁺ ion gradient maintenance at the level of the proximal and distal renal tubules (Sato et al. 1990) and also for CO_2 facilitated diffusion through alveolar epithelial cells (De Coursey 2000). Because the function of CA II in the brain is poorly understood at present, the mechanism of cerebral calcification in CA II deficiency is still unclear, but some suppositions consider it as an indirect effect of chronic systemic acidosis (Sly 1991). It should be remembered that at present, hCA II, in association with the HCO₃⁻/Cl⁻ exchanger, is seen as the major transporter for keeping the intra- and extracellular pH within its physiological limits (Rousselle and Heymann 2002). From this perspective, the CA deficiency syndrome can be essentially defined as a genetic disease of acid–base carriers/homeostasis (Alper 2002). No medical treatment for this disease is available to date. A possible pharmacological approach for such a treatment has been recently proposed by Scozzafava and Supuran (2000b) and Ilies et al. (2002). Thus, starting from the observation that in this genetic disease CA II is generally absent and CA I and CA IV are present in normal concentrations (Sato et al. 1990; Sly 1991), it has been suggested that the enhancement in the activity of the last two isoforms by exogenous modulators can compensate at least partially the functions of the absent isozyme. This hypothesis was based on the following considerations: (1) some of the physiological roles of CA I and CA IV are similar to those of CA II (Supuran and Puscas 1994), (2) isozyme I is a catalytically less effective form of CA (Coleman 1998) and is therefore very sensitive to activation with a large variety of natural or synthetic derivatives (Supuran et al. 1991; Ilies et al. 1997, 2002; Barboiu et al. 1997; Supuran and Scozzafava 1999; Briganti et al. 1999; Scozzafava et al. 2000b; Scozzafava and Supuran 2000a, 2000b, 2002b), and (3) the membrane-bound CA IV is susceptible to specific activation by positively charged CAAs (Ilies et al. 1997, 2002).

The very recent proposal (Sun and Alkon 2001, 2002) of the use of CAAs as therapeutic targets in cognitive disorders represents a new direction in this field. The following experimental conclusions strongly support this suggestion: (1) a significant decline of CA in the brain constantly accompanied the memory and learning alterations occurring in Alzheimer's disease or during aging (Meyer-Ruge et al. 1984); (2) the synaptic switch of the γ-aminobutyric acid (GABA)ergic responses from inhibitory to excitatory (that amplify input signals) depends on increased HCO_3^- conductance through the $GABA_A$ receptor–channel complex, and thereby directly depends on the activity of the CA present within the intracellular compartments of the hippocampus pyramidal cells (Sun and Alkon 2001); and (3) CAAs such as phenylalanine (Clare and Supuran 1994) or imidazole (Parkes and Coleman 1989) administered to experimental rats directly into the brain increase HCO_3^- concentrations in memory-related neural structures and thereby stimulated the different groups of GABA-releasing interneurons to change their function from filters to amplifiers, with a clear improvement of learning and memory abilities (Sun and Alkon 2001, 2002). Furthermore, in the absence of CAAs, the same stimuli were insufficient to trigger the previously mentioned synaptic switch (Sun and Alkon 2001), whereas CA inhibition suppressed it (Sun et al. 2001).

Such investigations (Sun and Alkon 2001, 2002; Sun et al. 2001) have shown that CA and its activators act as an effective gate-control that modulates the signal transfer through neural networks. CAAs can have pharmacological applications in the enhancement of synaptic efficacy, which in turn might constitute an excellent new approach for treating Alzheimer's disease and other conditions that need to achieve spatial learning and memory therapy (Sun and Alkon 2001; Scozzafava and Supuran 2002b). An even more selective approach has been further suggested: phenylalanine can be used in the majority of individuals who do not have a genetic lack of phenylalanine hydroxylase, whereas the development of more potent non-phenylalanine activators (such as derivatives of histamine or imidazole) can be applied to those with hydroxylase dysfunction (Sun and Alkon 2001).

CAAs derived from β-alanyl-histidine as lead molecule have been proposed very recently as precursors for another particular class of putative pharmacological agents (Scozzafava and Supuran 2002b). The clinical value of such agents will be related to the physiological roles of carnosine (i.e., antioxidant, free radical and aldehyde scavenger, or heavy metal ion complexing agent), with its prolonged metabolic survival *in vivo* and to the fact that its protective functions have already put this natural dipeptide into medical practice as a potential antiaging agent (Quinn et al. 1992; Hipkiss and Chana 1998; Hipkiss 1998).

12.5 CONCLUDING REMARKS

At present, the CA activation phenomenon seems to be clarified in great detail. Once the existence of CAAs was undoubtedly established, hundreds of novel compounds were synthesized and their activating potential evaluated against different CA isozymes and investigated for potential pharmacological applications. These compounds belong to well-defined structural classes, i.e., anions and amines, azoles, amino-azoles or amino acid derivatives. The largest majority of the reported strong activators are histamine or histidine derivatives, most of them possessing a peptidomimetic backbone. This focused approach has been motivated by (1) experimental evidence for efficient CA activation with biogenic amines and natural amino acids and (2) detailed x-ray analysis of the hCA II–histamine adduct, which has provided useful insights for the further design of novel CAAs.

Although belonging to diverse categories of organic derivatives, all the CAAs investigated share a common mechanism of action that is currently well understood at the molecular level by kinetic, spectroscopic and crystallographic studies.

Important SARs in all the CAA series (and also QSAR for a limited number of compounds) have been established and therefore some fundamental conclusions on an improved general strategy for CAAs design can be drawn. An efficient CAA molecule should incorporate two main structural elements: (1) at least one moiety able to shuttle protons between the enzyme active site and the external medium (generally an amino group or a NH imidazole ring), preferably with a pK_a value of 6 to 8, and (2) additional polar moieties that ensure a tight binding to the enzyme, of arylsulfonyl, carboxamido, ureido/thioureido or substituted pyridinium type. The substitution pattern of such a parent scaffold is an efficient tool for fine-tuning both the energy of binding of the activator molecule to the protein and the pK_a of the proton-transferring group, and hence for modulating the intensity of the activating properties. A certain type of *in vitro* or *ex vivo* isozyme specificity has also been identified for some of the various CAAs investigated.

CAAs are considered to be essential for investigating these widespread enzymes, which are deeply involved in major physiological processes in living organisms. Thus, CAAs are critical to (1) clarify some aspects of the catalytic mechanism, (2) explain the great differences in activity between the many CA isozymes, and (3) make it easier to understand the still-unknown physiological functions of certain isoforms. Different natural CAAs seen in the human body possess significant physiological roles. In addition, CAAs might also possess useful medical applications in the development of drugs and diagnostic agents for managing both the CA deficiency syndrome, for which no clinical treatment is available at present, and neurological disorders related to learning and memory impairments, such as Alzheimer's disease or aging. This field comprising potential pharmacological uses of CAAs is currently largely unexplored.

REFERENCES

Alberti, G., Bertini, I., Luchinat, C., and Scozzafava, A. (1981) A new class of inhibitors capable of binding both the acidic and alkaline forms of carbonic anhydrase. *Biochimica et Biophysica Acta* **668**, 16–26.

Alper, S.L. (2002) Genetic diseases of acid-base transporters. *Annual Review of Physiology* **64**, 899–923.

An, H., Tu, C., Duda, D., Montanez-Clemente, I., Math, K., Laipis, P.J., McKenna, R., and Silverman, D.N. (2002) Chemical rescue in catalysis by human carbonic anhydrases II and III. *Biochemistry* **41**, 3235–3242.

Barboiu, M., Supuran, C.T., Scozzafava, A., Briganti, F., Luca, C., Popescu, G., Cot, L., and Hovnanian, N. (1997) Supramolecular complexes of L-amino acids as efficient activators of the zinc enzyme carbonic anhydrase. *Liebigs Annalen der Chemie/Recueil* 1853–1859.

Bergenhem, N.C.H., Venta, P.J., Hopkins, P.J., Kim, H.J., and Tashian, R.E. (1992) Mutation creates an open reading frame within the 5' untranslated region of macaque erythrocyte carbonic anhydrase (CA) I mRNA that suppresses CA I expression and supports the scanning model for translation. *Proceedings of the National Academy of Sciences of the United States of America* **89**, 8798–8802.

Briganti, F., Iaconi, V., Mangani, S., Orioli, P., Scozzafava, A., Vernaglione, G., and Supuran, C.T. (1998) A ternary complex of carbonic anhydrase: X-ray crystallographic structure of the adduct of human carbonic anhydrase II with the activator phenylalanine and the inhibitor azide. *Inorganica Chimica Acta* **275/276**, 295–300.

Briganti, F., Mangani, S., Orioli, P., Scozzafava, A., Vernaglione, G., and Supuran, C.T. (1997a) Carbonic anhydrase activators: X-ray crystallographic and spectroscopic investigations for the interaction of isozymes I and II with histamine. *Biochemistry* **36**, 10384–10392.

Briganti, F., Scozzafava, A., and Supuran, C.T. (1997b) Novel carbonic anhydrase isozymes I, II and IV activators incorporating sulfonyl-histamino moieties. *Bioorganic and Medicinal Chemistry Letters* **9**, 2043–2048.

Catalan, J., Abboud, J.M., and Elguero, J. (1987) Basicity and acidity of azoles. *Advances in Heterocyclic Chemistry* **41**, 187–274.

Christianson, D.W., and Fierke, C.A. (1996) Carbonic anhydrase: Evolution of the zinc binding site by nature and by design. *Accounts of Chemical Research* **29**, 331–339.

Clare, B.W., and Supuran, C.T. (1994) Carbonic anhydrase activators. Part 3. Structure–activity correlations for a series of isozyme II activators. *Journal of Pharmaceutical Sciences* **83**, 768–773.

Coleman, J.E. (1998) Zinc enzymes. *Current Opinion in Chemical Biology* **2**, 222–234.

De Coursey, T.E. (2000) Hypothesis: Do voltage-gated H[+] channels in alveolar epithelial cells contribute to CO_2 elimination by the lung? *American Journal of Physiology — Cell Physiology* **278**, C1–C10.

Duda, D., Tu, C., Qian, M., Laipis, P.J., Agbandje-McKenna, M., Silverman, D.N., and McKenna, R. (2001) Structural and kinetic analysis of the chemical rescue of the proton transfer function of carbonic anhydrase II. *Biochemistry* **40**, 1741–1748.

Earnhardt, J.N., Wright, S.K., Qian, M., Tu, C., Laipis, P.J., Viola, R.E., and Silverman, D.N. (1999) Introduction of histidine analogs leads to enhanced proton transfer in carbonic anhydrase V. *Archives of Biochemistry and Biophysics* **361**, 264–270.

Engberg, P., Millqvist, E., Pohl, G., and Lindskog, S. (1985) Purification and some properties of carbonic anhydrase from bovine skeletal muscle. *Archives of Biochemistry and Biophysics* **241**, 628–638.

Eriksson, A.E., and Liljas, A. (1993) Refined structure of bovine carbonic anhydrase III at 2.0 Å resolution. *Proteins: Structure, Function and Genetics* **16**, 29–42.

Gao, J.M., Qiao, S., and Whitesides, G.M. (1995) Increasing binding constants of ligands to carbonic anhydrase by using "greasy tails." *Journal of Medicinal Chemistry* **38**, 2292–2301.

Hewett-Emmett, D. (2000) Evolution and distribution of the carbonic anhydrase gene families. In *The Carbonic Anhydrases — New Horizons,* Chegwidden, W.R., Edwards, Y., and Carter, N. (Eds.), Birkhäuser Verlag, Basel, pp. 29–76.

Hipkiss, A.R. (1998) Carnosine, a protective, anti-ageing peptide? *International Journal of Biochemistry and Cell Biology* **30**, 863–868.

Hipkiss, A.R., and Chana, H. (1998) Carnosine protects against methyl-glyoxal-mediated modifications. *Biochemical and Biophysical Research Communications* **248**, 28–32.

Igbo, I.N.A., Reigel, Jr. C.E., Greene, I.M., and Kenny, A.D. (1994) Effect of reserpine pretreatment on avian erythrocyte carbonic anhydrase activation by isoprotenerol. *Pharmacology* **49**, 112–120.

Ilies, M. (2002) Enzymes inhibitors and activators: Synthesis, characterization, and biological *activity*. Ph. D. Thesis, Polytechnic University, Bucharest, pp. 64–93.

Ilies, M., Banciu, M.D., Ilies, M.A., Scozzafava, A., Caproiu, M.T., and Supuran, C.T. (2002) Carbonic anhydrase activators: Design of high affinity isozymes I, II, and IV activators, incorporating tri-/tetrasubstituted-pyridinium-azole moieties. *Journal of Medicinal Chemistry* **45**, 504–510.

Ilies, M.A., Banciu, M.D., Ilies, M., Chiraleu, F., Briganti, F., Scozzafava, A., and Supuran, C.T. (1997) Carbonic anhydrase activators. Part 17. Synthesis and activation study of a series of 1-(1,2,4-triazole-(1H)-3-yl)-2,4,6-trisubstituted-pyridinium salts against isozymes I, II and IV. *European Journal of Medicinal Chemistry* **32**, 911–918.

Khalifah, R.G. (1971) The carbonic dioxide hydration activity of carbonic anhydrase. *Journal of Biological Chemistry* **246**, 2561–2573.

Leiner, M. (1940) Das Ferment Kohlensäureanhydrase in Tierkörper. *Naturwiss* **28**, 316–317.

Leiner, M., and Leiner, G. (1941) Die Aktivatoren der Kohlensäureanhydrase. *Naturwiss* **29**, 195–197.

Lindskog, S. (1997) Structure and mechanism of carbonic anhydrase. *Pharmacology and Therapeutics* **74**, 1–20.

Lindskog, S., and Silverman, D.N. (2000) The catalytic mechanism of mammalian carbonic anhydrases. In *The Carbonic Anhydrases — New Horizons,* Chegwidden, W.R., Edwards, Y., Carter, N. (Eds.), Birkhäuser Verlag, Basel, pp. 175–196 (and references cited therein).

Mallis, R.J., Poland, B.W., Chatterjee, T.K., Fisher, R.A., Darmawan, S., Honzatko, R.B., and Thomas, J.A. (2000) Crystal structure of S-glutathiolated carbonic anhydrase III. *FEBS Letters* **482**, 237–241.

Mann, T., and Keilin, D. (1940) Sulphanilamide as a specific inhibitor of carbonic anhydrase. *Nature* **146**, 164–165.

Mauksch, M., Bräuer, M., Wetson, J., and Anders, E. (2001) New insights into the mechanistic details of the carbonic anhydrase cycle as derived from the model system $[(NH_3)_3Zn(OH)^+/CO_2$: How does the H_2O/HCO_3^- replacement step occur? *Chembiochem: A European Journal of Chemical Biology* **2**, 190–198.

Meyer-Ruge, W., Iwangoff, P., and Reichlmeier, K. (1984) Neurochemical enzyme changes in Alzheimer's disease and Pick's disease. *Archives of Gerontology and Geriatrics* **3**, 161–165.

Murakami, H., and Sly, W.S. (1987) Purification and characterization of human salivary carbonic anhydrase. *Journal of Biological Chemistry* **262**, 1382–1388.

Nair, S.K., and Christianson, D.W. (1991) Unexpected pH-dependent conformation of His-64, the proton shuttle of carbonic anhydrase II. *Journal of American Chemical Society* **113**, 9455–9458.

Narumi, S., and Kanno, M. (1973) Effects of gastric acid stimulants on the activities of HCO_3^--stimulated, Mg^{2+}-dependent ATPase and carbonic anhydrase in rat gastric mucosa. *Biochimica et Biophysica Acta* **311**, 80–89.

Narumi, S., and Miyamoto, E. (1974) Activation and phosphorylation of carbonic anhydrase by adenosine-3′,5′-monophosphate-dependent protein kinase. *Biochimica et Biophysica Acta* **350**, 215–224.

Page, M.I., and Williams, A. (1989) Thermodynamics of enzymatic catalysis. In *Enzyme Mechanism*, Royal Society of Chemistry, London, pp. 1–13.

Paranawithana, S.R., Tu, C., Laipis, P.J., and Silverman, D.N. (1990) Enhancement of the catalytic activity of carbonic anhydrase III by phosphate. *Journal of Biological Chemistry* **265**, 22270–22274.

Parkes, J.L., and Coleman, P. (1989) Enhancement of carbonic anhydrase activity by erythrocyte membrane. *Archives of Biochemistry and Biophysics* **275**, 459–468.

Puscas, I. (1978) Histamine is a direct activator of gastric carbonic anhydrase. *Medicina Interna* **6**, 567–575.

Puscas, I., Coltau, M., Puscas, I.C., and Supuran, C.T. (1994) Carbonic anhydrase activators. Part 9. Kinetic parameters for isozyme II activation by histamine prove a noncompetitive type of interaction. *Revue Roumaine de Chimie* **39**, 114–119.

Puscas, I., Coltau, M., Supuran, C.T., Moisi, F., Lazoc, L., and Farcau, D. (1996) Carbonic anhydrase activators. Part 10. Kinetic studies of isozyme II activation with pentagastrin. *Revue Roumaine de Chimie* **41**, 159–163.

Puscas, I., Supuran, C.T., and Manole, G. (1990) Carbonic anhydrase inhibitors. Part 3. Relations between chemical structure of inhibitors and activators. *Revue Roumaine de Chimie* **35**, 683–689.

Quinn, P.J., Boldyrev, A.A., and Formazuyk, V.E. (1992) Carnosine: Its properties, functions and potential therapeutic applications. *Molecular Aspects of Medicine* **13**, 379–444.

Reithmeier, R.A. (2001) A membrane metabolon linking carbonic anydrase with chloride/bicarbonate anion exchangers. *Blood Cells, Molecules and Diseases* **27**, 85–89.

Roth, D.E., Venta, P.J., Tashian, R.E., and Sly; W.S. (1992) Molecular basis of human carbonic anhydrase II deficiency. *Proceedings of the National Academy of Sciences of the United States of America* **89**, 1804–1808.

Rousselle, A.-V., and Heymann, D. (2002) Osteoclastic acidification pathways during bone resorption. *Bone* **30**, 533–540.

Rowlett, R.S., Gargiulo, N.J., Santoli, F.A., Jackson, J.M., and Corbett, A.H. (1991) Activation and inhibition of bovine carbonic anhydrase III by dianions. *Journal of Biological Chemistry* **266**, 933–941.

Sato, S., Zhu, X.L., and Sly, W.S. (1990) Carbonic anhydrase isozymes IV and II in urinary membranes from carbonic anhydrase II-deficient patients. *Proceedings of the National Academy of Sciences of the United States of America*, **87**, 6073–6076.

Scozzafava, A., Briganti, F., Ilies, M.A., and Supuran, C.T. (2000a) Carbonic anhydrase inhibitors: Synthesis of membrane-impermeant low molecular weight sulfonamides possessing *in vivo* selectivity for the membrane-bound versus the cytosolic isozymes. *Journal of Medicinal Chemistry* **43**, 292–300.

Scozzafava, A., Iorga, B., and Supuran, C.T. (2000b) Carbonic anhydrase activators: synthesis of high affinity isozymes I, II and IV activators, derivatives of 4-(4-tosylureido-amino acyl)ethyl-1*H*-imidazole (histamine derivatives). *Journal of Enzyme Inhibition* **15**, 139–161.

Scozzafava, A., Menabuoni, L., Mincione, F., Briganti, F., Mincione, G., and Supuran, C.T. (1999) Carbonic anhydrase inhibitors: Synthesis of water-soluble, topically effective intraocular pressure lowering aromatic/heterocyclic sulfonamides containing cationic or anionic moieties — Is the tail more important than the ring? *Journal of Medicinal Chemistry* **42**, 2641–2650.

Scozzafava, A., and Supuran, C.T. (2000a) Carbonic anhydrase activators. Part 21. Novel activators of isozymes I, II and IV incorporating carboxamido and ureido histamine moieties. *European Journal of Medicinal Chemistry* **35**, 31–39.

Scozzafava, A., and Supuran, C.T. (2000b) Carbonic anhydrase activators. Part 24. High affinity isozymes I, II and IV activators, derivatives of 4-(4-chlorophenylsulfonylure-ido-amino acyl)ethyl-1*H*-imidazole. *European Journal of Pharmaceutical Sciences* **10**, 29–41.

Scozzafava, A., and Supuran, C.T. (2002a) Carbonic anhydrase activators: Human isozyme II is strongly activated by oligopeptides incorporating the carboxyterminal sequence of the bicarbonate anion exchanger AE1. *Bioorganic Medicinal Chemistry Letters* **12**, 1177–1180.

Scozzafava, A., and Supuran, C.T. (2002b) Carbonic anhydrase activators: high affinity isozymes I, II, and IV activators, incorporating a β-alanyl-histidine scaffold. *Journal of Medicinal Chemistry*, **45**, 284–291.

Shelton, J.B., and Chegwidden, W.R. (1988) Activation of carbonic anhydrase III by phosphate. *Biochemical Society Transactions* **16**, 853–854.

Shelton, J.B., and Chegwidden, W.R. (1996) Modification of carbonic anhydrase III activity by phosphates and phosphorylated metabolites. *Comparative Biochemistry and Physiology* **114A**, 283–289.

Silverman, D.N., Tu, C., and Wynns, G.C. (1978) Proton transfer between hemoglobin and the carbonic anhydrase active site. *Journal of Biological Chemistry* **253**, 2563–2567.

Sly, W.S. (1991) Carbonic anhydrase II deficiency syndrome: clinical delineation, interpretation and implications. In *The Carbonic Anhydrases*, Dodgson, S.J., Tashian, R.E., Gros, G., and Carter, N.D. (Eds.), Plenum Press, New York, pp. 183–196 (and references cited therein).

Sly, W.S., Hewett-Emmett, D., Whyte, M.P., Yu, Y-SL., and Tashian, R.E. (1983) Carbonic anhydrase II deficiency identified as the primary defect in the autosomal recessive syndrome of osteopetrosis with renal tubular acidosis and cerebral calcification. *Proceedings of the National Academy of Sciences of the United States of America* **80**, 2752–2756.

Sly, W.S., and Hu, P.Y. (1995a) Human carbonic anhydrases and carbonic anhydrase deficiencies. *Annual Review of Biochemistry* **64**, 375–401.

Sly, W.S., and Hu, P.Y. (1995b) The carbonic anhydrase II deficiency syndrome: osteopetrosis with renal tubular acidosis and cerebral calcification. In *The Metabolic Basis of Inherited Disease*, 7th ed., Beaudet, A., Sly, W.S., and Valle, D. (Eds.), McGraw-Hill, New York, pp. 4113–4124.

Soda, H., Yukizane, S., Koga, Y., Aramaki, S., and Kato, H. (1996) A point mutation in exon 3 (His 107 Tyr) in two unrelated Japanese patients with carbonic anhydrase II deficiency and central nervous system involvement. *Human Genetics* **97**, 435–437.

Stams, T., and Christianson, D.W. (2000) X-ray crystallographic studies of mammalian carbonic anhydrase isozymes. In *The Carbonic Anhydrases — New Horizons*, Chegwidden, W.R., Edwards, Y., and Carter, N. (Eds.), Birkhäuser Verlag, Basel, pp. 159–174 (and references cited therein).

Sun, M-K., and Alkon, D.L. (2001) Pharmacological enhancement of synaptic efficacy, spatial learning and memory through carbonic anhydrase activation in rats. *Journal of Pharmacology and Experimental Therapeutics* **297**, 961–967.

Sun, M.-K., and Alkon, D.L. (2002) Carbonic anhydrase gating of attention: memory therapy and enhancement. *Trends in Pharmacological Sciences* **23**, 83–89.

Sun, M.K., Dahl, D., and Alkon, D.L. (2001) Heterosynaptic transformation of GABAergic gating in the hippocampus and effects of carbonic anhydrase inhibition. *Journal of Pharmacology and Experimental Therapeutics* **296**, 811–817.

Supuran, C.T. (1991) Design of carbonic anhydrase inhibitors and activators. Ph. D. Thesis, Polytechnic University, Bucharest, pp. 1–148.

Supuran, C.T. (1992) Carbonic anhydrase activators. Part 4. A general mechanism of action for activators of isozymes I, II and III. *Revue Roumaine de Chimie* **37**, 411–421.

Supuran, C.T., and Balaban, A.T. (1994) Carbonic anhydrase activators. Part 8. pK_a-activation relationship in a series of amino acid derivatives activators of isozyme II. *Revue Roumaine de Chimie* **39**, 107–113.

Supuran, C.T., Balaban, A.T., Cabildo, P., Claramunt, R.M., Lavandera, J.L., and Elguero, J. (1993a) Carbonic anhydrase activators. Part 7. Isozyme II activation with bis-azolyl-methanes, -ethanes and related azoles. *Biological and Pharmaceutical Bulletin* **16**, 1236–1239.

Supuran, C.T., Barboiu, M., Luca, C., Pop, E., Brewster, M.E., and Dinculescu, A. (1996a) Carbonic anhydrase activators. Part 14. Synthesis of mono- and bis-pyridinium salt derivatives of 2-amino-5-(2-aminoethyl)- and 2-amino-5-(3-aminopropyl)-1,3,4-thiadiazole, and their interaction with isozyme II. *European Journal of Medicinal Chemistry* **31**, 597–606.

Supuran, C.T., Claramunt, R.M., Lavandera, J.L., and Elguero, J. (1996b) Carbonic anhydrase activators. Part 15. A kinetic study of the interaction of bovine isozyme II with pyrazoles, bis- and tris-azolylmethanes. *Biological and Pharmaceutical Bulletin* **19**, 1417–1422.

Supuran, C.T., Dinculescu, A., and Balaban, A.T. (1993b) Carbonic anhydrase activators. Part 5. CA II activation by 2,4,6-trisubstituted pyridinium cations with 1-(-aminoalkyl)side chains. *Revue Roumaine de Chimie* **38**, 343–349.

Supuran, C.T., Dinculescu, A., Manole, G., Savan, F., Puscas, I., and Balaban, A.T. (1991) Carbonic anhydrase activators. Part 2. Amino acids and some of their derivatives may act as potent enzyme activators. *Revue Roumaine de Chimie* **36**, 937–946.

Supuran, C.T., and Puscas, I. (1994) Carbonic anhydrase activators. In Carbonic Anhydrase and Modulation of Physiologic and Pathologic Processes in the Organism, Puscas, I. (Ed.), Helicon, Timisoara, Romania, pp. 113–145.

Supuran, C.T., and Scozzafava, A. (1999) Carbonic anhydrase activators. Part 23. Aminoacyl/dipeptidyl histamine derivatives bind with high affinity to isozymes I, II and IV and act as efficient activators. *Bioorganic Medicinal Chemistry* **7**, 2397–2406.

Supuran, C.T., and Scozzafava, A. (2000a) Activation of carbonic anhydrase isozymes. In *The Carbonic Anhydrases — New Horizons*, Chegwidden, W.R., Edwards, Y., and Carter, N. (Eds.), Birkhäuser Verlag, Basel, pp. 197–219.

Supuran, C.T., and Scozzafava, A. (2000b) Carbonic anhydrase inhibitors and their therapeutic potential. *Expert Opinion on Therapeutic Patents* **10**, 575–600.

Supuran, C.T., and Scozzafava, A. (2000c) Carbonic anhydrase activators: Synthesis of high affinity isozymes I, II and IV activators, derivatives of 4-(arylsulfonylureido-amino acyl)ethyl-1H-imidazole. *Journal of Enzyme Inhibition* **15**, 471–486.

Supuran, C.T., and Scozzafava, A. (2001) Carbonic anhydrase inhibitors. *Current Medicinal Chemistry — Immunologic, Endocrine and Metabolic Agents* **1**, 61–97.

Supuran, C.T., Scozzafava, A., and Casini, A. (2003) Carbonic anhydrase inhibitors. *Medicinal Research Reviews* **23**, 146–189.

Supuran, C.T., Scozzafava, A., Ilies, M.A., Iorga, B., Cristea, T., Briganti, F., Chiraleu, F., and Banciu, M.D. (1998) Carbonic anhydrase inhibitors. Part 53. Synthesis of substituted-pyridinium derivatives of aromatic sulfonamides: The first non polymeric membrane-impermeable inhibitors with selectivity for isozyme IV. *European Journal of Medicinal Chemistry* **33**, 577–594.

Tanis, R.J., Ferrell, R.E., and Tashian, R.E. (1973) Substitution of lysine for threonine at position 100 in human carbonic anhydrase Id Michigan. *Biochemical and Biophysical Research Communications* **51**, 699–703.

Tu, C., Couton, J.M., Van Heeke, G., Richards, N.G.J., and Silverman, D.N. (1993) Kinetic analysis of a mutant (His[107] Tyr) responsible for human carbonic anhydrase deficiency syndrome. *Journal of Biological Chemistry* **268**, 4775–4779.

Tu, C., Paranawithana, S.R., Jewell, D.A., Tanhauser, S.M., LoGrasso, P.V., Wynns, G.C., Laipis, P.J., and Silverman, D.N. (1990) Buffer enhancement of proton transfer in catalysis of human carbonic anhydrase III. *Biochemistry* **29**, 6400–6405.

Tu, C.K., Sanyal, G., Wynns, G.C., and Silverman, D.N. (1983) The pH dependence of the hydration of carbon dioxide catalyzed by carbonic anhydrase III from skeletal muscle of the cat. Steady state and equilibrium studies. *Journal of Biological Chemistry* **258**, 8867–8871.

Tu, C., Wynns, G.C., and Silverman, D.N. (1981) Inhibition by cupric ions of oxygen-18 exchange catalyzed by human carbonic anhydrase II. Relation to the interaction between carbonic anhydrase and hemoglobin. *Journal of Biological Chemistry* **256**, 9466–9471.

Venta, P.J. (2000) Inherited deficiencies and activity variants of the mammalian carbonic anhydrases. In *The Carbonic Anhydrases — New Horizons*, Chegwidden, W.R., Edwards, Y., and Carter, N. (Eds.), Birkhäuser Verlag, Basel, pp. 403–412.

Venta, P.J., Welty, R.J., Johnson, T.M., Sly, W.S., and Tashian, R.E. (1991) Carbonic anhydrase II deficiency syndrome in a Belgian family is caused by a point mutation at an invariant histidine residue (107 His Tyr): Complete structure of the normal human CA II gene. *American Journal of Human Genetics* **49**, 1082–1090.

Vince, J.W., and Reithmeier, R.A. (2000) Identification of the carbonic anhydrase II binding site in the Cl$^-$/HCO$_3^-$ anion exchanger AE1. *Biochemistry* **39**, 5527–5533.

Index

A

Acatalytic CAs, 25–43
  evolutionary analysis and subcellular
      localization of CA-RPs, 30
  expression of CA-RPs, 31–35
      developmental expression, 34
      immunohistochemical localization in
          brain, 32–34
      mRNA expression, 31–32
      oncogenic expression, 35
  functional prosecution of CA-RPs, 35–38
      interaction of CA II with bicarbonate
          transporter, 37–38
      ligand binding to CA-RP domain of
          RPTPβ, 36–37
  future direction of functional analysis of
      CA-RPs, 38–39
  molecular properties of CA-RPs, 26–29
      active-site residues and catalytic activity,
          28–29
      cDNA cloning, 26–28
Acetazolamide, 11, 37, 47, 52, 67, 121, 135, 272
  acidic character of, 185
  behavior exhibited by, 199
  chelating properties of, 183
  complexes, 186
  coordination ways of, 191
  donor atoms, 185
  –hCAII complex, 198
  high-altitude illness and, 292
  IR bands for, 188
  ligand behavior of, 195
  test, 137
  treatment of edema using, 289
  treatment of retinal detachment with, 251
  use of as antiglaucoma drug, 243
  use of as diuretic drug, 184
Acetobacterium woodi, 17
Acid/base homeostasis, isozymes involved in, 19
Adrenaline, activation of CAs by, 343
AE, see Anion exchanger
Affinity grid, calculation of, 54
AIDS, treatment of with amprenavir, 68
Alcon Laboratories, 109, 246
Aldehyde hydration, 1
Alimentary tract

CAIs in, 287
  distribution of CA isozymes in, 285
  role of CAs in, 284
Aliphatic sulfonamides, 134, 135, 150, 172
Altitude sickness, 150
Alzheimer's disease, 318, 345
Amber atomic charges, 53
AMBER molecular mechanics force field, 174
Ambidentate, behavior of acetazolamide as, 190
Amino acid derivatives, CA activation with, 338
Aminothiadiazole sulfonamides, 167
Amprenavir, 68
Anabaena variabilis, 17
Aniline, 216
Anion exchanger (AE), 37
Anionic inhibitors
  apparent affinity constants of, 212
  CA inhibition mechanism by, 7
  inhibition constants of, 213
Anorexia, 247
Anthranilate, 216
Antibacterial drug, 183
Antibiotics, possibility of developing
          CA-inhibitor-based, 16
Antibodies
  anti-CA, 308
  monoclonal, 32
Anticancer agent
  acidic pH and uptake of, 270
  novel, 125
Anticonvulsant drugs, 86, 132
Antiepileptics, 19, 51, 131
Antiglaucoma agents, 19, 93
Antiglaucoma CAIs, design of, 68
Antiglaucoma drugs, 11, 67, 70
Antiglaucoma sulfonamides, 122
Antitumor sulfonamides, 125
Apparent affinity constant, 211, 214
Aquatic photosynthetic organisms, 18
Aromatic sulfonamides, 69
  benzenoid, 151
  CA II inhibition with, 74
  Schiff bases of, 75, 121
Arylsulfonyl isocyanates, reaction of amino
          acids/dipeptides with, 336
Atom–atom polarizabilities, 165, 174
Autacoids, 343

AutoDock, 52, 54, 57, 58
Autoimmune diseases, CAs and pathogenesis of,
    308

B

Baeyer–Piccard reaction conditions, 328
Benzene-carboxamide derivatives, inhibition data
    for, 78–79
Benzene derivatives, CA inhibitory activity in,
    162
Benzenedisulfonamides, 157–159
Benzenesulfonamide, 70, 71, 152
    connectivity index QSAR, 165
    derivatives, electronic correlates, 151
Benzenoid aromatics, 151
Benzimidazoles, 80
Benzolamide
    behavior exhibited by, 199
    chelating properties of, 183
    complexes, 195, 196
    coordination behavior of, 195
    dideprotonated, 200
Benzothiadiazine diuretics, 67, 89
Benzothiazole, 80
Bicarbonate
    dehydration, hCA III activity in, 343
    secretion, 244
    transport activity, push–pull mechanism, 268
    transporter
        CA-RPs and, 38
        interaction of CA II with, 37
Bicyclic sulfonamides, 95
Bidentate ligands, sulfonamides behaving as, 200
Binding energy, 167–168, 174
Biosynthetic reactions, 2
Bis-sulfonamides, 89
Bitopic protein, CA XII as, 265
Bladder cancer, CA IX in, 262
Blindness, 251
Bone-targeted sulfonamides, development of, 137
Brain
    distribution of CA isozymes in, 290
    enzyme, compounds acting as specific
        inhibitors for, 70
    tumors, neurons surrounding, 290
Branhamella catarrhalis, 16
Breast cancer, 35
Brinzolamide, 11, 47, 52, 56, 68, 247, 272
    common problem for, 111
    congeners, 109
    substitution of, 112
α-Bromoketones, 88
Burns, treatment of, 202

C

CA, see Carbonic anhydrase
α-CA, 2
    reactions catalyzed by, 4
    tetrahedral coordination geometry, 15
    x-ray crystal structures, 3
    zinc hydroxide mechanism for, 13
β-CA, 2
    Pisum sativum, 14
    x-ray structures of, 12
γ-CA, 2
CA I
    affinity of, 118
    expression of in skin, 305
    Michigan 1, 9
    physiological function of, 8
    slow-reacting isozyme, 92
CA II
    activation data, 331
    –activator adducts, 317
    activity, external peptides enhancing, 268
    –bicarbonate complex, structure of, 226
    –bromide adduct, 225
    diffusion control, 5
    expression of in skin, 305
    inhibition by thiadiazoles, 173
    interaction of with bicarbonate transporter, 37
    monosubstituted benzenesulfonamide
        inhibitors of, 151
    proton transfer process, 5
    rapid-reacting isozyme, 92
    serum antibodies reactive to, 308
    –topiramate
        complex, 293
        crystallographic study of, 230
CA III
    catalytic efficiency for CO₂ hydration, 342
    inhibition of, 121
CA IV
    active site of, 120
    inhibition of by thiadiazolines, 173
    inhibitory properties, 132
    membrane-anchored, 212
    -specific inhibitors, 330
CA V, activators of histidine analogs, 342
CA IX, see also Cancer-related anhydrase
        isozymes
    expression pattern, 258
    gastric carcinomas and, 259
    gastric mucosa, 122
    involvement of in cell adhesion, 267
    as marker of tumor hypoxia, 255
    original detection of, 257

protein, expression of in MDCK cells, 269
spectrum of cancers expressing, 259
CA XII, identification of, 263, *see also* Cancer-
related anhydrase isozymes
CA XIV, expressions of in neurons, 291
CAAs, *see* Carbonic anhydrase activators
Cadmium CA, 16
CAI, *see* Carbonic anhydrase inhibitor
Callinectes sapidus, 122
CAM plants, *see* Crassulacean acid metabolism
plants
Cancer, *see also specific types*
development of, 283
emerging role of CAs in, 271
genes, control of, 260
hallmark of, 256
invasion, reaction-diffusion model of, 270
Cancer-related anhydrase isozymes, 255–281
CA inhibition as approach to anticancer
therapy, 270–272
CA IX as marker of tumor hypoxia, 260–263
catalytic properties and proposed roles of CA
IX and CA XII, 265–269
identification, distribution and regulation of
CA XII, 263–265
identification, structure and distribution of CA
IX, 257–259
influence of extracellular pH on tumor
progression and possible implication
of CA activity, 270
inhibition profile of CA IX, 272–273
regulation of CA IX expression, 259–260
CAP, *see* Carbonic anhydrase-like protein
Carbonic anhydrase (CA), 2, 25, 45, 149
attractiveness of research on, 283
bicarbonate interaction with, 226
binding of sugar sulfamates to, 51
cadmium, 16
catalytic mechanism of, 210
cytosolic forms, 2
deficiency syndrome, 344, 346
emerging role of in cancer, 271
extracellular, 30
first epithelial tissue found to contain, 284
γ-class, 15
gene families, 2
genetic deficiency, 318
Helicobacter pylori, 17
–histamine complex dissociation, 322
inhibition, 209
data
benzothiophene-2-sulfonamide
derivatives, 98
4-isothiocyanatobenzenesulfonamide,
82–83

docking results for, 55
drugs developed for, 272
mechanism, 7
inhibitory activity(ies)
benzene derivatives, 162
chemical structure and, 81
frontier orbital energies and, 162
monohalogenoderivatives, 135
intracellular, 30
involvement of in secretion of electrolytes, 19
isoforms, 15, 303
isolated from *N. gonorrhoeae* (NGCA), 17
isozyme(s)
difference between cancer-related, 267
distribution of in brain, 290
expression of in human skin, 303
pharmaceutical research and, 46
point mutation within active site of, 10
secreted, 2
susceptibility to anions, 212
-like protein (CAP), 309
mechanistic studies of, 183
membrane-associated, 121, 288
membrane-bound isozymes, 2
mitochondrial form, 2
poor substrate for, 222
sulfanilamide as specific inhibitor of, 11
Thalassiosira weissflogii, 16
unpurified blood, 90
unsubstituted aromatic sulfonamides and, 67
zonisamide inhibition of, 307
Carbonic anhydrase activators (CAAs), 317–352
complex, first x-ray structure obtained for, 319
connections of CAAs with physiopathology,
343–346
first membrane-impermeant, 330
histamine-derived, 336
mechanism of action of CAAs, 319–325
prototypical, 329
QSAR of, 328
research, 318
synthesis of, 341
types, 325–342
amine derivatives, 328–329
amino acid derivatives, 338–342
anions, 342–343
azole derivatives, 329–333
background, 326–328
histamine derivatives, 333–338
Carbonic anhydrase inhibitor (CAI), 6, 45, 67,
270, 318
alimentary tract, 287
anchoring groups for designing, 230
anticonvulsant activity, 68
antiglaucoma, 68, 136

antitumor properties, 68
availability of, 283
benzothiadiazine diuretics and, 67
bicyclic sulfonamides as, 95
binding to erythrocytes of, 74
classification of, 213–215
complexes, generation of x-ray structures of, 46
compounds effective as, 91
design of membrane-impermeant, 122
future prospects of, 136
gastric mucosa CA-specific, 132
heterocyclic sulfonamide, 183
hydroxamates as, 229
kidney, 289
nervous system, 291
skin cancer, 310
sulfonamide, 10, 202
systemic sulfonamide, 11
topical antiglaucoma, 229
treatment of sleep apnea with, 47
unsubstituted aromatic/heterocyclic sulfonamides acting as, 184
α-Carbonic anhydrase isoforms, isolation of in higher vertebrates, 1
Carbonic anhydrase-related proteins (CA-RPs), 2, 25, *see also* Acatalytic CAs
cDNA sequences, 28
developmental expression of, 34
gene knockout of, 39
immunohistochemical localization of, 33
molecular properties of, 27
mRNA expression, 26, 31, 32
protein–protein interactions, 36
CA-RP, *see* Carbonic anhydrase-related proteins
CA-RP VIII cDNA, encoding of amino acid residues, 27
CA-RP X
difference of expression in humans and mice, 34
expression of in myelin sheath, 32
CA-RP XI expression, murine, 34
Catalytic and inhibition mechanisms, distribution and physiological roles, 1–23
catalytic and inhibition mechanisms of CAs, 3–16
α-CA, 3–12
β-CA, 12–15
γ-CA, 15–16
cadmium CA, 16
distribution of CAs, 16–18
physiological functions of CAs, 18–20
Catecholamines, activation of CAs by, 343

cDNA
cloning, 26
MN, isolation and sequence analysis of, 257
Cell
adhesion, involvement of CA IX in, 267
–cell adhesion disruption, 263
invasion rate, 271
Cerebrospinal fluid (CSF), 289
formation of, 289
role of CA activity in production of, 291
shunting, 291
Cerebrovasodilator agents, 86, 132
Cervical tumors, CA IX expression, 262
Chenodeoxycholic acid, 133
Cherry hemangioma, 307, 308
Chlamydomonas reinhardtii, 18
Chloroform–buffer partition coefficients, 87
N-Chlorosuccinimide (NCS), 97, 100
Chlorosulfonyl isocyanate, 88
Cholic acid, 133
Ciliary processes enzyme, 117
CNDO approximation, *see* Complete neglect of differential overlap approximation
CO_2
fixation, 18
homeostasis, 2
hydration
activity, 2, 25
CA III catalytic efficiency for, 342
reversible, 318
Cognitive disorders, use of CAAs as therapeutic targets in, 345
Colon, role of CAs in, 285
Colorectal tumors, aberrant expression of CA IX in, 259
CoMFA, *see* Comparative Molecular Field Analysis
Comparative Molecular Field Analysis (CoMFA), 50
Comparative Molecular Similarity Indices Analysis (CoMSIA), 50
Complete neglect of differential overlap (CNDO) approximation, 328
Computer-aided drug design, 133
CoMSIA, *see* Comparative Molecular Similarity Indices Analysis
Confocal fluorescence microscopy, 30
Congestive heart failure, treatment of edema induced by, 289
Contact allergy, 247
Contactin, 25
Cornea
inadequate drug penetration through, 246
permeability, 95

COSMO technique, 153
Crassulacean acid metabolism (CAM) plants, 18
Creatine, 113
CSF, *see* Cerebrospinal fluid
Cyanobacteria, 18
Cyclic unsaturated anhydrides, 103
Cyclodextrins, association of ophthalmologic
    drugs with,249
Cystoid macular edema, 251
Cytoplasmic proteins, 30

D

Dementia, 247
Deoxycholic acid, 133
Depression, 247
Dermatology, carbonic anhydrase inhibitors in,
    303–315
    CAIs and skin cancer, 310–311
    CAs in skin, 305–306
    role of CAs in skin, 306–307
    role of CAs in skin diseases, 307–309
    role of CAs in skin tumors, 309–310
Dichlorophenamide, 11, 47, 68, 90, 272
    treatment of edema using, 289
    use of as antiglaucoma drug, 243
Dideprotonated species, 185
Diethylenetriaminopentaacetic acid dianhydride,
    336
Diisopropylcarbodiimide (DIPC), 335
DIPC, *see* Diisopropylcarbodiimide
Dipole moment, 165, 175
Disulfonamides, activities and structures of, 161
Diuretics, 19, 47, 89, 184
Docking
    results, CA inhibitors, 55
    studies, 58
Dopamine, activation of CAs by, 343
Dorzolamide, 11, 47, 52, 68, 104, 247, 272
Drug design(s)
    aims of CAA, 325
    computer-aided, 133
    docking approaches, 51
    tools for, 49
Drug diagnostic tools, 319
Duodenal ulcers, 20

E

EI, *see* Enzyme–inhibitor
Electrolyte(s)
    involvement of CAs in secretion of, 19
    secretion, 2

Electron
    -attracting groups, 121
    microscopy, 307
    transport processes, Photosystem II-mediated,
        18
Electronic spectroscopy, 216
Electrostatic potential, 163
Enzyme
    binding mode of cyanamide to, 223
    distortion energies, 165
    excessive activity, 318, 319
    inhibitors, antimalarial effect of, 18
Enzyme–activator
    adducts, 334, 340
    complex, stability of, 325, 339
Enzyme–inhibitor (EI)
    adducts, x-ray diffraction experiments, 215
    complexes, 134
    interaction, 211
Epileptic seizures, treatment of, 292
Erythrocyte(s)
    CA inhibition, 73
    selective radiolabeling of, 70
Erythropoesis, 260
Escherichia coli, 13, 17, 28
Essential tremor, treatment of with CAIs, 47
Ethenesulfonylchloride, 88
Ethoxozolamide (HEZA), 11, 47, 68, 200, 272
    treatment of edema using, 289
    use of as antiglaucoma drug, 243
Extracellular pH, influence of on tumor
    progression, 270

F

Frontier orbital energies, CA inhibitory activity
    and, 162

G

GABA, 113
    receptor–channel complex, 345
    responses, bicarbonate-ion-induced, 291
Gallbladder, role of CAs in, 286
Gasteiger–Marsili charges, 54
Gastric acid
    production, 133
    secretion, inhibition of, 287
Gastric carcinomas, CA IX and, 259
Gastric ulcers, 20
Gastroduodenal bicarbonate, 284
Gastroduodenal ulcers, treatment of, 47, 83, 149

Gastroenterology, neurology, and nephrology, role
        of carbonic anhydrase in, 283–301
    CAIs in alimentary tract, 287
    CAIs in kidney, 289
    CAIs in nervous system, 291–293
    role of CAs in alimentary tract, 284–287
        colon, 285–286
        pancreas, liver and gallbladder, 286–287
        upper alimentary tract, 284–285
    role of CAs in kidney, 287–289
    role of CAs in nervous system, 289–291
Gastrointestinal carcinoma, 35
Glaucoma
    animal models of, 202, 250
    attacks, control of acute, 245
    management of, 83
    topically active inhibitors in, 84
    treatment of, 47, 149
Gluconeogenesis, 2
Glucose transport, 260
Glutamic acid residues, 27
Glutathione-S-transferase (GST), 39
Glycolysis, 260, 270
Glycosylphosphatidylinositol (GPI), 288
GPI, see Glycosylphosphatidylinositol
Green fluorescent protein, 30
Grid-based automated docking program, 52
GST, see Glutathione-S-transferase

H

hCA II–histamine
    adduct, x-ray analysis of, 346
    complex, 324
    interaction, 334
hCA II–urea adduct, 223
Helicobacter pylori, 17
Heteroaromatics, 165, 168
Heterocyclic sulfonamides, 80, 81, 164
HEZA, see Ethoxozolamide
HIF, see Hypoxia-inducible transcription factor
High-altitude illness, drug for managing, 292
High-ceiling diuretics, 89, 184
High-throughput screening (HTS), 50
Histamine
    activation, 333, 343
    CAA behavior of, 326
Histidine cluster, 5
HIV infection, treatment of with amprenavir, 68
HRE, see Hypoxia response element
HTS, see High-throughput screening

Huckel calculations, 193
Hydrocephalus, treatment of with CAIs, 47
Hydrochloric acid production, 258
Hydrochlorothiazide, 197
Hydrogen bond donor, 162, 218, 221
Hydrolytic processes, 1
Hydroxamates, 126
Hypoxia-inducible transcription factor (HIF), 260
    -binding sequence, 265
    target, CA IX as, 260
    transcription complex, 260
Hypoxia response element (HRE), 260, 261, 264
Hypoxic markers, secreted, 263

I

Imidazole, 216, 343
Indisulam, 125
Inhibition
    kinetics, 210
    mechanism, 211
Inhibitors, see also Carbonic anhydrase inhibitors
    anionic, 7
    nonsulfonamide, 228
    sulfonamide, 7
In situ hybridization, 32
Intramolecular proton transfer, inhibition of, 229
Intraocular pressure (IOP), 77, 93
    CAIs and, 93
    drugs effective in reducing, 246
    lowering of, 103, 115, 118, 202
    prolonged reduction of, 243
IOP, see Intraocular pressure
Isozyme(s)
    bovine, 202
    CA inhibition against, 72
    cytosolic, 76
    gene, missing, 344
    inhibition of, 120
    involvement of in acid/base homeostasis, 19
    membrane-anchored, 30
    mitochondrial, 213
    physiological function of, 19
    QSAR differences, 178
    rapid-reacting, 92
    secretory, 30
    slow-reacting, 92
    -specific inhibitors, 68, 118
    specificity, differences in, 333
    transmembrane, 30, 35

K

Ketone(s)
    activities found for, 105
    conversion of lithioderivatives to, 99
Kidney
    CAIs in, 289
    role of CAs in, 287
Knowledge-based pair potentials, 58

L

Lamarckian genetic algorithm, 55
Ligand(s)
    behavior, acetazolamide, 185, 189, 195
    bidentate, sulfonamides behaving as, 200
    deprotonated methazolamide, 194
    monodentate, 192–193, 221
    oxygen-complexing, 225
    placement, 53
    screening approaches, structure-based, 50
Lipid solubility, 115
Lipophilicity, 160, 174
Lithioderivatives, conversion of to ketones, 99
Lithocholic acid, 133
Liver, role of CAs in, 286
Local dipole index, 179
Lung cancer, 35, 262

M

Macular degeneration, CAIs in, 251
Macular edema, CAIs in, 251
Maleic anhydride, 103
Matrix metalloproteinases (MMPs), 68, 130, 231,
        232
Membrane-associated CAs, selective inhibitors
        for, 121
Memory therapy, 318, 345
Merck, thienothiopyran sulfonamides developed
        by, 110
Merck group, 95, 103
Merck, Sharp & Dome, 107
Mercury-containing compounds, 229
Metabolic dysfunctions, 137
Metal complexes of heterocyclic sulfonamides,
        183–207
    acetazolamide complexes, 184–192
    applications of metal complexes of
        sulfonamides in therapy, 202–203
    benzolamide complexes, 195–200

complexes containing other sulfonamide
        CAIs, 200–202
    methazolamide complexes, 192–195
Metalloprotease inhibitors, 68
Metalloprotein inhibition, 209
Methanesulfonamide, 135
Methanobacterium thermoautotrophicum, 12, 13
Methanosarcina thermophila, 15
Methazolamide, 11, 47, 84, 125, 272
    chelating properties of, 183
    complexes, 192
    treatment of edema using, 289
    use of as antiglaucoma drug, 243
Methyloxycarbonylazide, 89
Microvascular density (MVD), 262
Mitochondrial isozyme, 213
Mitoxantrone, 270
MMPs, *see* Matrix metalloproteinases
MN antigen, 257
Monoclonal antibodies, 32
Monohalogenoderivatives, CA inhibitory
        activities, 135
Motion sickness, 150
Mountain sickness, prevention of acute, 307
Mulliken population analysis, 154
Multiple binding modes observed in x-ray
        structures of carbonic anhydrase
        inhibitor complexes, 45–65
    binding of sugar sulfamates to CAs, 51
    docking results for CA inhibitors, 55
    observed docking conformations of
        topiramate and RWJ-37947, 57
Mutation rate, 55
MVD, *see* Microvascular density

N

β-Naphthyl acetate hydrolyses, activation
        mechanism for, 9
Nasopharyngeal carcinoma, hypoxic profile, 262
National Cancer Institute, 271
National Institutes of Health (NIH), 125, 310
NCS, see *N*-Chlorosuccinimide
Neisseria, 16, 17
Neoangiogenesis, 256, 260, 261
Nephrolithiasis, 247
Nephrology, *see* Gastroenterology, neurology, and
        nephrology, role of carbonic
        anhydrase in
Nervous system
    CAIs in, 291
    role of CAs in, 289
Neurological disorders, 20, 47, 149

Neurology, *see* Gastroenterology, neurology, and nephrology, role of carbonic anhydrase in

NGCA, *see* CA isolated from *N. gonorrhoeae*

NIH, *see* National Institutes of Health

Nipecotic acid, 77

Nitrogen, nonprotonated, 218

NMR spectroscopy, 51

Nonphotosynthesizing prokaryotes, 19

Nonsmall cell lung carcinomas, CA IX and, 259

Nonsulfonamide carbonic anhydrase inhibitors, 209–241

    inhibition of CAs by anions, 210–228

    other types of nonsulfonamide inhibitors, 228–232

        inhibitors of proton shuttle, 228–229

        organic sulfamates and hydroxamates as CAIs, 229–233

Noradrenaline, activation of CAs by, 343

Northern blot analysis, mRNA expression of CA-RPs by, 31

Nuclear magnetic resonance spectroscopy, 216

Nucleotide synthesis, 125

O

Ocular hypertension, 244

Ocular pathologies, CAIs in, 251

Ocular pigment, 103

Olygohydrosis, 307

Ophthalmology, clinical applications of carbonic anhydrase inhibitors in, 243–254

    CAIs in macular edema, macular degeneration and related ocular pathologies, 251

    combination antiglaucoma therapy of CAIs with other pharmacological agents, 247–250

    sulfonamides in treatment of glaucoma, 244–247

Orbital magnitude, 163

Organic sulfamates, 229

Osteoclasts, bone resorption in, 19

Osteoporosis, 20

Oxidation product, 91

Oxygen-complexing ligands, 225

P

Paclitaxel, 270

Pallas software, 156

Pancreas, role of CAs in, 286

Panniculitis, 305

Parkinson's disease, treatment of with CAIs, 47

PCC, *see* Pyridinium chlorochromate

PDB, *see* Protein Data Bank

Pearson correlation coefficient, 163

PEP carboxylase, *see* Phosphoenolpyruvate carboxylase

Peptic ulcers, use of CAIs for therapy of, 287

Perfluoroderivatives, 113, 135

PFMZ, van der Waals contacts with, 10

Pharmaceutical research, CA isozymes and, 46

Phosgene, 88

Phosphoenolpyruvate (PEP) carboxylase, 18

Phosphonates, 126

Photosystem II-mediated electron transport processes, 18

Pisum sativum, 12, 14

Plasmodium, 17, 18, 138

Polarizability, 159, 160, 173

Polyaminopolycarboxylic acid tails, 113

Polyhalogenoacetamido-thiadiazoles, 93

Polysaccharide, pseudoplastic properties, 249

Porphyridium purpureum, 12, 13, 14

Prokaryotes, nonphotosynthesizing, 19

Prontosil, 183

Protein(s)

    bitopic, CA XII as, 265

    CA IX, expression of in MDCK cells, 269

    CA XII, 264

    carbonic anhydrase-like, 309

    cytoplasmic, 30

    glycosylphosphatidylinositol-linked membrane-associated, 288

    green fluorescent, 30

    kinase, cAMP-dependent, 326

    –ligand interaction geometries, 51

    neuronal cell-surface, 25

    –protein interactions, 35–36, 267

    recombinant

        CA-RP VIII, 29

        Escherichia coli, 28

    tumor-associated, 35

    tyrosine phosphatases (PTPs), 26, 36

    von Hippel–Lindau tumor suppressor, 259, 261

    zinc-bound water in native, 220

Protein Data Bank (PDB), 46, 52

Proteus, 17

Proton

    acceptance, 331–332

    release channel, rapid, 323

    shuttle, 210

        inhibitors of, 228

        moiety, 338

        residue, 120, 319

    transfer

pathway, active-site zinc, 15
process, CA II, 5
reaction, 4
Pseudomonas, 16
PTPs, *see* Protein tyrosine phosphatases
Pyrazole, activating properties, 343
Pyrrole-carboxamido-containing compounds, 118

Q

QSAR, *see* Quantitative structure–activity
        relationship
QSAR studies of sulfonamide carbonic anhydrase
        inhibitors, 149–182
   aliphatics, 175–178
   benzenoid aromatics, 151–166
      direct binding studies, 165–166
      electronic correlates, 151–160
      lipophilicity, 160–162
      quantum measures other than charge,
         162–163
      steric effects, area, volume and
         polarizability, 160
      topological indices, 163–165
   heteroaromatics, 166–175
      charge and dipole moment, 166–174
      direct binding, 174–175
      lipophilicity, 174
      quantum measures other than charge, 174
      steric variables and polarizability, 174
      topological indices, 174
Quantitative structure–activity relationship
        (QSAR), 49, 150, 327, 339
Quantum similarity, definition of, 163

R

RCCs, *see* Renal cell carcinomas
Receptor protein tyrosine phosphatases (RPTP),
        26, 256–257
   ligand binding to CA-RP domain of, 36
   molecular structure of, 37
Recombinant protein
   CA-RP VIII, 29
   Escherichia coli, 28
Red cell isozyme, 8
Renal acidification, impaired, 288
Renal cell carcinomas (RCCs), 256, 260
Renal physiology, role of sulfonamide CAIs in
        understanding, 137
Retinitis pigmentosa, 251
Rhodospirillum rubrum, 17
Ritter reaction, 105, 109

RPTP, *see* Receptor protein tyrosine phosphatases
RT-PCR, 34, 289

S

Saliva, CA present in, 247
Salivary glands, most prominent isozyme of, 284
Sarcosine, 113
SARs, *see* Structure–activity relationships
Schiff bases, 101, 102, 104
   aromatic sulfonamides, 121
   sulfanilamide, 155
SDS-PAGE, 28
Self-similarity, definition of, 163
Serotonin, activation of CAs by, 343
Serratia, 17
Sjögren's syndrome, 308
Skin, *see also* Dermatology, carbonic anhydrase
        inhibitors in
   cancer, CAIs and, 310
   diseases, role of CAs in, 307
   expression of CA isozymes in, 303, 305
   role of CAs in, 306
   tumors, role of CAs in, 309
Sleep apnea, treatment of with CAIs, 47
Sodium bicarbonate cotransporter, 38
Solvation energy, 159
Staphylococcus, 16
Steric effects, 160
Steric variables, 174
Streptococcus, 17
Structural genomics, progress in, 50
Structure–activity relationships (SARs), 327, 330,
        338, 346
Sugar sulfamates, binding of to CAs, 51
Sulfamates, organic, 229
Sulfanilamide
   derivatives, synthesis of, 183–184
   pharmacological agents developed from, 69
   reaction of with potassium thiocyanate, 91
   Schiff bases, 155
Sulfides, oxidation of, 71
Sulfonamide(s)
   aliphatic, 134, 135, 150, 172
   aminothiadiazole, 167
   anticonvulsant effects of, 49
   antiepileptic, 131
   antiglaucoma, 117, 122
   antitumor, 125
   aromatic, 69, 121
   -avid enzyme, 256
   bicyclic, 95
   bone-targeted, 137
   CAI(s)

antifungal properties, 202
association of with cyclodextrins, 250
CA active site and, 10
cationic, 122
charges, 156
deprotonated, 7
derivatives, inhibition data for, 107, 111
first clinically used topically acting, 104
heteroaromatic, 168
heterocyclic, 80, 81, 164
IC50 and pKa data for, 103
inhibitors
    CA inhibition mechanism by, 7
    organ-selective, 136
    potency of, 211
    x-ray crystallographic structures for
        adducts of, 6
metal complexes of, 202, 203
mixed group of, 169–171
moiety(ies)
    hydrogen bond, 10
    modified, 128
parent, 112
QSAR calculations, 153
second-generation topically acting, 112
topical, 93
unsaturated primary, 134
unsubstituted, 150
water-soluble metal complexes of, 201
Sulfonamide carbonic anhydrase inhibitors,
        development of, 67–147
aliphatic sulfonamides, 134–136
antiepileptic sulfonamides and other
        miscellaneous inhibitors, 131–134
antitumor sulfonamides, 125–126
classical inhibitors, 69–93
    aromatic sulfonamides, 69–80
    bis-sulfonamides, 89–93
    heterocyclic sulfonamides, 80–89
future prospects of CAIs, 136–138
isozyme-specific inhibitors, 118–121
    isozyme I, 118–120
    isozyme III, 121
    isozyme IV, 120–121
selective inhibitors for membrane-associated
        CAs, 121–124
sulfonamides with modified moieties and
        other zinc-binding groups, 126–131
topical sulfonamides as antiglaucoma agents,
        93–118
    ring approach, 93–112
    tail approach, 112–118
Sulfonylated amino acid hydroxamates, 130
SUPERQUAD program, 185
Systemic sulfonamide CAIs, 11

T

Tail approach, 112, 114
new derivatives prepared by, 115
sulfonamide CAI obtained by, 115
variant of, 122
Thalassiosira weissflogii, 16
Thiadiazine diuretics, 184
Thiadiazoles, 172, 173, 216
Thienothiopyran sulfonamides, 107, 110
Thiocyanate adduct, 6
Thiols, 126
Thiophene-2-sulfonamide, derivatives of, 86
Thiophenols, alkylation of substituted, 97
Thiosulfonic acid, 127
Tissue integrity, CA IX and, 258
Topical antiglaucoma drugs, 272
Topical sulfonamides, 93
Topiramate, 46, 56, 131
anticonvulsant effects of, 49
antiepileptic activity of, 49
docking conformations, 57
replacement of diisopropylidene moiety of, 50
side-effects of, 49
treatment of epileptic seizures using, 292
use of as antiepileptic drug, 51
Topological indices, 163s, 174
Topotecan, 270
Transmembrane isozyme, overexpression of in
        tumors, 35
Trichloroacetaldehyde, 216
Trifluoroacetohydroxamic acid, 231
Trifluoromethazolamide, 84
Trigonal-bipyramidal adducts, anion binding as, 6
Tumor
    -associated protein, 35
    cell proliferation, compromised, 256
    growth inhibition data, 311
    hypoxia, CA IX as marker of, 255, 260
    metabolism
        changes in, 255
        performance of, 268
    progression, influence of extracellular pH on,
        270
    suppressor protein, von Hippel–Lindau, 259,
        261
Tumorigenesis, relationships of CAs with, 306

U

Unsubstituted sulfonamides, 150
Urea synthesis, 2
Ureate, zinc-bound, 223
Ursodeoxycholic acid, 133
Uterine carcinoma, 35